T0215940

Lecture Notes in Mathematics

Edited by A. Dold, F. Takens and B. Teissier

Editorial Policy
for the publication of monographs

1. Lecture Notes aim to report new developments in all areas of mathematics – quickly, informally and at a high level. Monograph manuscripts should be reasonably self-contained and rounded off. Thus they may, and often will, present not only results of the author but also related work by other people. They may be based on specialized lecture courses. Furthermore, the manuscripts should provide sufficient motivation, examples and applications. This clearly distinguishes Lecture Notes from journal articles or technical reports which normally are very concise. Articles intended for a journal but too long to be accepted by most journals, usually do not have this "lecture notes" character. For similar reasons it is unusual for doctoral theses to be accepted for the Lecture Notes series.

2. Manuscripts should be submitted (preferably in duplicate) either to one of the series editors or to Springer-Verlag, Heidelberg. In general, manuscripts will be sent out to 2 external referees for evaluation. If a decision cannot yet be reached on the basis of the first 2 reports, further referees may be contacted: the author will be informed of this. A final decision to publish can be made only on the basis of the complete manuscript, however a refereeing process leading to a preliminary decision can be based on a pre-final or incomplete manuscript. The strict minimum amount of material that will be considered should include a detailed outline describing the planned contents of each chapter, a bibliography and several sample chapters.
Authors should be aware that incomplete or insufficiently close to final manuscripts almost always result in longer refereeing times and nevertheless unclear referees' recommendations, making further refereeing of a final draft necessary.
Authors should also be aware that parallel submission of their manuscript to another publisher while under consideration for LNM will in general lead to immediate rejection.

3. Manuscripts should in general be submitted in English.
Final manuscripts should contain at least 100 pages of mathematical text and should include
- a table of contents;
- an informative introduction, with adequate motivation and perhaps some historical remarks: it should be accessible to a reader not intimately familiar with the topic treated;
- a subject index: as a rule this is genuinely helpful for the reader.

Continued on back inside cover

Lecture Notes in Mathematics 1645

Editors:
A. Dold, Heidelberg
F. Takens, Groningen

Springer
Berlin
Heidelberg
New York
Barcelona
Budapest
Hong Kong
London
Milan
Paris
Santa Clara
Singapore
Tokyo

Hendrik W. Broer George B. Huitema
Mikhail B. Sevryuk

Quasi-Periodic Motions in Families of Dynamical Systems

Order amidst Chaos

Springer

Authors

Hendrik W. Broer
Department of Mathematics
University of Groningen
P.O. Box 800
NL-9700 AV Groningen, The Netherlands
e-mail: h.w.broer@math.rug.nl

Mikhail B. Sevryuk
Institute of Energy Problems
of Chemical Physics
Lenin prospect 38, Bldg. 2
117829 Moscow, Russia
e-mail: rusin@iepcp.msk.ru

George B. Huitema
KPN Research
P.O. Box 15000
NL-9700 CD Groningen, The Netherlands
e-mail: g.b.huitema@research.kpn.nl

Cataloging-in-Publication Data applied for

Die Deutsche Bibliothek - CIP-Einheitsaufnahme

Broer, Hendrik W.:
Quasi-periodic motions in families of dynamical systems : order amidst chaos / Hendrik W. Broer ; George B. Huitema ; Mikhail B. Sevryuk. - Berlin ; Heidelberg ; New York ; Barcelona ; Budapest ; Hong Kong ; London ; Milan ; Paris ; Santa Clara ; Singapore ; Tokyo : Springer, 1996
(Lecture notes in mathematics ; 1645)
ISBN 3-540-62025-7
NE: Huitema, George B.:; Sevrjuk, Michail B.:; GT

Mathematics Subject Classification (1991): 58F27, 58F30, 34C50, 70H05, 34D30, 34C20, 11K60, 34D20, 70K30, 34C23

ISSN 0075-8434
ISBN 3-540-62025-7 Springer-Verlag Berlin Heidelberg New York

First Reprint 2002

This work is subject to copyright. All rights are reserved, whether the whole or part of the material is concerned, specifically the rights of translation, reprinting, re-use of illustrations, recitation, broadcasting, reproduction on microfilms or in any other way, and storage in data banks. Duplication of this publication or parts thereof is permitted only under the provisions of the German Copyright Law of September 9, 1965, in its current version, and permission for use must always be obtained from Springer-Verlag. Violations are liable for prosecution under the German Copyright Law.

Springer-Verlag Berlin Heidelberg New York
a part of Springer Science+Business Media
© Springer-Verlag Berlin Heidelberg 1996
Printed in Germany

The use of general descriptive names, registered names, trademarks, etc. in this publication does not imply, even in the absence of a specific statement, that such names are exempt from the relevant protective laws and regulations and therefore free for general use.

Typesetting: Camera-ready TEX output by the author
SPIN: 10997413 46/3111 – 5 4 3 2 – Printed on acid-free paper

Preface

This book is devoted to the phenomenon of quasi-periodic motion in dynamical systems. Such a motion in the phase space densely fills up an invariant torus. This phenomenon is most familiar from Hamiltonian dynamics. Hamiltonian systems are well known for their use in modelling the dynamics related to frictionless mechanics, including the planetary and lunar motions. In this context the general picture appears to be as follows. On the one hand, Hamiltonian systems occur that are in complete order: these are the integrable systems where all motion is confined to invariant tori. On the other hand, systems exist that are entirely chaotic on each energy level. In between we know systems that, being sufficiently small perturbations of integrable ones, exhibit coexistence of order (invariant tori carrying quasi-periodic dynamics) and chaos (the so called stochastic layers). The Kolmogorov–Arnol'd–Moser (KAM) theory on quasi-periodic motions tells us that the occurrence of such motions is *open* within the class of all Hamiltonian systems: in other words, it is a phenomenon persistent under small Hamiltonian perturbations. Moreover, generally, for any such system the union of quasi-periodic tori in the phase space is a nowhere dense set of positive Lebesgue measure, a so called Cantor family. This fact implies that open classes of Hamiltonian systems exist that are not ergodic.

The main aim of the book is to study the changes in this picture when other classes of systems – or contexts – are considered. Examples of such contexts are the class of reversible systems, of volume preserving systems or the class of all systems, often referred to as "dissipative". In all these cases, we are interested in the occurrence of quasi-periodic motions, or tori, persistent under small perturbations within the class in question. By an application of the KAM theory it turns out that in certain cases, in order to have this persistence, the systems are required to depend on *external* parameters. An example of such a situation is the dissipative class, where quasi-periodic attractors are found. These attracting quasi-periodic tori are isolated in the phase space and they are only persistent when at least one parameter is present. In that case, generally, the set of parameters for which such an attractor occurs has positive Lebesgue measure. Quasi-periodic attractors are well known to be a transient stage in a bifurcation route from order to chaos.

The KAM theory is a powerful instrument for the investigation of this problem in a broad sense, describing the organization of invariant tori as Cantor families of positive Lebesgue (Hausdorff) measure. It yields a unifying approach for all cases, leading to a formulation with a minimal number of parameters. In this book, we discuss various aspects of the KAM theory. However, there are still many problems of the theory outside the scope of the present text. Some of these will be briefly indicated.

We proceed in giving an outline of the text. In introductory Chapter 1 we present a more precise formulation of our main problem illustrating this with many examples. Here we also define the contexts to which we apply our approach throughout. These include two different reversible contexts and, in the Hamiltonian setting next to the isotropic case, also the coisotropic one. The Chapters 2 and 3 form the "kernel" of the book. In Chapter 2 first we formulate the conjugacy or stability theory. Depending on the context, we introduce a suitable number of unfolding parameters which stabilize the systems within their context for small perturbations. This stability only holds on Cantor families of invariant tori with Diophantine frequencies (or KAM tori), the corresponding conjugacies being smooth in the sense of Whitney. This approach first leaves us with families that depend on a great many parameters, but the discussion continues by systematically reducing the number of parameters to a minimum, where still sufficiently many tori are left, in the sense of Hausdorff measure. Main tool here is the Diophantine approximation lemma in the form close to that of V.I. Bakhtin. Chapter 3 subsequently discusses the theory on the continuation of analytic tori due to A.D. Bruno.

Next, in Chapter 4, we discuss the organization of the Cantor families of tori as they occur in our various contexts, including estimates of the appropriate Hausdorff measure. We also present some considerations regarding the dynamics in the "resonance zones", i.e., in the complement of the KAM tori (including Nekhoroshev estimates on solutions near those tori).

Chapter 5 presents conclusive remarks on the subject. Correspondences and differences between the cases of vector fields and diffeomorphisms are discussed. Also we show that, generally, the KAM tori accumulate very much on each other, which can be concisely formulated in terms of Lebesgue density points.

Chapter 6 consists of appendices. In the first of these, the stability theorem is proven in one of its simplest forms. Other appendices fully describe the Bruno theory and the Diophantine approximation lemma.

The style of the book makes it suitable for both experts and beginners regarding the KAM theory. On the one hand, it presents an up to date and therefore quite advanced overview of the theory. On the other hand, it contains an elementary introduction to Whitney differentiability and a complete proof of the simplest stability theorem in this respect. By this and the other details of the appendices, the text is largely self-contained. Also it contains an extended bibliography (which does not, of course, claim to be complete).

The authors are grateful to V.I. Arnol'd, B.L.J. Braaksma, S.-N. Chow, S.A. Dovbysh, J.K. Hale, M.R. Herman, I. Hoveijn, Yu.S. Il'yashenko, H.H. de Jong, S.B. Kuksin, J.S.W. Lamb, V.F. Lazutkin, M.V. Matveev, J.K. Moser, A.I. Neĭshtadt, G.R.W. Quispel, J. Pöschel, F. Takens, and D.V. Treshchëv (in alphabetical order) for long-time support, fruitful discussions, pointing out references, meticulous reading of some technical details or, generally, encouragement. Also the first author acknowledges hospitality from the Georgia Institute of Technology during the preparation of the manuscript. The third author thanks the Groningen University for its hospitality.

H.W.B., G.B.H. & M.B.S.
Groningen, September 1996

Notation

Although, as a rule, we explain each notation where it first appears in the book, some notations used frequently in the sequel are collected here. Some basic sets are denoted by the "open font" (or "blackboard bold") characters. The symbols $\mathbb{Z}$, $\mathbb{Q}$, $\mathbb{R}$, and $\mathbb{C}$ require no comments, $\mathbb{R}_+$ will denote the set of *non-negative* real numbers, $\mathbb{Z}_+$ is the set of *non-negative* integers, and $\mathbb{N} = \mathbb{Z}_+ \setminus \{0\}$ is the set of positive integers. The symbol $\mathbb{S}^n$ denotes the unit n-dimensional sphere in $\mathbb{R}^{n+1}$, and $\mathbb{T}^n = (\mathbb{S}^1)^n = (\mathbb{R}/2\pi\mathbb{Z})^n$ is the standard n-torus. Also, $\mathbb{RP}^n$ is the n-dimensional real projective space, and $\Pi : \mathbb{R}^{n+1} \setminus \{0\} \to \mathbb{RP}^n$ denotes the natural projection. Note that each of the spaces $\mathbb{T}^0$ and $\mathbb{RP}^0$ is a point, while $\mathbb{S}^0$ is the disjoint union of two points. Note also that $\mathbb{RP}^1 \simeq \mathbb{S}^1 = \mathbb{T}^1$.

The symbols $\mathcal{O}_n(a)$ will denote a neighborhood of a point $a \in \mathbb{R}^n$. In particular, $\mathcal{O}_n(0)$ is a neighborhood of the origin in $\mathbb{R}^n$. We write $\mathcal{O}(a)$ instead of $\mathcal{O}_1(a)$ for $a \in \mathbb{R}$.

By the angle brackets, we will denote the standard inner product of two vectors, so that

$$\langle a, b\rangle = \textstyle\sum_{i=1}^n a_i b_i \quad \text{for} \quad a \in \mathbb{R}^n, b \in \mathbb{R}^n.$$

The symbols $|a|$ and $\|a\|$ will denote the l_1-norm and l_2-norm (Euclidean norm), respectively, of vector a (unless when stated otherwise):

$$|a| = \textstyle\sum_{i=1}^n |a_i| \quad \text{and} \quad \|a\|^2 = \textstyle\sum_{i=1}^n |a_i|^2 \quad \text{for} \quad a \in \mathbb{C}^n.$$

For the Landau symbols, we write $O_m(u)$ instead of $O(|u|^m)$ [and $O(u)$ instead of $O_1(u) = O(|u|)$] for $m \in \mathbb{N}$ and any (scalar or vector) independent variable u. By $D_u F$ we will sometimes denote the Jacobi matrix $\partial F/\partial u$. The relation $a := b$ will mean that equality $a = b$ is the definition of quantity a. By $\log u$ we denote $\log_e u$ (which is often designated by $\ln u$ elsewhere). The boundary of manifold or set M is denoted by ∂M. The interior of set M is denoted by $\mathrm{int}(M)$ and the closure of set M by $\mathrm{cl}(M)$ or $\overline{M}$. The symbols $\mathit{diag}(a_1, a_2, \ldots, a_n)$ mean the diagonal $n \times n$ matrix with diagonal entries $a_1, a_2, \ldots, a_n$. The matrix transposed to A is denoted by A^t. The dot means differentiation with respect to time: $\dot{x} := dx/dt$ and $\ddot{x} := d^2x/dt^2$. The average of a function over $\mathbb{T}^n$ will be sometimes denoted by the square brackets $[\cdot]$. Mark $\square$ means the end of the proof.

The notations $[a, b]$, $[a, b[$, $]a, b]$, and $]a, b[$ for $-\infty \le a < b \le +\infty$ mean respectively the intervals $\{x : a \le x \le b\}$, $\{x : a \le x < b\}$, $\{x : a < x \le b\}$, and $\{x : a < x < b\}$. For example, $\mathbb{R}_+ = [0, +\infty[$. Given $x \in \mathbb{R}$, the integral part of x is denoted by $\mathrm{Entier}(x) := \max\{m \in \mathbb{Z} : m \le x\} = \max(\mathbb{Z} \cap]-\infty, x])$. If $\mathrm{Entier}(x) = \ell$ then $\ell \le x < \ell + 1$.

The n-dimensional Hausdorff measure in $\mathbb{R}^N$ for $N \ge n$ (see Federer [114] or Morgan [242]) will be denoted by meas_n (elsewhere usually denoted by $\mathcal{H}^n$). For $N = n$ the measure meas_n coincides with the standard Lebesgue measure $\mathcal{L}^n$ in $\mathbb{R}^n$.

The term "analytic" will always refer to mappings between *real* manifolds (equipped with an analytic structure), whereas holomorphic functions $f : D \to (\mathbb{C}/2\pi\mathbb{Z})^{N_1} \times \mathbb{C}^{N_2}$, $D \subset (\mathbb{C}/2\pi\mathbb{Z})^{n_1} \times \mathbb{C}^{n_2}$, that are real-valued for real arguments will be called "real analytic".

... there are so many ways to deal with formulas.

Donald E. Knuth. *The TEXbook.* Addison-Wesley, 1986.

There are some formulas that can't be handled easily ...

Leslie Lamport. LATEX. *A Document Preparation System.* Addison-Wesley, 1986.

Contents

Chapter 1

Introduction and examples

1.1 A preliminary setting of the problem

This book investigates the occurrence of quasi-periodic motions in dynamical systems with special emphasis on the persistence of these motions under small perturbations of the system. Quasi-periodic motions densely fill up invariant tori, therefore this study can be regarded as a part of a more general theory of invariant manifolds. The existence, persistence and other properties of invariant manifolds play a fundamental rôle in the analysis of nonlinear dynamical systems [67, 115, 158, 356]. In this book we confine ourselves with finite dimensional systems. For the theory of quasi-periodic motions in infinite dimensional dynamical systems, the reader is recommended to consult, e.g., [185, 186, 279–281] and references therein.

The perturbations we will consider, although small in an appropriate sense, will be arbitrary. However, it is important to specify whether the whole perturbation problem, for example, sits in the Hamiltonian context, or has to respect a certain symmetry, or is subject to no restriction whatsoever. Such a class of vector fields to be examined is often referred to as a "*context*" of the problem [60, 62, 162]. In many of these cases one needs parameters to achieve persistent occurrence of quasi-periodicity, where the specific rôle of the parameters depends on the context at hand.

One reason for the need of parameters is the following. In the perturbation analysis of quasi-periodic tori we follow a specific torus through the perturbation. In this "continuation process" the frequencies of the torus are kept constant. This means that these frequencies somehow have to be treated as parameters of the system. In the classical Hamiltonian context with Lagrangian tori, these frequency-parameters can be accounted for by the action variables, granted some nondegeneracy. However, in the context of general (or "dissipative") systems this is not possible and parameters have to be explicitly present in the setting. Therefore in the title of this book we speak of "families of dynamical systems".

The main problem will be what is the *minimal* number of parameters needed in order to have persistence of quasi-periodicity. We will discuss the *organization* of the tori in families parametrized over Cantor sets of positive Lebesgue measure in a Whitney-smooth manner. A related problem is how to apply the theory to examples where a number of parameters is available.

Several types of examples of dynamical systems with quasi-periodicity will appear in the sequel. Among these are oscillators with weak forcing, either periodic or quasi-periodic, or with weak couplings between them. Another class of applications is given by Bifurcation Theory: subordinate to some degenerate bifurcations quasi-periodic motion shows up in a persistent way.

1.1.1 Definitions

A torus with parallel dynamics. Consider a smooth vector field X on a manifold M with an invariant n-torus T. We say that X on T induces *parallel* (or *conditionally periodic*, or *Kronecker*, or *linear*) motion, evolution, dynamics, or flow, if there exists a diffeomorphism $T \to \mathrm{T}^n$ transforming the restriction $X|_T$ to a constant vector field $\sum_{i=1}^n \omega_i \partial/\partial x_i$ on the standard n-torus $\mathrm{T}^n := (\mathrm{S}^1)^n = (\mathrm{R}/2\pi\mathrm{Z})^n$ with angular coordinates $x_1, x_2, \dots, x_n$ modulo 2π. In a more familiar notation, this vector field determines the system $\dot{x}_i = \omega_i$, $1 \le i \le n$, of differential equations. The numbers $\omega_1, \omega_2, \dots, \omega_n$ are called *(internal) frequencies* of the motion (evolution, dynamics, or flow) on T, but also of the invariant torus T itself.

Remark 1. The *frequency vector* $\omega = (\omega_1, \omega_2, \dots, \omega_n) \in \mathrm{R}^n$ is determined uniquely up to changes of the form $\omega \mapsto A\omega$, where $A \in GL(n, \mathrm{Z})$, i.e., A is an $n \times n$ matrix with integer entries and determinant ± 1.

Remark 2. Invariant tori with parallel dynamics are of great importance in the theory of dynamical systems which stems, in the long run, from the fact that *any finite dimensional connected and compact abelian Lie group is a torus* [2, 50, 236]. A close more geometric statement is that the factor group R^N/Γ of R^N by a discrete subgroup Γ is T^N provided that R^N/Γ is compact [13].

More generally, any finite dimensional connected abelian Lie group is the product $\mathrm{T}^n \times \mathrm{R}^m$ and the factor group of R^N by any discrete subgroup is $\mathrm{T}^n \times \mathrm{R}^{N-n}$ for some $0 \le n \le N$. The latter statement is the key one in the proof of the Liouville–Arnol'd theorem on completely integrable Hamiltonian systems (see Theorem 1.2 in § 1.3.2 below).

The dynamical properties of an invariant torus with linear flow are very sensitive to the number-theoretical properties of its frequency vector.

A quasi-periodic torus. A parallel motion on an invariant n-torus T with frequency vector ω is called *quasi-periodic* or *nonresonant* if the frequencies $\omega_1, \omega_2, \dots, \omega_n$ are rationally independent, i.e., if for all $k \in \mathrm{Z}^n \setminus \{0\}$ one has $\langle \omega, k \rangle := \sum_{i=1}^n \omega_i k_i \neq 0$. In this case the torus T itself also is said to be quasi-periodic. Otherwise an invariant torus T with parallel dynamics is called *resonant*. For example, the 2-torus of Figure 1.1 with parallel dynamics is quasi-periodic if and only if the ratio of the corresponding frequencies ω_1 and ω_2 is irrational.

Quasi-periodic tori are densely filled up by each of the orbits (or solution curves) contained therein. The whole motion then is ergodic [17]. However, quasi-periodic dynamics is not chaotic, since by the parallelity there is no sensitive dependence on the initial conditions.

The resonant tori are foliated by invariant subtori of smaller dimension. In all the contexts to be met below, a Kupka–Smale theorem holds (cf. [260, 265]), generically for-

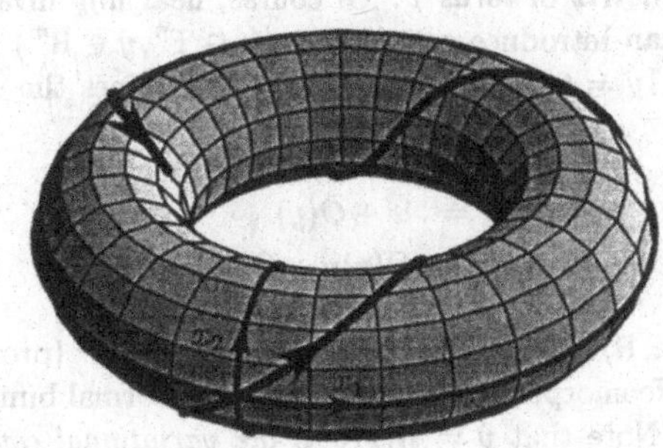

Figure 1.1: Evolution or solution curve of a constant vector field $\omega_1\partial/\partial x_1 + \omega_2\partial/\partial x_2$ on the two-dimensional torus T^2.

bidding the existence of resonant tori. For a more detailed discussion, see one of the examples in the next section. As a consequence, all the invariant tori with parallel flow of a *generic* dynamical system are quasi-periodic. Moreover, it turns out that most of the invariant tori with parallel dynamics in the phase space of a generic vector field satisfy stronger nonresonance conditions, as we shall introduce now.

A Diophantine torus. Our study is a part of the Kolmogorov–Arnol'd–Moser (KAM) theory (named after its founders A.N. Kolmogorov [179, 180], V.I. Arnol'd [4, 5, 8], and J. Moser [243, 245]) which generally establishes the existence and persistence of quasi-periodic tori in dynamical systems. In this theory the frequencies of quasi-periodic tori are not only rationally independent, but have to meet the following, stronger nonresonance condition. We say that an invariant n-torus T with parallel dynamics is *Diophantine* if for some constants $\tau > 0$ and $\gamma > 0$ the corresponding frequency vector ω satisfies the following infinite system of inequalities:

$$|\langle\omega, k\rangle| \geq \gamma|k|^{-\tau} \tag{1.1}$$

for all $k \in \mathbb{Z}^n \setminus \{0\}$, where $|k| := \sum_{i=1}^n |k_i|$. Clearly Diophantine tori are quasi-periodic, but not *vice versa*. For $\tau > n-1$ the set of all frequency vectors $\omega \in \mathbb{R}^n$ that are Diophantine in the above sense has positive Lebesgue measure [337].

Example 1.1 Each equilibrium point of a vector field is a Diophantine invariant 0-torus. Each S-periodic trajectory of a vector field is a Diophantine invariant 1-torus with frequency $2\pi/S$.

A Floquet torus. An invariant n-torus T with parallel dynamics of a vector field X on an $(n+m)$-dimensional manifold is called *Floquet* if near T, one can introduce coordinates $(x \in \mathbb{T}^n, y \in \mathbb{R}^m)$ in which the torus T itself gets the equation $\{y = 0\}$ while the field X determines the system of differential equations of the so called Floquet form

$$\begin{aligned} \dot{x} &= \omega + O(y) \\ \dot{y} &= \Omega y + O_2(y) \end{aligned} \tag{1.2}$$

with $\Omega \in gl(m, \mathrm{R})$ independent of $x \in \mathrm{T}^n$, compare [60, 62, 162]. If this is the case, matrix Ω is called the *Floquet matrix* of torus T. Of course, near *any* invariant n-torus T with parallel dynamics, one can introduce coordinates ($x \in \mathrm{T}^n, y \in \mathrm{R}^m$) in which the torus T itself gets the equation $\{y = 0\}$ while the field X determines the system of differential equations

$$\begin{aligned} \dot{x} &= \omega + O(y) \\ \dot{y} &= \Omega(x)y + O_2(y) \end{aligned} \tag{1.3}$$

with $\Omega = \Omega(x) \in gl(m, \mathrm{R})$ depending smoothly on $x \in \mathrm{T}^n$ (provided that a certain neighborhood of T is diffeomorphic to $\mathrm{T}^n \times \mathrm{R}^m$, i.e., the normal bundle of T in the phase space is trivial[1] [157]). Note that $\dot{y} = \Omega(x)y$ is *the variational equation* along T. The torus T is Floquet if matrix Ω can be made independent of the point on T (*reduced* to a constant) by an appropriate choice of local coordinates.

To any Floquet invariant torus with parallel dynamics, one associates the so called *normal frequencies*, i.e., the *positive imaginary parts* of the eigenvalues of its Floquet matrix. The Diophantine tori that the KAM theory can deal with in general also have to be Floquet [12, § 26], for some exceptions see, e.g., [137, 138, 364] (these papers concern the so called hyperbolic lower-dimensional tori in Hamiltonian systems, cf. the remark at the end of § 2.3.4) and Chapter 3 below as well. Moreover, the internal $\omega_1, \omega_2, \ldots, \omega_n$ and normal $\omega_1^N, \omega_2^N, \ldots, \omega_r^N$ frequencies of these tori ($0 \le r \le m$) should satisfy further Diophantine conditions

$$|\langle \omega, k \rangle + \langle \omega^N, \ell \rangle| \ge \gamma |k|^{-\tau} \tag{1.4}$$

for all $k \in \mathrm{Z}^n \setminus \{0\}$, $\ell \in \mathrm{Z}^r$, $|\ell| \le 2$ with some constants $\tau > 0$ and $\gamma > 0$, as we will see below in Section 2.1. Note that for $\ell = 0$, inequalities (1.4) are just the standard Diophantine inequalities (1.1).

Remark 1. The Floquet matrix is determined up to similarity, i.e., up to changes of the form $\Omega \mapsto A^{-1}\Omega A$, where $A \in GL(m, \mathrm{R})$.

Remark 2. Recall that the *Floquet multipliers* of a periodic trajectory are the eigenvalues of its monodromy operator [12, § 34] (sometimes these are also called *characteristic multipliers* [1, Sect. 7.1]). If in (1.2) $n = 1$, then the period and monodromy operator of closed trajectory $\{y = 0\}$ are equal respectively to $2\pi/\omega$ and $e^{2\pi\Omega/\omega}$. Consequently, if the eigenvalues of the Floquet matrix of an S-periodic trajectory are $\lambda_1, \ldots, \lambda_m$ then its Floquet multipliers are $e^{S\lambda_1}, \ldots, e^{S\lambda_m}$.

Remark 3. An equilibrium point is always Floquet. A periodic trajectory is Floquet if and only if its monodromy operator has a real logarithm (the Floquet theorem, see Arnol'd [12, § 26]). The question is now prompted under what conditions the system of differential equations (1.3) near an invariant torus with parallel dynamics can be reduced to the Floquet form (1.2). It is known, see [5, 12], that for $m = 1$ and arbitrary n reducibility to the Floquet form does take place for *Diophantine* tori. This problem will be treated below in detail (see § 1.5.1).

[1] This is not always the case. No neighborhood of the central circle on the Möbius strip is diffeomorphic to the cylinder $\mathrm{S}^1 \times \mathrm{R}$

For $m \geq 2$, non-pathological examples exist where reducibility does not hold. So, in general reducibility involves an assumption, which turns out to be satisfied typically when sufficiently many parameters are present. For more details here see, e.g., [5, 42, 62, 64, 148, 149, 162, 167, 168, 170, 171, 184].

In examples, reducibility often holds due to a normalization (averaging), see, e.g., Arnol'd [12] and Broer & Vegter [56, 66]. For Hamiltonian vector fields, reducibility is often implied by the presence of sufficiently many additional integrals in involution [184, 256].

1.1.2 Contexts

We next point briefly at the contexts to be explored, for detailed definitions and more up-to-date references see Section 1.3 below. The most important contexts are the Hamiltonian, the volume preserving and the reversible ones, as well as the general "dissipative" one. A Lie algebra of systems is often involved in the definition of the context, this being a natural way to express the "preservation of a structure".

In the sequel, several refinements of these contexts will be considered. Also the central question mentioned before, regarding the minimal number of parameters necessary to obtain the persistence of Diophantine invariant tori, will be addressed.

The Hamiltonian context. With respect to quasi-periodicity the most well known context is Hamiltonian (conservative), defined by the preservation of a symplectic form. This context is notorious for its strong relation to mechanics [1, 5, 13, 17, 130, 182, 183, 201, 335]. The classical result roughly states that in this context, it is a typical property to have many so called Lagrangian quasi-periodic invariant tori, seen from the measure-theoretical point of view. This implies that non-ergodicity is a typical property as well [219].

This "classical" KAM theory was initiated by Kolmogorov [179] in 1954 (see also [180] and [14, 235] as well) and further developed by Arnol'd [4, 5], Moser [243], and many others. For details we refer to, e.g., [13, 20, 30, 91–93, 103, 122, 129, 130, 133, 201, 248, 277, 278, 282, 306, 307, 335]. For a review and a large bibliography, also see Bost [43]. A further refinement of this theory is given in the sequel.

The dissipative context. Another important context is the "dissipative" one, where no structure at all is present. Here the notion of a quasi-periodic attractor comes up as a quasi-periodic torus that is isolated in the phase space. Although quasi-periodicity itself is not considered to be chaotic, this type of dynamics is a possible transient stage between order and chaos, cf. Ruelle & Takens [296, 297], see also [32]. In the dissipative context, parameters are needed in order to have the persistence of quasi-periodic motions. In a suitable class of families of dissipative systems, it is typical to have many parameter values with a quasi-periodic motion, again in the sense of the measure theory.

This part of the KAM theory was first developed by Moser [244, 246] and later on taken up by Broer, Huitema & Takens [62, 162]. Also these results will be explored further below.

The volume preserving context. The volume preserving context shows up, for example, when describing the velocity field of an incompressible fluid. With respect to quasi-periodicity, it has been studied by Moser [244, 246] and later on by Broer & Braaksma [46, 53]. It turns out that the case of codimension 1 Diophantine tori is much related to the above Hamiltonian one, while the case of codimension no less than 2 is very similar to the general dissipative one. This theory was also taken up in [62, 162] and will be presented and extended in the sequel.

The reversible context. Reversible systems are compatible with some involution G, that takes motions to motions while reversing the time parametrization. This concept often comes up in physical systems and is also known for some great similarities with the Hamiltonian case (in particular, regarding quasi-periodicity), see a list of such similarities in [320]. The reversible KAM theory was initiated by Moser [244, 246, 248] and Bibikov & Pliss [35, 36, 38, 40] and later on extended by Scheurle [310–312], Parasyuk [262], Pöschel [278], Arnol'd & Sevryuk [11, 22, 315, 316, 318], and others. Recently the theory has taken a lot of interest, leading to new developments that we will come back to later.

1.2 Occurrence of quasi-periodicity

This section is devoted to a few simple examples of dynamical systems, sometimes depending on parameters, where quasi-periodicity occurs in a persistent way. These examples are situated in various contexts and motivate the introduction of parameters. Also, a first idea is given of the Whitney-smoothness of families of Diophantine invariant tori. The examples given here all are in the world of oscillators, with forcings (both periodic and quasi-periodic), couplings, etc.

1.2.1 Quasi-periodic attractors

The first examples are within the dissipative context, involving quasi-periodic attractors, cf. Broer, Dumortier, van Strien & Takens [58, Ch. 4] and Broer [57] (see also Bogolyubov, Mitropol'skiĭ & Samoĭlenko [42]). We shall see that for the persistence of these attractors, the systems in question (vector fields) have to depend on "external" parameters. For simplicity we restrict ourselves to the case of 2-tori, where the discussion leads to circle maps, compare [12, § 11].

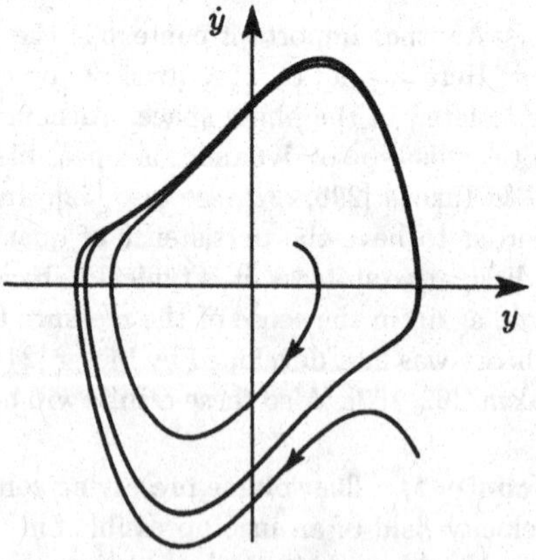

Figure 1.2: A hyperbolic periodic attractor (limit cycle) of the free oscillator.

Preliminaries

We give two examples, both being based on a nonlinear oscillator

$$\ddot{y} + c\dot{y} + ay + f(y, \dot{y}) = 0$$

with $y \in \mathrm{R}$, which is assumed to have a periodic attractor. Here a and c are real constants. For example, one may think of the "Van der Pol" form $f(y, \dot{y}) = by^2\dot{y}$ [58, Ch. 1], b being a real constant. In Figure 1.2, the corresponding phase portrait is shown in the $(y, \dot{y})$-plane.

Periodic forcing. In the first example we force this oscillator periodically, which leads to the following equation of motion:

$$\ddot{y} + c\dot{y} + ay + f(y, \dot{y}) = \epsilon g(y, \dot{y}, t) \tag{1.5}$$

with $g(y, \dot{y}, t + 2\pi) \equiv g(y, \dot{y}, t)$. For simplicity we assume the functions f and g to be analytic in all the arguments; ϵ is a small parameter "controlling" the forcing. In this example we consider the three-dimensional extended phase space $\mathrm{R}^2 \times \mathrm{S}^1$ with coordinates $(y, \dot{y}, t \bmod 2\pi)$, where we obtain a vector field determining the system of differential equations

$$\begin{aligned} \dot{y} &= z \\ \dot{z} &= -ay - cz - f(y, z) + \epsilon g(y, z, t) \\ \dot{t} &= 1. \end{aligned}$$

Coupling. In the second example we consider two "Van der Pol"-type oscillators, as before, with a coupling:

$$\begin{aligned} \ddot{y}_1 + c_1\dot{y}_1 + a_1y_1 + f_1(y_1, \dot{y}_1) &= \epsilon g_1(y_1, y_2, \dot{y}_1, \dot{y}_2) \\ \ddot{y}_2 + c_2\dot{y}_2 + a_2y_2 + f_2(y_2, \dot{y}_2) &= \epsilon g_2(y_1, y_2, \dot{y}_1, \dot{y}_2). \end{aligned} \tag{1.6}$$

Here $y_j \in \mathrm{R}$ while a_j, c_j are real constants for $j = 1, 2$. The functions f_j, g_j $(j = 1, 2)$ are again assumed to be analytic and ϵ is a small parameter. The phase space here is therefore R^4 with coordinates $(y_1, y_2, \dot{y}_1, \dot{y}_2)$, on which we obtain a vector field determining the system of differential equations

$$\begin{aligned} \dot{y}_j &= z_j \\ \dot{z}_j &= -a_jy_j - c_jz_j - f_j(y_j, z_j) + \epsilon g_j(y_1, y_2, z_1, z_2) \end{aligned}$$

for $j = 1, 2$.

A preliminary perturbation theory: The torus as an invariant manifold

The torus as an invariant manifold. First we consider the "unperturbed" case $\epsilon = 0$, where in both examples the situation is simple. Indeed, in the three- (respectively four-dimensional) phase space we find an attracting invariant 2-torus with suitable coordinates $(x_1, x_2$ modulo $2\pi)$, in which the differential equations afforded by the restricted vector field have the constant form $\dot{x}_1 = \omega_1$, $\dot{x}_2 = \omega_2$. From hyperbolicity of each limit cycle it follows that normally to this torus, the attraction already can be seen from the linear

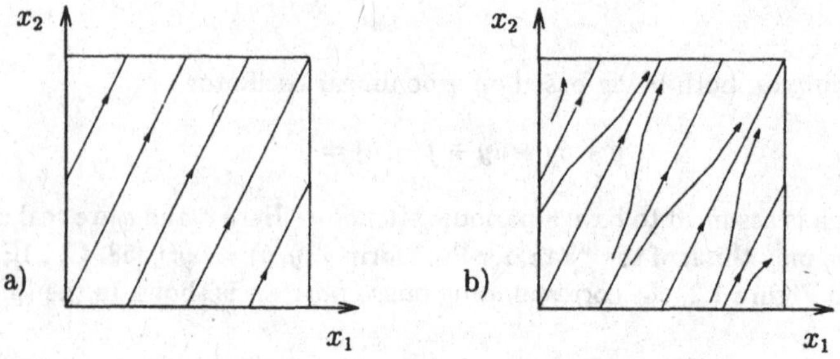

Figure 1.3: Perturbation from a parallel a) to a "phase-locked" b) 2-torus in the resonant case $\omega_1 : \omega_2 = 1 : 2 \Leftrightarrow \alpha = 2$.

terms: the invariant torus is normally hyperbolic (see [67, 115, 158, 356]). Consequently, this torus, as an invariant manifold, persists if ϵ varies near 0 in R.

However, the linear dynamics in the torus is not necessarily persistent under small perturbations: it may even become topologically different. To see this, consider (constant) perturbations of the form $\dot{x}_1 = \omega_1 + \epsilon_1$, $\dot{x}_2 = \omega_2 + \epsilon_2$. For (ϵ_1, ϵ_2) varying near $(0,0)$ in R^2, one meets both quasi-periodicity and resonance: in the latter case all trajectories are periodic (with the same period). These two situations are clearly different since in the former case the trajectories are not periodic.

Parameters. The present approach looks for smooth *equivalences* between unperturbed and perturbed tori, i.e., for diffeomorphisms which take evolutions into evolutions and which preserve the direction of the time parametrization, though not necessarily this time parametrization itself. The smoothness (or analyticity) of the equivalences puts this into the framework of the classical perturbation theory.

However, within the class of constant vector fields on the torus, the frequency ratio is an invariant for (even topological) equivalences. To construct a perturbation theory for at least parallel motions on the torus, we wish to keep the control on the frequency ratio under perturbations. This motivates the introduction of external parameters. From now on we assume the systems (1.5) and (1.6) to depend on parameters in such a way that for $\epsilon = 0$ the Jacobi matrix of the frequency ratio $\alpha = \omega_2/\omega_1$, as a function of the parameters, has maximal rank 1. One could use, e.g., the coefficient a, respectively the ratio of the coefficients a_1/a_2, as a parameter. We can assume that α is bounded, otherwise we replace α by α^{-1}.

The Kupka–Smale theorem. Still in the resonant case where α is rational, the dynamics on the torus may change drastically under a perturbation to $\epsilon \neq 0$. In fact there exist arbitrarily small perturbations which lead to the so called *phase-locking*: for an example see Figure 1.3. This dynamics definitely is not parallel. Moreover it is preserved under small(er) perturbations. The situation illustrates the so called Kupka–Smale theorem, cf. Peixoto [265] or Palis & de Melo [260, Ch. 3], which asserts that generically continua of periodic evolutions of the same period do not occur (see also [1, Sect. 7.3] and [58, Ch. 2]). Similar theorems hold in the parameter dependent case, compare Sotomayor [336] or Brunovský [76, 77]. The analytic case has been specifically analyzed by

Broer & Tangerman [65]. This gives a heuristic motivation for restricting our study of the persistence of parallel flows to the case of quasi-periodic tori, i.e., with α irrational. See, e.g., [12, 39, 308] for a detailed discussion on "phase-locked" invariant tori in dissipative systems.

Remark. There are known analogues of the Kupka–Smale theorem for other contexts as well, see, e.g., [293] for the conservative case and [105] for the reversible case (a particular example of so called convex billiards is studied in [200]).

Reduction to circle maps

After passage to the invariant (center) manifold, compare [115, 158], we are left with a family of vector fields on the torus T^2. Although such a reduction may have caused some loss of differentiability, for simplicity we assume this family to be still analytic. This restriction is not very essential since with a modification of the approach below, one obtains similar results in the C^k-category with $k \leq \infty$ sufficiently large (cf. [151, 243, 277, 278, 306] as well as [62, 162]). For $\epsilon = 0$ the members of the family in question are constant vector fields with a varying frequency ratio α. In fact, again for simplicity, we further restrict to the case where α itself is the parameter.

Consider the circle $\{(x_1, x_2) \in T^2 \ : \ x_1 = 0\}$, which for $|\epsilon|$ small intersects all orbits of each vector field in our family transversally. This enables us to define the *Poincaré return map* A^α_ϵ as follows. To each point $(0, x_2)$, we assign the point $(2\pi, A^\alpha_\epsilon(x_2))$ where the orbit starting from $(0, x_2)$ hits the circle at its first return. If we write $x := x_2$, then $\{A^\alpha_\epsilon\}_\alpha$ can be seen as a family of diffeomorphisms of the circle $S^1 = T^1 = R/2\pi Z$ which has the form

$$A^\alpha_\epsilon : x \mapsto x + 2\pi\alpha + \epsilon a(x, \alpha, \epsilon) \mod 2\pi.$$

Notice that for $\epsilon = 0$ the map A^α_ϵ reduces to the rigid rotation $R^\alpha : x \mapsto x + 2\pi\alpha \mod 2\pi$. Also observe that, without loss of generality, we may confine ourselves with the case $0 \leq \alpha \leq 1$. The problem of finding equivalences between the vector fields on T^2 mentioned above now translates to finding *conjugacies* between the corresponding Poincaré maps on S^1. For completeness, this last problem will be sketched below. The mathematics closely follows Arnol'd [12, § 11], although both settings differ regarding the rôle of the parameters.

For any orientation preserving homeomorphism of S^1, the rotation (or winding) number is defined as the $2\pi^{\text{th}}$ part of its average rotation, see [12, § 11]. In the case of the rigid rotation R^α, the rotation number exactly is α. An important property of the rotation number is its invariance under (even topological) conjugacies, compare the above discussion about the invariance of the frequency ratio.

A further perturbation theory: Conjugacies on the circle

Formal solutions of the conjugacy equation. We now look for analytic conjugacies on S^1 between the families R^α and A^α_ϵ where α is irrational. The invariance of the rotation number implies that a *parameter shift* may be needed here. In order to deal with the parameter dependence in general, we pass to the cylinder $S^1 \times [0, 1]$, where we consider "vertical" maps $R, A_\epsilon : S^1 \times [0, 1] \to S^1 \times [0, 1]$, defined by

$$R(x, \alpha) := (R^\alpha(x), \alpha) \quad \text{and} \quad A_\epsilon(x, \alpha) := (A^\alpha_\epsilon(x), \alpha).$$

The desired conjugacy then is a mapping $\Phi_\epsilon : S^1 \times [0,1] \to S^1 \times [0,1]$ of the form

$$\Phi_\epsilon(x,\alpha) = (x + \epsilon U(x,\alpha,\epsilon), \alpha + \epsilon\sigma(\alpha,\epsilon))$$

which makes the following diagram commute:

$$\begin{array}{ccc} S^1 \times [0,1] & \xrightarrow{A_\epsilon} & S^1 \times [0,1] \\ \uparrow \Phi_\epsilon & & \uparrow \Phi_\epsilon \\ S^1 \times [0,1] & \xrightarrow{R} & S^1 \times [0,1]. \end{array}$$

This means that $A_\epsilon \circ \Phi_\epsilon = \Phi_\epsilon \circ R$, or, equivalently, that the mappings U and σ satisfy the nonlinear equation

$$U(x + 2\pi\alpha, \alpha, \epsilon) - U(x,\alpha,\epsilon) = 2\pi\sigma(\alpha,\epsilon) + a(x + \epsilon U(x,\alpha,\epsilon), \alpha + \epsilon\sigma(\alpha,\epsilon), \epsilon). \qquad (1.7)$$

Let us proceed by comparing the Taylor series expansions in ϵ of both sides of this equation. We only consider the coefficients of the 0th power in ϵ, since the procedure for higher order coefficients is similar. So we write

$$\begin{aligned} U(x,\alpha,\epsilon) &= U_0(x,\alpha) + O(\epsilon), \\ \sigma(\alpha,\epsilon) &= \sigma_0(\alpha) + O(\epsilon), \\ a(x,\alpha,\epsilon) &= a_0(x,\alpha) + O(\epsilon). \end{aligned}$$

Substituting these expressions in equation (1.7) we get

$$U_0(x + 2\pi\alpha, \alpha) - U_0(x,\alpha) = 2\pi\sigma_0(\alpha) + a_0(x,\alpha) \qquad (1.8)$$

often called *the homological equation*, cf. [12, §§ 12,13,22,25–28]. We wish to solve (1.8) with respect to U_0 and σ_0 by using Fourier expansions in x. The value of σ_0 is obtained immediately:

$$\sigma_0(\alpha) = -\frac{1}{2\pi}[a_0(\cdot,\alpha)] = -\frac{a_{00}(\alpha)}{2\pi},$$

where $[a_0(\cdot,\alpha)]$ denotes the mean value of a_0 over S^1 (i.e., the 0th Fourier coefficient). If we then put

$$a_0(x,\alpha) = \sum_{k=-\infty}^{\infty} a_{0k}(\alpha)e^{ikx} \quad \text{and} \quad U_0(x,\alpha) = \sum_{k=-\infty}^{\infty} U_{0k}(\alpha)e^{ikx}$$

it follows that, for $k \in Z \setminus \{0\}$,

$$U_{0k}(\alpha) = \frac{a_{0k}(\alpha)}{e^{2\pi ik\alpha} - 1},$$

while the function $U_{00}(\alpha)$ is arbitrary.

First observe that in general the condition that α is irrational is necessary and sufficient for the existence of a formal solution U_0. Recall that this amounts to the nonresonance of the frequencies ω_1 and ω_2.

But even then the denominators $e^{2\pi ik\alpha} - 1$ are not bounded away from zero, which makes the convergence of the Fourier series for U_0 problematic. This is the well known *small divisors problem* which can be solved by a further restriction of α by Diophantine conditions: for some constants $\tau > 2$, $\gamma > 0$ and for all integers p, q with $q > 0$ we require that

$$\left| \alpha - \frac{p}{q} \right| \geq \frac{\gamma}{q^\tau}. \tag{1.9}$$

This condition is closely related to the above condition (1.1) on the frequencies ω_1, ω_2.

Let us sketch how this runs. If α is as above then a short calculation shows that for all $k \in \mathbb{Z} \setminus \{0\}$ one has $|e^{2\pi i k\alpha} - 1| \geq 4\gamma |k|^{1-\tau}$. For such α the growth of the coefficients U_{0k} as $|k| \to +\infty$ is therefore controlled. In fact, for analytic a_0 the coefficients a_{0k} decay exponentially as $|k| \to +\infty$; this is the well known Paley–Wiener lemma (Lemma 1.8), for a proof see § 1.5.2. Then the coefficients U_{0k} also have exponential decay as $|k| \to +\infty$ and the Fourier series of U_0 converges to an analytic function.

In this way one could proceed with the higher order terms in ϵ and try to prove the convergence, so having solved the conjugacy equation (1.7), compare the Poincaré–Lindstedt method in the periodic case [20, 275]. However, the theory often uses a Newton-like algorithm to show that, for $|\epsilon|$ small and α satisfying (1.9) the desired conjugacy Φ_ϵ exists and is analytic in x and ϵ. For more details in a slightly different setting we again refer to Arnol'd [12, §§ 11,12]. In the sequel (see Section 6.1) a more general proof will be given for vector fields instead of maps.

The small divisors difficulties are in fact central in the KAM theory. The latter has been defined as "a collection of ideas of how to approach certain problems in perturbation theory connected with small divisors" (Pöschel [282]).

A Cantor set of positive measure. Next let us consider the set $[0,1]_\gamma$ of α's in $[0,1]$ that satisfy the condition (1.9) for τ fixed. This set has the following properties:

(i) $[0,1]_\gamma$ is the union of a countable set and a *Cantor set*;

(ii) the Lebesgue measure of $[0,1] \setminus [0,1]_\gamma$ is of the order of γ as $\gamma \downarrow 0$.

Property (i) follows from the Cantor–Bendixson theorem, cf. Hausdorff [147], which implies that a closed set in $\mathbb{R}$ is the union of a countable set and a perfect set. Since clearly $[0,1]_\gamma$ is totally disconnected (the dense set of rationals lies in its complement), the perfect set must be a Cantor set. Property (ii) can be shown by observing how $[0,1]_\gamma$ is obtained from $[0,1]$: in fact for each $q = 2, 3, \ldots$ around the rationals p/q with $1 \leq p \leq q-1$ a symmetric open interval of length $2\gamma q^{-\tau}$ is deleted, compare Figure 1.4. One should also delete the intervals $[0, \gamma[$ and $]1-\gamma, 1]$. Consequently,

$$meas_1([0,1] \setminus [0,1]_\gamma) \leq 2\gamma + \sum_{q=2}^{\infty} (q-1)\frac{2\gamma}{q^\tau} < 2\gamma \sum_{q=1}^{\infty} \frac{1}{q^{\tau-1}}$$

where the series converges since $\tau > 2$. For a similar discussion regarding the Diophantine condition (1.1), see § 1.5.2 below.

Whitney-smoothness. Cantor sets are nowhere dense and therefore have no interior points. Nevertheless, the dependence of Φ_ϵ on the parameter α as it varies over the nowhere dense set $[0,1]_\gamma$ is smooth in a certain sense, compare Herman [151]. In fact, this dependence is of Whitney class C^∞, which means that Φ_ϵ can be *extended* to a C^∞-diffeomorphism of $S^1 \times [0,1]$ (cf. Figure 1.5). This extension is not unique and outside

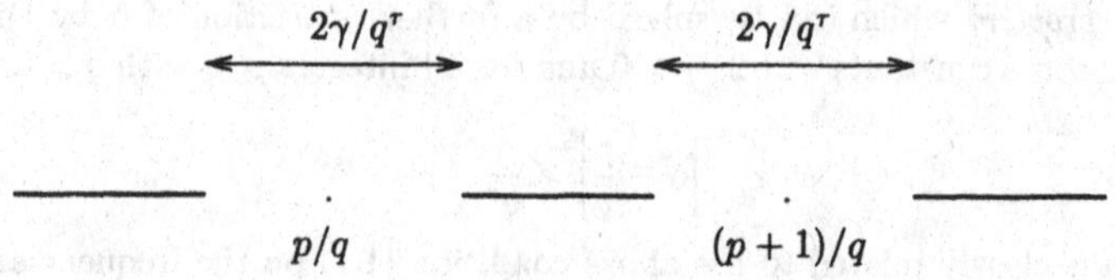

Figure 1.4: Construction of the set $[0,1]_\gamma$.

$S^1 \times [0,1]_\gamma$ in general it *loses* the conjugation property. See Sections 1.3 and 2.2 for further discussion.

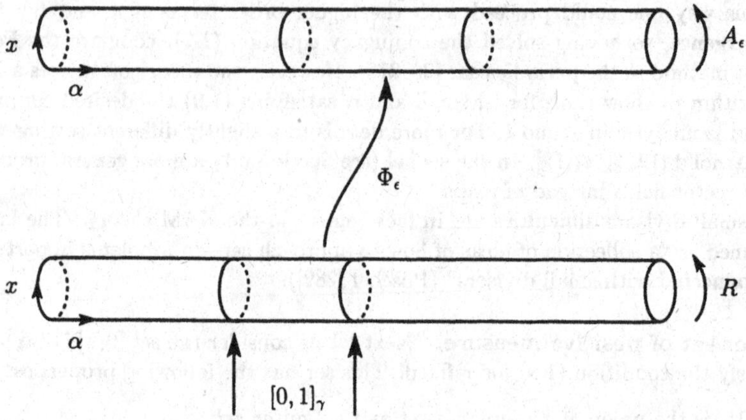

Figure 1.5: Φ_ϵ conjugates R to A_ϵ, restricted to $S^1 \times [0,1]_\gamma$.

Conclusion

We now return to the original problem regarding the weakly forced oscillator, respectively the weakly coupled oscillators. From the above we conclude the following, for a general definition of the terminology referring to Section 2.2.

For $|\epsilon|$ sufficiently small, in the product of the phase space and the parameter (α) space, there exists a Whitney-smooth family of Diophantine invariant 2-tori, parametrized over a parameter set of large measure, the tori being quasi-periodic attractors. From Arnol'd [8] it already follows that for all $|\epsilon|$ small, in the family $\{A_\epsilon^\alpha\}_\alpha$ the persistence holds of an uncountable number of quasi-periodic circles. The Whitney-smoothness in α was added by Herman [151], see above. An independent proof of this fact is again given in Section 6.1.

It is this smoothness that allows of the above measure-theoretical conclusions. Indeed, a C^1-diffeomorphism preserves positive measure, which generally is not the case for a homeomorphism.

Remark 1. We observe that the above result on the existence of a conjugacy can be phrased in terms of *(structural) stability* (later on we shall specify a suitable topology to be used on the space of analytic systems, see § 6.1.2). Indeed, for all sufficiently small perturbations of the family of rigid rotations $\{R^\alpha\}_\alpha$ there exists a family of smooth conjugacies (and a reparametrization) to this family $\{R^\alpha\}_\alpha$, provided that $\alpha \in [0,1]_\gamma$. This implies that in the product of the phase space of the forced oscillator or coupled oscillators and the α-space, the unperturbed family of quasi-periodic invariant tori, in an analogous way, is stable with respect to smooth equivalences. These two types of stability will be called respectively *quasi-periodic* and *weak quasi-periodic stability*, see Huitema [162], Broer, Huitema & Takens [62], and Broer & Huitema [60]. Below in the next chapter we shall come back to this in a wider setting.

Remark 2. If we take the ambient space into account, somewhat more can be said.

(a) Doing the finitely differentiable KAM theory in the *center manifold*, one finds smooth Diophantine tori. Since by the parallelity these are r_*-normally hyperbolic for every $r_* \in \mathrm{N}$, by uniqueness it follows that they are even of class C^∞ (cf. [62, 158, 162, 356]). The corresponding subset of parameter values is nowhere dense, so it is topologically small. Nevertheless this set has positive Lebesgue measure. If one restricts to the case where the perturbed Diophantine tori are Floquet, one can even establish analyticity, cf. [62, 162] and § 2.3.1 below. On the other hand, the continuation theory (Chapter 3) implies analyticity of the perturbed Diophantine tori that are not necessarily Floquet.

(b) What of the rest of the parameter space? Here, generically, one meets Kupka–Smale ("phase-lock") tori, that are only finitely differentiable, see, e.g., [12, 347]. Residual parameter sets exhibiting this kind of dynamics are topologically large (and at least dense).

For a broader mathematical background on the subject of (Lebesgue) measure and (Baire) category, we refer to Oxtoby [258], Morgan [242], and Federer [114].

1.2.2 Quasi-periodic motions in conservative examples

The above examples can both be mimicked in the Hamiltonian (conservative) context, which may be somewhat more familiar. Here, however, no parameters are needed for the persistent occurrence of quasi-periodic motions. Moreover, these motions fill a subset of the phase space of positive measure.

Forcing. To start with, consider the periodically forced pendulum without friction, with the equation of motion given by

$$\ddot{y} = -\omega^2 \sin y + \epsilon \cos t.$$

In the extended phase space $\mathrm{S}^1 \times \mathrm{R} \times \mathrm{S}^1$ with coordinates $(y \bmod 2\pi, \dot{y}, t \bmod 2\pi)$ this gives the system of equations

$$\begin{aligned} \dot{y} &= z \\ \dot{z} &= -\omega^2 \sin y + \epsilon \cos t \\ \dot{t} &= 1. \end{aligned}$$

As before, ϵ is a small parameter. We study this perturbation problem by its Poincaré map $A_\epsilon : S^1 \times R \to S^1 \times R$, which again is the return map of the section $t \equiv 0 \mod 2\pi$. This map is area preserving, which here expresses the conservative character of the problem.

For $\epsilon = 0$ the system is autonomous: A_0 is just the time 2π flow map of the pendulum. Restricting to the region with oscillatory evolution, we pass to action-angle variables $(\eta, \xi) \in R \times S^1$, see, e.g., [1, 13]. In these coordinates the (autonomous) pendulum takes the form

$$\begin{aligned} \dot{\xi} &= \alpha(\eta) \\ \dot{\eta} &= 0 \end{aligned}$$

with a certain function α and one therefore has

$$A_0(\xi, \eta) = (\xi + 2\pi\alpha(\eta), \eta).$$

One can show that the derivative $d\alpha/d\eta$ nowhere vanishes, for which reason A_0 is called *a* (pure) *twist map.* So we end up with the following perturbation problem:

$$A_\epsilon(\xi, \eta) = A_0(\xi, \eta) + O(\epsilon)$$

as $\epsilon \to 0$, which is completely contained in the area preserving setting.

The dynamics of A_0 is quite simple: the oscillatory motions of the pendulum give rise to a cylinder of invariant circles with rotation numbers $\alpha(\eta)$. Compared with the above dissipative situation, now the rôle of the parameter is played by the action variable η. Again the question is which of the invariant circles $\{\eta = const\}$ are persistent for $\epsilon \neq 0$ small. Without going into the details we mention that the present result is completely analogous: the circles with Diophantine rotation numbers survive and their union has positive measure in the phase cylinder, being organized as a Whitney-C^∞ family. These surviving circles give corresponding Diophantine invariant 2-tori for the flow. For details compare, e.g., [78, 150, 199, 243, 247, 248, 298, 299, 301, 335].

Coupling. The second example consists in two pendula

$$\begin{aligned} \ddot{y}_1 &= -\omega_1^2 \sin y_1 + \epsilon \frac{\partial U(y_1, y_2)}{\partial y_1} \\ \ddot{y}_2 &= -\omega_2^2 \sin y_2 + \epsilon \frac{\partial U(y_1, y_2)}{\partial y_2} \end{aligned}$$

with a conservative coupling (U being a potential). Putting $z_j = \dot{y}_j$ this leads to the system of equations in R^4

$$\begin{aligned} \dot{y}_j &= z_j \\ \dot{z}_j &= -\omega_j^2 \sin y_j + \epsilon \frac{\partial U(y_1, y_2)}{\partial y_j} \end{aligned}$$

($j = 1,\ 2$), which is Hamiltonian with Hamilton function

$$H_\epsilon(y_1, z_1, y_2, z_2) = \tfrac{1}{2} z_1^2 + \tfrac{1}{2} z_2^2 - \omega_1^2 \cos y_1 - \omega_2^2 \cos y_2 - \epsilon U(y_1, y_2).$$

Since H_ϵ is a (first) integral of this system, we restrict to the three-dimensional energy level surfaces and arrive at the (isoenergetic) Poincaré return map A_ϵ with respect to some transversal section of dimension two. Again this map preserves an appropriate area form, see Proposition 5.7 in § 5.1.7 below. Although the computations here are more cumbersome than in the previous case, a similar perturbation result holds.

Remark. The latter example easily generalizes to n weakly coupled oscillators and further to the general case of Lagrangian invariant tori in nearly integrable Hamiltonian systems. The present discussion then is in the setting of the so called isoenergetic KAM theorem, compare, e.g., Arnol'd [5, 13], Broer & Huitema [59, 162], or Delshams & Gutiérrez [102–104] (see also § 4.2.3 below).

1.2.3 Quasi-periodic responses

Another class of examples involving parameters is obtained as follows. Consider a nonlinear oscillator with quasi-periodic forcing, given by the second order differential equation

$$\ddot{y} + c\dot{y} + ay = f(y, \dot{y}, t, a, c)$$

with y, $t \in \mathbb{R}$, where a and c (subject to $a > 0$ and $c^2 < 4a$) serve as parameters (for the condition $c^2 < 4a$, see § 1.5.1 below). Moreover, one assumes that f is quasi-periodic in t, i.e., $f(y, \dot{y}, t, a, c) = F(y, \dot{y}, t\omega_1, t\omega_2, \ldots, t\omega_n, a, c)$ for some function $F : \mathbb{R}^2 \times \mathbb{T}^n \times \mathbb{R}^2 \to \mathbb{R}$ and a fixed nonresonant frequency vector $\omega \in \mathbb{R}^n$.

The problem then is to find a ***response solution***, i.e., a solution that is quasi-periodic with these same frequencies $\omega_1, \omega_2, \ldots, \omega_n$. This leads to a problem as before. Indeed, we rewrite the second order equation above as a system of equations on $\mathbb{T}^n \times \mathbb{R}^2$:

$$\begin{aligned}
\dot{x} &= \omega \\
\dot{y} &= z \\
\dot{z} &= -ay - cz + F(y, z, x, a, c),
\end{aligned}$$

where $x = (x_1, x_2, \ldots, x_n)$ modulo 2π. Now suppose that the vector field determining this system of differential equations has an invariant n-torus which is a graph $(y, z) = (y(x, a, c), z(x, a, c))$ over $\mathbb{T}^n$ with dynamics $\dot{x} = \omega$. Evidently this torus would give a family of response solutions as required above, so the problem now is to look for such a torus. Since it is natural to assume this torus to be close to $\mathbb{T}^n \times \{0\}$ for small f, this again can be seen as a perturbation problem. It turns out that for $c \approx 0$, even stronger nonresonance (Diophantine) conditions have to be imposed on the frequencies, cf. § 1.1.1.

As stated here the problem sits within the general dissipative context, for which reason two parameters a and c are "needed" for the persistence of a solution, the damping c controlling the normal hyperbolicity of the torus. The result roughly is that for small f a Whitney-C^∞ family of invariant tori survives, parametrized by c and a. This family contains analytic subfamilies being curves parametrized by c. In the a-direction the family has the structure of a Cantor set.

Below in § 1.5.1 the corresponding linear problem (where f does not depend on y and $\dot{y}$) will be studied to some detail.

Remark. For $|c|$ large, the problem was treated by Stoker [339], just using hyperbolicity. For $c \approx 0$ the small divisors enter the problem, compare Friedman [117], Moser [244], and

Braaksma & Broer [47]. It turns out that the above nowhere dense family of tori can be "fattened" using hyperbolicity [47, 117], cf. also Section 4.3 below.

As a modification of the above, consider the quasi-periodically forced oscillator

$$\ddot{y} + ay = \epsilon f(t, y, \dot{y}),$$

which is reversible in the sense that $f(-t, y, -\dot{y}) \equiv f(t, y, \dot{y})$, see § 1.1.2 above. The reversing involution here is given by $G : (y, \dot{y}) \mapsto (y, -\dot{y})$. We now can formulate the same problem as before, but notice that we have one parameter less, or in other words, we have that $c = 0$. Still the corresponding problem can be solved as a consequence of the reversibility. It follows that for small $|\epsilon|$ a Whitney-C^∞ family of invariant tori survives, parametrized by a (compare, e.g., Moser [244, 246] and Broer & Huitema [60]).

1.3 A further setting of the problem

Now we proceed towards a more detailed description of the set-up. The above conservative examples suggest looking for certain families of quasi-periodic tori rather than individual tori. On the other hand, the examples with attractors suggest seeking quasi-periodic tori of systems depending on external parameters. Besides, the properties of such invariant tori are sensitive to what structures on the phase space (such as the area form or reversibility, cf. the above response problem) the system is assumed to preserve or to be compatible with. How to unify all these examples?

1.3.1 The main problem

In answer to this, consider a real finite dimensional connected manifold M endowed with some structure $\mathfrak{S}$ (e.g., a tensor field, the action of a Lie group, etc.). We will assume both the manifold M itself and the structure $\mathfrak{S}$ to be analytic. Our general aim is to study l-parameter families of Floquet Diophantine invariant n-tori in s-parameter families of vector fields on M, compatible with the structure $\mathfrak{S}$, in a sense to be made precise. The tori we look for are also supposed to be compatible with the structure $\mathfrak{S}$, again in a sense to be specified.

Thus, we will encounter mainly two different kinds of objects: families of vector fields and families of quasi-periodic invariant tori. An s-parameter family $\{X^\mu\}_\mu$ of vector fields on M is just a vector field on M depending on a parameter $\mu \in P \subset \mathbb{R}^s$ (P being some open domain). Sometimes it will be convenient to regard such a family of vector fields as a "vertical" vector field on the product $M \times P$.

An l-parameter family of quasi-periodic invariant n-tori of $\{X^\mu\}_\mu$ is a more complicated object, for its precise definition see Section 2.2. For the time being, one can think of a certain set of n-tori in $M \times P$ such that the $(n+l)$-dimensional Hausdorff measure of the union of all the tori is finite (neither 0 nor $+\infty$), while each torus lies in one of the "fibers" $M \times \{\mu_0\}$, is invariant under the flow of field X^{μ_0} and carries quasi-periodic dynamics.

To fix thoughts, let C^r, $r \in \mathbb{N} \cup \{\infty, \omega\}$, be some smoothness class (where ω stands for analyticity). By $\mathfrak{X}^r_s$ we denote the space of all s-parameter families of C^r-smooth vector fields on M compatible with the structure $\mathfrak{S}$, the dependence of these fields on the s-dimensional parameter also being of class C^r. For $r \le \infty$ we equip the space $\mathfrak{X}^r_s$ with the weak or strong C^r topology, cf. Hirsch [157] (the difference between these two standard topologies is entirely irrelevant for our purposes). In the analytic case we use the topology induced by the compact-open topology on the space of holomorphic extensions, cf. Broer & Tangerman [65] and § 6.1.2 below.

Our concern will be with properties that define open sets in these topologies. Such properties often are called *persistent* and the corresponding families *typical* (cf. Broer, Dumortier, van Strien & Takens [58, Ch. 2]).

From now on, for simplicity, we shall confine ourselves with the analytic category $r = \omega$, i.e., we will consider families of vector fields that depend both on the phase space variables and the parameters in an analytic manner. All our results have straightforward analogues in the other smoothness classes, cf. [62, 162]. The C^∞ and finitely differentiable KAM theory has been developed in, e.g., [42, 150, 151, 197–201, 243, 245, 273, 277, 278, 298, 299, 301, 306, 307, 340, 364].

We are now able to further specify the main problem addressed in this book, compare the beginning of Section 1.1:

For what values of $n \in \mathbb{Z}_+$ and $l \in \mathbb{Z}_+$, does a *typical* family of vector fields in $\mathfrak{X}^r_s$ possess l-parameter families of Floquet Diophantine invariant n-tori compatible with the structure $\mathfrak{S}$? What are the properties of these families of tori in terms of Whitney-smoothness and Lebesgue (or Hausdorff) measure?

The best known way to study quasi-periodic tori for typical families of vector fields is to perturb *integrable* families, containing vector fields equivariant with respect to a free action of the torus group. Integrable systems are quite exceptional, but their perturbations, no matter how small their size is, are already typical.

If we succeed in proving that *any* sufficiently small perturbation of an integrable family $\{X_0^\mu\}_\mu$ admits an l-parameter family of Floquet Diophantine invariant n-tori,[2] we may already conclude that such families of tori occur typically in $\mathfrak{X}^r_s$, namely, in a neighborhood of $\{X_0^\mu\}_\mu$. If desired, we can impose some further restrictions on the unperturbed integrable family $\{X_0^\mu\}_\mu$ (e.g., some nondegeneracy and nonresonance conditions). In fact, we can choose the unperturbed family as special as we please, without affecting at all our final conclusion on the typicality.

Remark. Throughout this book, the parameter is assumed to vary in an *open* domain of the Euclidean space. However, more general parameter domains are also considered sometimes in the KAM theory.

The general picture of the influence a perturbation has upon an integrable family of dynamical systems can now be summarized as follows:

a) When the perturbation is zero, the evolution of the systems is completely regular, and there are many quasi-periodic motions which are well organized (into analytic families).

b) When the perturbation is other than zero but sufficiently small, the systems still possess many quasi-periodic motions which are however worse organized than in the unperturbed systems (namely, into Cantor families). Moreover, chaotic motions

[2] in which case $\{X_0^\mu\}_\mu$ is sometimes said to be *KAM-stable* (cf. [184, 327])

appear. As the perturbation grows, the set constituted by quasi-periodic solutions is shrinking while the region of solutions of another type (e.g., chaotic) is being expanded.

c) Finally, when the perturbation becomes sufficiently large, quasi-periodic motions (at least of the same topological type as the unperturbed ones) disappear, and the systems may get very chaotic.

We will be interested exclusively in the transition a) $\rightsquigarrow$ b) and will study mainly the persistence of quasi-periodicity rather than the onset of chaos.

1.3.2 Contexts revisited: definitions

As indicated before, the structures $\mathfrak{S}$ we will be interested in are a volume element, a symplectic structure, and an involution. Accordingly, we will distinguish four main "contexts" of our theory, see [60, 62, 162]:

(i) the dissipative context, where no structure on the phase space is present;

(ii) the volume preserving context, where the structure on M is a volume element, i.e., a nowhere vanishing differential form of maximal degree;

(iii) the Hamiltonian context, where the structure on M is a symplectic structure, i.e., a closed nondegenerate differential 2-form;

(iv) the reversible context, where the structure on M is an involution $G : M \to M$ (i.e., a mapping whose square is the identity transformation).

The volume preserving context. In the volume preserving context, the vector fields in question are globally divergence-free. Recall that if σ is a volume element on an N-dimensional manifold M then the *divergence* $\operatorname{div}X$ of a vector field X on M is a real-valued function on M defined as

$$d(i_X\sigma) = (\operatorname{div}X)\sigma.$$

Here $i_X\sigma$ is the $(N-1)$-form whose value at the vectors $X_1, \ldots, X_{N-1}$ is equal to the value of σ at the vectors $X, X_1, \ldots, X_{N-1}$. Thus, divergence-free vector fields X are those for which the form $i_X\sigma$ is closed. A divergence-free vector field X is said to be *globally divergence-free* if the form $i_X\sigma$ is not only closed but even exact, cf. [1]. Invariant tori in the volume preserving context are assumed to satisfy no further conditions.

The Hamiltonian context. In the Hamiltonian context, the vector fields in question are Hamiltonian. Recall that if the symplectic structure is ω^2, then *Hamiltonian* vector fields X are by definition those for which the 1-form $i_X\omega^2$ is exact, hence being minus the differential of the corresponding Hamilton function H:

$$dH(\xi) = -i_X\omega^2(\xi) = \omega^2(\xi, X) \quad \forall\xi.$$

The invariant n-tori with parallel dynamics, to be studied here, are assumed to be either isotropic (for $n \leq N$) or coisotropic (for $n \geq N$), where N is the number of degrees of freedom (so that the phase space M is of dimension $2N$). We recall these definitions. First, for tangent vectors ξ, $\eta \in T_uM$ at a point $u \in M$ we say that $\xi \perp \eta$ (skew-orthogonal) if $\omega^2(\xi, \eta) = 0$. Then, a submanifold L of manifold M is *isotropic* [*coisotropic*] if the tangent space T_uL to L at each point $u \in L$ lies in its skew-orthogonal complement $(T_uL)^\perp$ [respectively, contains $(T_uL)^\perp$], cf. [1, 18, 201]. For any isotropic submanifold L, one has $\dim L \le N$, while for any coisotropic submanifold L, one has $\dim L \ge N$. Any submanifold of dimension 0 or 1 is isotropic, while any submanifold of codimension 0 or 1 is coisotropic. For N-dimensional submanifolds the concepts of isotropicity and coisotropicity coincide, and an N-dimensional submanifold $L \subset M$ for which $(T_uL)^\perp = T_uL$ at each point $u \in L$ is called *Lagrangian*.

The most important class of integrable Hamiltonian vector fields is constituted by so called *completely integrable* fields described by the celebrated Liouville–Arnol'd theorem. Before formulating the latter, recall that the *Poisson bracket* (H_1, H_2) of two Hamilton functions H_1 and H_2 on a symplectic manifold is defined as

$$(H_1, H_2) := dH_1(X_2) = \omega^2(X_2, X_1) = -dH_2(X_1),$$

where X_1 and X_2 are the corresponding Hamiltonian vector fields and ω^2 is the symplectic structure.

Theorem 1.2 (see [1, 5, 6, 13, 17, 18, 20, 58, 182, 183, 201]) *Let N smooth functions $H_i : M \to \mathrm{R}$, $1 \le i \le N$, on a $2N$-dimensional symplectic manifold M be in involution,*[3] *i.e., let their pairwise Poisson brackets vanish identically: $(H_i, H_j) \equiv 0$. Denote the corresponding Hamiltonian vector fields by X_i, $1 \le i \le N$, and consider a common level surface of the functions H_i:*

$$M_h := \{u \in M \ : \ H_i(u) = h_i \quad \text{for all} \quad i = 1, \ldots, N\}$$

($h \in \mathrm{R}^N$). Suppose that M_h is not empty and consider a connected component M_h^{cc} of M_h. Assume that the following two conditions are satisfied:

1) the N vector fields X_i, $1 \le i \le N$, are linearly independent at any point $u \in M_h^{cc}$,

2) any solution of each Hamiltonian differential equation $\dot u = X_i(u)$ with the initial condition $u(0) \in M_h^{cc}$ is defined over the whole real line R (this condition is met automatically if M_h^{cc} is compact).

Then the following holds.

a) M_h^{cc} is a smooth Lagrangian submanifold of M (invariant under the flows of vector fields X_i).

b) M_h^{cc} is diffeomorphic to $\mathrm{T}^n \times \mathrm{R}^{N-n}$ for some $0 \le n \le N$ (if M_h^{cc} is compact then $n = N$). Moreover, there exist coordinates ($\varphi \in \mathrm{T}^n, w \in \mathrm{R}^{N-n}$) on M_h^{cc} in which the Hamiltonian differential equations $\dot u = X_i(u)$ restricted to M_h^{cc} take the form

$$\dot\varphi = \omega^{(i)} = \mathrm{const} \in \mathrm{R}^n, \quad \dot w = C^{(i)} = \mathrm{const} \in \mathrm{R}^{N-n}, \quad 1 \le i \le N.$$

c) If M_h^{cc} is compact (and therefore $n = N$) then in some neighborhood of M_h^{cc}, one can choose coordinates ($x \in \mathrm{T}^N, y \in \mathrm{R}^N$) such that y depends on H_i only: $y = y(H_1, \ldots, H_N)$ [and M_h^{cc} takes the form $\{y = \mathrm{const}\}$], $\varphi = x|_{M_h^{cc}}$, and the symplectic structure ω^2 gets the

[3] This term is not to be confused with mappings $G : M \to M$ such that $G^2 = id$.

form $\omega^2 = \sum_{i=1}^N dy_i \wedge dx_i$. *The Hamiltonian differential equations* $\dot{u} = X_i(u)$ *in coordinates* (x, y) *take the form*

$$\dot{x} = \frac{\partial H_i(y)}{\partial y}, \quad \dot{y} = 0, \quad 1 \le i \le N.$$

d) *If* H_i *are analytic, so are* M_h^{cc} *and the coordinates* (w, φ) *and* (y, x).

e) *The coordinates* (w, φ) *and* (y, x) *[and, consequently, the solutions of equations* $\dot{u} = X_i(u)$*] can be found by "quadratures", i.e., by a finite number of "algebraic operations" (including evaluating a function specified implicitly), differentiations, and integrations (integration means evaluating the integral of a known function).*[4]

Under the conditions of this theorem, the Hamiltonian systems governed by the Hamilton functions H_i are said to be *completely integrable*, or *integrable in the sense of Liouville.* If M_h^{cc} is compact, they are indeed equivariant in a neighborhood of M_h^{cc} with respect to a free action of T^N. The coordinates (y, x) introduced in item c) of the theorem are called *the action-angle variables* ($y_1, \ldots, y_N$ are the action variables and $x_1, \ldots, x_N$ the angle variables).

Remark. If M_h^{cc} is compact then M_h^{cc} itself and all the nearby connected components of common level surfaces of the functions H_i are given by $\{y = const\}$ and exemplify Lagrangian invariant tori (of each vector field X_i) with parallel dynamics and frequency vectors $\partial H_i(y)/\partial y$. So, a whole neighborhood of M_h^{cc} in the phase space M is smoothly foliated into Lagrangian invariant tori with parallel dynamics (called *Liouville tori*).

For some generalizations of Theorem 1.2, see [20, 182, 183, 252, 256, 285].

The reversible context. In the reversible context, the vector fields X in question are *reversible* with respect to the fixed involution G of the phase space M. This means that G transforms field X into the opposite field $-X$, to be precise:

$$(D_u G)X(u) \equiv -X(G(u)).$$

In other words, a vector field X [locally determining the differential equation $\dot{u} = X(u)$] is said to be reversible with respect to G if $G(u(-t))$ is a solution of that equation whenever $u(t)$ is. This definition is applicable to any diffeomorphism $G : M \to M$ but we will only consider the involutive case $G^2 = id$. Besides, n-tori with parallel dynamics in the reversible context are assumed to be invariant not only under the flow of the field itself, but also under the reversing involution G. It turns out that on any of such tori with a nonresonant frequency vector, one can choose a coordinate system $x \in T^n$ which normalize both the dynamics and the involution.

Lemma 1.3 (Sevryuk [315]) *If the differential equation* $\dot{x} = \omega$ *on the torus* T^n *with nonresonant frequency vector* ω *is reversible with respect to a diffeomorphism* $G : T^n \to T^n$ *then this diffeomorphism is an involution and has the form* $G(x) = c - x$ *for some constant* $c \in T^n$.

[4] It is this property that the term "integrable" descends from.

Proof. Choose an arbitrary point $x_0 \in \mathrm{T}^n$. Then $G(x_0 + \omega t) = G(x_0) - \omega t$ for all $t \in \mathrm{R}$. In other words, $G(x) = G(x_0) + x_0 - x$ for all $x \in \mathrm{T}^n$ of the form $x_0 + \omega t$. But ω is nonresonant, and points of this form are everywhere dense in T^n. Therefore $G(x) = c - x$ for all the points $x \in \mathrm{T}^n$ with $c = G(x_0) + x_0$. □

Remark 1. Note that the coordinate change $\xi = x - c/2$ retains the differential equation (which takes the form $\dot{\xi} = \omega$) but puts the involution into the standard form $G : \xi \mapsto -\xi$.

Remark 2. The nonresonance condition on ω in Lemma 1.3 is very essential. For instance, the system $\dot{x}_1 = 1$, $\dot{x}_2 = 0$ on T^2 is reversed by any diffeomorphism of the form $G : (x_1, x_2) \mapsto (a(x_2) - x_1, b(x_2))$.

Recall that the action of any compact Lie group on an arbitrary manifold M around a fixed point $u \in M$ is linear in an appropriate local coordinate system centered at u (this is the contents of the Bochner theorem, cf. [50, 236]). Since an involution generates a group Z_2 of two elements, around a fixed point it is always conjugate to its linear part. Consequently, the fixed points of any analytic involution $G : M \to M$ constitute an analytic submanifold $\operatorname{Fix} G \subset M$. However, this submanifold can be disconnected and its connected components can be of different dimensions, even in the case where the manifold M itself is connected, see Bredon [50] or Quispel & Sevryuk [287, 328]. An involution $G : M \to M$ of a $(q_+ + q_-)$-dimensional manifold M is said to be of type (q_-, q_+) if its submanifold $\operatorname{Fix} G$ of fixed points is non-empty and all the connected components of $\operatorname{Fix} G$ are of dimension q_+.

Lemma 1.3 implies that the restriction of an involution G to a quasi-periodic n-torus T of a G-reversible vector field is always of type $(n, 0)$, and that the set $(\operatorname{Fix} G) \cap T$ consists of 2^n isolated points. Hence, if the involution G itself is of type (q_-, q_+) then $q_- \geq n$.

The reader is referred to works [22, 49, 51, 105, 106, 189–195, 248, 284, 287, 289–292, 315, 320, 326] for a survey of the main properties of reversible dynamical systems, examples, physical applications and recent generalizations (the papers [287, 292, 320, 326] and thesis [190] also contain an extended bibliography).

1.3.3 Integrability

Next let us consider in detail the integrable set-up in the above four contexts. In other words, we will deal with s-parameter families $\{X^\mu\}_\mu$ of vector fields on M under the assumption that these vector fields as well as structure $\mathfrak{S}$ are equivariant with respect to a free action of the standard n-torus T^n. We also assume that the structure $\mathfrak{S}$ is "linear" in a certain sense, although, due to normalization theorems like the Darboux or the Bochner theorem (see [1, 13, 18, 50, 201, 233, 236, 354]), this often leads to no additional restrictions. The parameter μ varies over an open domain $P \subset \mathrm{R}^s$, and we will always assume that $0 \in P$ and that the field X^0 possesses an invariant n-torus T with parallel dynamics. This torus is one of the orbits of the action of T^n on M. For an extensive coördinate free discussion of this in terms of normal bundles, see Broer, Huitema & Takens [60, 62, 162].

The dissipative (n, p, s) context. In this case we consider invariant n-tori of an s-parameter integrable family of vector fields $X = \{X^\mu\}_\mu$ on an $(n+p)$-dimensional manifold. A certain neighborhood of the n-torus T is diffeomorphic to $\mathrm{T}^n \times \mathrm{R}^p$, and one can introduce coordinates ($x \in \mathrm{T}^n, w \in \mathrm{R}^p$) near T in which the torus T itself gets the equation $\{w = 0\}$ and the family $X = \{X^\mu\}_\mu$ of vector fields obtains the form

$$X^\mu(x,w) = \varpi(w,\mu)\frac{\partial}{\partial x} + W(w,\mu)\frac{\partial}{\partial w} \tag{1.10}$$

where $W(0,0) = 0$, cf. [62, 162]. The equivariance just means that these vector fields do not depend on x. Generically the Jacobi matrix $D_w W(0,0)$ is non-singular, and the equation $W(w,\mu) = 0$ can be solved with respect to w as $w = a(\mu)$, $a(0) = 0$. Having introduced the new coordinate $z = w - a(\mu)$, we arrive at the family of vector fields

$$[\omega(\mu) + O(z)]\frac{\partial}{\partial x} + [\Omega(\mu)z + O_2(z)]\frac{\partial}{\partial z}, \tag{1.11}$$

where z varies over a neighborhood of $0 \in \mathrm{R}^p$ and where the matrix $\Omega(\mu)$ is non-singular for all μ. We then conclude that for each value of μ, the n-torus $\{z = 0\}$ is invariant under the flow of X^μ and carries parallel dynamics with frequency vector $\omega(\mu)$. Moreover, this torus is of Floquet type with Floquet matrix $\Omega(\mu)$, cf. (1.2). Observe that for the occurrence of such analytic s-parameter families of Floquet invariant n-tori with parallel dynamics, no restrictions on the non-negative integers n, p, s are needed.

The volume preserving (n,p,s) context. Next we consider invariant n-tori of s-parameter integrable families of globally divergence-free vector fields on an $(n+p)$-dimensional manifold. In this context, we have the same family (1.10) of vector fields which, however, are supposed now to be globally divergence-free with respect to volume element $dx \wedge dw$. First of all, observe that this requirement excludes the case $p = 0$ since the constant vector field $\omega\partial/\partial x$ on the torus T^n with volume element dx is always divergence-free but never *globally* divergence-free (except for the trivial case $\omega = 0$). For $p = 1$, one can easily verify that vector fields (1.10) are divergence-free if and only if W does not depend on w, and they are globally divergence-free if and only if $W \equiv 0$. For $p > 1$, the notion of being divergence-free and that of being globally divergence-free for vector fields on $M = \mathrm{T}^n \times \mathrm{R}^p$ are equivalent (because the cohomology $H^{n+p-1}(M, \mathrm{R}) = 0$ in this case [221]), and fields (1.10) are divergence-free (and globally divergence-free) with respect to $dx \wedge dw$ if and only if the vector fields $W(w,\mu)\partial/\partial w$ are divergence-free (and globally divergence-free) with respect to dw (i.e., $\mathrm{Tr}\, D_w W(w,\mu) \equiv 0$). The two cases $p = 1$ and $p > 1$ therefore turn out to be drastically different in the volume preserving context, cf. [62, 162].

If $p = 1$, we arrive at the family of vector fields

$$\omega(y,\mu)\frac{\partial}{\partial x} \tag{1.12}$$

where $y \in \mathrm{R}$ (to achieve consistence with the notations in the Hamiltonian and reversible contexts discussed below, we prefer to write y instead of w and ω instead of ϖ). For each value of μ, each torus $\{y = const\}$ is invariant under the flow of X^μ and carries parallel dynamics with frequency vector $\omega(y,\mu)$. So, we obtain an $(s+1)$-parameter analytic family of invariant n-tori with parallel dynamics.

If $p > 1$, we proceed in exactly the same way as in the dissipative context. Thus we obtain a family (1.11) of vector fields and, again, an s-parameter analytic family of Floquet invariant n-tori $\{z = 0\}$ with parallel dynamics. The only difference is that now $\Omega(\mu) \in sl(p, \mathrm{R})$ for each μ (i.e., $\mathrm{Tr}\,\Omega(\mu) \equiv 0$).

The Hamiltonian isotropic (n,p,s) context. This context concerns isotropic invariant n-tori of s-parameter families of Hamiltonian vector fields with $n+p$ degrees of freedom ($n \geq 0$, $p \geq 0$, $n+p \geq 1$). *The codimension of the tori in the phase space is equal to* $n+2p$. We suppose that field X^0 possesses an isotropic invariant n-torus T and near this torus, there exists a coordinate system ($\varphi \in \mathbb{T}^n, w \in \mathbb{R}^{n+2p}$) in which T takes the form $\{w=0\}$, the Hamilton function does not depend on φ and the symplectic structure has constant coefficients. It is not hard to verify that after an appropriate coordinate change $\varphi = x + Ay + Bz$, $w = Cy + Dz$, where $x \in \mathbb{T}^n$, $y \in \mathbb{R}^n$, $z \in \mathbb{R}^{2p}$, and A, B, C, D are constant matrices of suitable sizes, this symplectic structure will get the form

$$\omega^2 = \sum_{i=1}^{n} dy_i \wedge dx_i + \sum_{j=1}^{p} dz_j \wedge dz_{j+p}. \tag{1.13}$$

In fact, the assumption of constant coefficients is not really necessary for such a normalization of the symplectic structure because the latter can always be reduced to form (1.13) around an isotropic torus (the generalized Darboux theorem [1, 354]). The Hamiltonian $H^\mu(y,z)$ determines the family of vector fields $X = \{X^\mu\}_\mu$ of the form

$$X^\mu(x,y,z) = \frac{\partial H}{\partial y}\frac{\partial}{\partial x} + \left(J\frac{\partial H}{\partial z}\right)\frac{\partial}{\partial z}$$

with

$$J = \begin{pmatrix} 0 & -I \\ I & 0 \end{pmatrix}, \qquad I = \mathrm{diag}(1,\ldots,1) \in SL(p,\mathbb{R}). \tag{1.14}$$

Since the torus $\{y=0, z=0\}$ is invariant under the flow of X^0, the origin $z=0$ is a critical point of the function $\mathbb{R}^{2p} \to \mathbb{R}$, $z \mapsto H^0(0,z)$, i.e., $D_z H^0(0,0) = 0$. Generically this critical point is nondegenerate, i.e., $\det \partial^2 H^0(0,0)/\partial z^2 \neq 0$, in which case the equation $D_z H^\mu(y,z) = 0$ determines an analytic surface $z = Z(y,\mu)$, $Z(0,0) = 0$. We obtain an $(n+s)$-parameter analytic family of isotropic invariant n-tori with parallel dynamics for X^μ. Namely, for each value of μ, each torus $\{y = const, z = Z(y,\mu)\}$ is isotropic, invariant under the flow of X^μ and carries parallel dynamics with frequency vector $\partial H^\mu(y, Z(y,\mu))/\partial y$.

We may assume that $Z(y,\mu) \equiv 0$, otherwise one has to introduce the new coordinates

$$\begin{aligned} x_i^{\text{new}} &:= x_i + \sum_{j=1}^{p} \frac{\partial Z_j(y,\mu)}{\partial y_i}\left[z_{j+p} - \frac{Z_{j+p}(y,\mu)}{2}\right] - \sum_{j=1}^{p} \frac{\partial Z_{j+p}(y,\mu)}{\partial y_i}\left[z_j - \frac{Z_j(y,\mu)}{2}\right] \\ & \quad (1 \leq i \leq n), \\ z^{\text{new}} &:= z - Z(y,\mu) \end{aligned} \tag{1.15}$$

and verify easily that in the new coordinate system $(x^{\text{new}}, y, z^{\text{new}})$, the symplectic structure ω^2 retains its form (1.13):

$$\omega^2 = \sum_{i=1}^{n} dy_i \wedge dx_i^{\text{new}} + \sum_{j=1}^{p} dz_j^{\text{new}} \wedge dz_{j+p}^{\text{new}},$$

while the equality $z = Z(y,\mu)$ turns to $z^{\text{new}} = 0$. In fact, for each fixed value of μ, the canonical transformation $(x,y,z) \mapsto (x^{\text{new}}, y, z^{\text{new}})$ given by (1.15) is just the time one flow map of the Hamiltonian vector field governed by the Hamilton function

$$\mathfrak{H} = \mathfrak{H}^\mu(y,z) = \sum_{j=1}^{p}[z_{j+p}Z_j(y,\mu) - z_jZ_{j+p}(y,\mu)].$$

The Hamiltonian differential equations associated to $\mathfrak{H}$ are

$$\begin{aligned}
\dot{x}_i &= \frac{\partial \mathfrak{H}}{\partial y_i} = \sum_{j=1}^{p}\left[\frac{\partial Z_j(y,\mu)}{\partial y_i}z_{j+p} - \frac{\partial Z_{j+p}(y,\mu)}{\partial y_i}z_j\right]\\
\dot{y}_i &= -\frac{\partial \mathfrak{H}}{\partial x_i} \equiv 0\\
\dot{z}_j &= -\frac{\partial \mathfrak{H}}{\partial z_{j+p}} = -Z_j(y,\mu)\\
\dot{z}_{j+p} &= \frac{\partial \mathfrak{H}}{\partial z_j} = -Z_{j+p}(y,\mu)
\end{aligned}$$

($1 \le i \le n$, $1 \le j \le p$), so that the corresponding flow map $(x,y,z) \overset{t}{\mapsto} (x',y',z')$ is

$$\begin{aligned}
x'_i &= x_i + \sum_{j=1}^{p}\frac{\partial Z_j(y,\mu)}{\partial y_i}\left[tz_{j+p} - \frac{t^2}{2}Z_{j+p}(y,\mu)\right] - \sum_{j=1}^{p}\frac{\partial Z_{j+p}(y,\mu)}{\partial y_i}\left[tz_j - \frac{t^2}{2}Z_j(y,\mu)\right]\\
&\quad (1 \le i \le n),\\
y' &= y,\\
z' &= z - tZ(y,\mu),
\end{aligned}$$

and for $t = 1$ we obtain (1.15).

Thus, in the sequel we will always suppose that $Z(y,\mu) \equiv 0$, in which case the Hamiltonian has the form

$$H = H^\mu(y,z) = F(y,\mu) + \tfrac{1}{2}\langle z, K(y,\mu)z\rangle + O_3(z) \tag{1.16}$$

(where the $2p \times 2p$ Hessian matrix $K(y,\mu)$ is symmetric for all values of y and μ). The corresponding family of vector fields has the form

$$X^\mu(x,y,z) = [\omega(y,\mu) + O_2(z)]\frac{\partial}{\partial x} + [\Omega(y,\mu)z + O_2(z)]\frac{\partial}{\partial z} \tag{1.17}$$

where $\omega = \partial F/\partial y$ and $\Omega = JK$. Matrices of the form JK with symmetric K are said to be Hamiltonian, or *infinitesimally symplectic* [1]. For each value of μ, each torus $\{y = const, z = 0\}$ is isotropic, invariant under the flow of X^μ and carries parallel dynamics with frequency vector $\omega(y,\mu)$. Moreover, each such torus is Floquet.

Remark 1. The eigenvalues of a Hamiltonian matrix occur in pairs $(\lambda, -\lambda)$, the root spaces corresponding to λ and $-\lambda$ having the same Jordan structure. Moreover, the number of nilpotent Jordan blocks of any given odd order is even (see [120, 161, 178]). The spectrum of a generic $2p \times 2p$ Hamiltonian matrix has the form

$$\pm\delta_1, \ldots, \pm\delta_{N_1}, \quad \pm i\varepsilon_1, \ldots, \pm i\varepsilon_{N_2}, \quad \pm\alpha_1 \pm i\beta_1, \ldots, \pm\alpha_{N_3} \pm i\beta_{N_3} \tag{1.18}$$

where $\delta_j > 0$, $\varepsilon_j > 0$, $\alpha_j > 0$, $\beta_j > 0$ and $N_1 + N_2 + 2N_3 = p$.

Remark 2. In the Hamiltonian context, the Floquet nature of isotropic invariant tori with parallel dynamics is often ensured by the presence of sufficiently many additional integrals in involution [184, 256].

Remark 3. For $p = 0$, vector fields (1.17) take the form $X^\mu = (\partial F/\partial y)\partial/\partial x$ and are completely integrable in the Liouville sense (see Theorem 1.2).

The Hamiltonian coisotropic (n, p, s) context. Here we are concerned with coisotropic invariant n-tori of s-parameter families of Hamiltonian vector fields with $n-p$ degrees of freedom ($n \geq 3$, $0 < p < n/2$). *The codimension of the tori in the phase space is equal to* $n-2p$. The case $p = n/2$ is excluded because invariant tori in the Hamiltonian context lie in the level hypersurfaces of the Hamilton function (= energy) and have therefore positive codimension. For n odd and $p = (n-1)/2$, each invariant torus is a connected component of an energy level hypersurface. We consider the space $M = \mathrm{T}^n \times \mathrm{R}^{n-2p}$ with coordinates (x, y) and a symplectic structure ω^2 with constant coefficients, each torus $\{y = const\}$ being coisotropic with respect to ω^2. Identify the tangent spaces $T_{x_0}\mathrm{T}^n = U \cong \mathrm{R}^n$ to T^n at all the points $x_0 \in \mathrm{T}^n$ and the tangent spaces $T_{y_0}\mathrm{R}^{n-2p} = V \cong \mathrm{R}^{n-2p}$ to R^{n-2p} at all the points $y_0 \in \mathrm{R}^{n-2p}$. The structure ω^2 can be treated as a nondegenerate skew-symmetric bilinear form on $U \oplus V$ such that the $(n-2p)$-dimensional skew-orthogonal complement $U^\perp$ of plane U lies in U. We will denote this bilinear form by the same symbols ω^2. The bilinear form $\omega^2(\mathbf{v}, \mathbf{u})$, where $\mathbf{u} \in U^\perp$ and $\mathbf{v} \in V$, is nondegenerate (otherwise the whole form ω^2 would be degenerate).

As usual by V^* we denote the dual of V, then defining a linear mapping $\mathbf{f} : V^* \to U^\perp$ by

$$\omega^2(\mathbf{v}, \mathbf{f}(\xi)) = \xi(\mathbf{v}) \quad \forall \mathbf{v} \in V$$

for $\xi \in V^*$. Consider a family

$$H = H^\mu(y) \tag{1.19}$$

of integrable Hamiltonians on M and the corresponding family of Hamiltonian vector fields $X = \{X^\mu\}_\mu$.

Lemma 1.4 *For each value of μ and at each point of M*

$$X = \mathbf{f}\,(dH|_V) \in U^\perp.$$

Proof. Let $\mathbf{f}\,(dH|_V) = \mathbf{a}$. Then $\forall \mathbf{u} \in U$ $\forall \mathbf{v} \in V$

$$dH(\mathbf{u} + \mathbf{v}) = dH(\mathbf{v}) = \omega^2(\mathbf{v}, \mathbf{a}) = \omega^2(\mathbf{u} + \mathbf{v}, \mathbf{a})$$

(the first equality in this chain follows from $\partial H/\partial x \equiv 0$ and the last one, from $\mathbf{a} \in U^\perp$). Consequently, $\mathbf{a} = X$. □

Thus, all the tori $\{y = const\}$ are invariant under the flow of X^μ for each μ and carry parallel dynamics with frequency vectors $\omega(y, \mu) = \mathbf{f}\,(dH^\mu(y)|_V)$ (after the identification of U and R^n). So, we obtain an $(n-2p+s)$-parameter analytic family of coisotropic invariant n-tori with parallel dynamics (in fact, the space $M \times P$ is foliated into those tori). On the other hand, all the frequency vectors of the tori lie in the fixed $(n-2p)$-dimensional subspace $U^\perp$ of R^n (any vector in this subspace can be realized as the frequency vector for a suitable Hamiltonian). The subspace $U^\perp$ is determined by the symplectic structure ω^2 only and does not depend on the Hamilton function H. The behavior of Hamiltonian systems on M is very sensitive to the arithmetical properties of the arrangement of the plane $U^\perp$ with respect to the lattice $\mathbf{Z}^n$. For instance, it is possible that all the vectors in $U^\perp$ are resonant (e.g., if $U^\perp$ is one of the coordinate planes). The alternative possibility for $1 \le p < (n-1)/2$ is that both resonant vectors and nonresonant ones constitute everywhere dense subsets of $U^\perp$. For $p = (n-1)/2$ [of course, in this case n is odd] the alternative possibility is that all the nonzero vectors in $U^\perp$ are *nonresonant* [$U^\perp$ is a straight line for $p = (n-1)/2$]. We see that the cases $1 \le p < (n-1)/2$ and $p = (n-1)/2$ within the Hamiltonian coisotropic (n, p, s) context are rather different.

As the arithmetical properties of the space $U^\perp$ are different for different forms ω^2, there is no universal normal form for the symplectic structure in the Hamiltonian coisotropic context [like (1.13) in the Hamiltonian isotropic context].

Coisotropic invariant tori of Hamiltonian systems are encountered, e.g., in the quasi-classical theory of motion of a conduction electron [264]. It turns out that the motion of a conduction electron in an electric and magnetic field can proceed along four-dimensional coisotropic invariant tori in the six-dimensional phase space $T^*\mathbf{R}^3$. Here $n = 4$, $p = 1$. For homogeneous fields, an open domain in the phase space is smoothly foliated into such tori.

Remark. It is worthwhile to note that throughout this book, we always suppose that the structure $\mathfrak{S}$ on the phase space M is independent of the parameter μ labeling the vector fields. This assumption is usually not essential but in the Hamiltonian coisotropic context, it is severe greatly simplifying the dynamics and the perturbation theory. Indeed, if the structure ω^2 depends on μ then the space $U^\perp$ of allowed frequency vectors also depends on μ. Thus, in this case the integrable (and consequently nonintegrable) dynamics on M for different values of μ would be generically quite different (however, the sets of the μ values corresponding to various types of dynamics are determined in terms of ${\omega^2}_\mu$ and do not depend on a particular Hamiltonian).

The reversible context

Now we consider invariant n-tori of s-parameter families of reversible vector fields. Here the phase space is $M = \mathbf{T}^n \times \mathbf{R}^{u+v}$ with coordinates (x, w^+, w^-) where $x \in \mathbf{T}^n$, $w^+ \in L^+ = \mathbf{R}^u$, $w^- \in L^- = \mathbf{R}^v$, and where the involution G has the form

$$G : (x, w^+, w^-) \mapsto (-x, w^+, -w^-), \tag{1.20}$$

being of type $(n+v, u)$. An integrable s-parameter family $X = \{X^\mu\}_\mu$ of G-reversible vector fields on M then has the form

$$X^\mu(w^+, w^-, \mu) = \varpi(w^+, w^-, \mu)\frac{\partial}{\partial x} + W^+(w^+, w^-, \mu)\frac{\partial}{\partial w^+} + W^-(w^+, w^-, \mu)\frac{\partial}{\partial w^-} \tag{1.21}$$

where the functions ϖ and W^- are even in w^- whereas the function W^+ is odd in w^- [and therefore $W^+(w^+, 0, \mu) \equiv 0$]. We suppose that the torus $\{w^+ = 0, w^- = 0\}$ is invariant under the flow of field X^0, i.e., $W^-(0,0,0) = 0$. An n-torus $\{w^+ = w_0^+, w^- = w_0^-\}$ is invariant under involution (1.20) if and only if $w_0^- = 0$, and for $w_0^- = 0$ it is invariant under the flow of X^{μ_0} if and only if $W^-(w_0^+, 0, \mu_0) = 0$. Consider the mapping

$$\mathbf{R}^{u+s} \to \mathbf{R}^v, \quad (w^+, \mu) \mapsto W^-(w^+, 0, \mu). \tag{1.22}$$

The preimage of zero under this mapping is generically empty for $u + s < v$. This means that for $u + s < v$ a generic s-parameter family of integrable G-reversible vector fields on M admits no invariant n-tori: even if some particular family possesses such tori, the latter can be destroyed by an arbitrarily small perturbation of the family (within the integrable realm!). Now consider the case $u + s \geq v$ which, in turn, splits into two quite different subcases: $u \geq v$ and $s \geq v - u > 0$. These will be referred to as the *reversible context 1* and the *reversible context 2*, respectively. In the first subcase, we will write $v = p$, $u = m + p$. In the second subcase, we will write $u = p$, $v = m + p$ $(1 \leq m \leq s)$. We shall now consider these subcases separately.

Reversible (n, m, p, s) context 1. This context concerns invariant n-tori of s-parameter families of vector fields reversible with respect to involutions of type $(n + p, m + p)$ with $m \geq 0$, $p \geq 0$, $s \geq 0$. Here $w^+ \in L^+ = \mathbb{R}^{m+p}$, $w^- \in L^- = \mathbb{R}^p$. For each fixed value of μ, one can generically introduce a new coordinate system (y, z^+) in $L^+ = \mathbb{R}^{m+p}$ $(y \in \mathbb{R}^m$, $z^+ \in \mathbb{R}^p)$ via the formula $w^+ = Q(y, z^+, \mu)$ in such a way that $Q(0, 0, 0) = 0$ and $W^-(Q(y, z^+, \mu), 0, \mu) \equiv z^+$. We will also write z^- instead of w^-. Then involution (1.20), the family of differential equations on M determined by vector fields (1.21), and mapping (1.22) take the form

$$G : (x, y, z^+, z^-) \mapsto (-x, y, z^+, -z^-), \tag{1.23}$$

$$\begin{aligned} \dot{x} &= \omega(y, \mu) + O(z^+, z^-), & \dot{z}^+ &= O(z^-) = a(y, \mu)z^- + O_2(z^+, z^-), \\ \dot{y} &= O(z^-), & \dot{z}^- &= z^+ + O_2(z^-), \end{aligned} \tag{1.24}$$

and

$$\mathbb{R}^{m+p+s} \to \mathbb{R}^p, \quad (y, z^+, \mu) \mapsto z^+, \tag{1.25}$$

respectively. For each value of μ, each torus $\{y = const, z^+ = z^- = 0\}$ is invariant under both involution G and the flow of X^μ and carries parallel dynamics with frequency vector $\omega(y, \mu)$. We thus obtain an $(m + s)$-parameter analytic family of invariant n-tori with parallel dynamics for X^μ.

In the sequel, we will prefer to consider involutions G and families of G-reversible differential equations of a slightly more general form

$$G : (x, y, z) \mapsto (-x, y, Rz) \tag{1.26}$$

and

$$\dot{x} = \omega(y, \mu) + O(z), \quad \dot{y} = O(z), \quad \dot{z} = \Omega(y, \mu)z + O_2(z), \tag{1.27}$$

where $z \in \mathbb{R}^{2p}$, R is an arbitrary fixed involutive $2p \times 2p$ matrix whose 1- and (-1)-eigenspaces are p-dimensional, and $\Omega(y, \mu)R + R\Omega(y, \mu) \equiv 0$, cf. [60]. Matrices Ω satisfying the latter equality are said to be *infinitesimally reversible* with respect to R [315, 321]. Expressions (1.23) and (1.24) correspond to

$$R = \begin{pmatrix} I & 0 \\ 0 & -I \end{pmatrix}, \quad \Omega(y, \mu) = \begin{pmatrix} 0 & a(y, \mu) \\ I & 0 \end{pmatrix}$$

[I being defined in (1.14)].

Remark. The eigenvalues of an infinitesimally reversible matrix occur in pairs $(\lambda, -\lambda)$, the root spaces corresponding to λ and $-\lambda$ having the same Jordan structure (see [161,

321, 333]). The spectrum of a generic $2p \times 2p$ infinitesimally reversible matrix has the form (1.18) provided that the 1- and (-1)-eigenspaces of the reversing involutive matrix are p-dimensional.

Reversible (n, m, p, s) context 2. This context concerns invariant n-tori of s-parameter families of vector fields reversible with respect to involutions of type $(n+m+p, p)$ with $m \geq 1$, $p \geq 0$, $s \geq m$. Here $w^+ \in L^+ = \mathbb{R}^p$, $w^- \in L^- = \mathbb{R}^{m+p}$. Generically one can split the coordinates $\mu_1, \ldots, \mu_s$ in the parameter space $\mathbb{R}^s$ into two groups $\mu = (\mu^1, \mu^2)$, $\mu^1 \in \mathbb{R}^m$, $\mu^2 \in \mathbb{R}^{s-m}$ in such a way that for each fixed value of μ^2, the mapping

$$\mathbb{R}^{p+m} \to \mathbb{R}^{m+p}, \quad (w^+, \mu^1) \mapsto W^-(w^+, 0, \mu^1, \mu^2) \tag{1.28}$$

is a local diffeomorphism near the point $(w^+ = 0, \mu^1 = 0)$, the preimage of zero being $(w^+ = \xi(\mu^2), \mu^1 = \zeta(\mu^2))$, $\xi(0) = 0$, $\zeta(0) = 0$. Shifting the variables w^+ and μ^1, if necessary, one can achieve $\xi \equiv 0$, $\zeta \equiv 0$. For each μ^2, introduce the new coordinate system (z^-, y) in $L^- = \mathbb{R}^{m+p}$ ($z^- \in \mathbb{R}^p$, $y \in \mathbb{R}^m$) via the formula

$$w^- = Q(z^-, y, \mu^2) = \frac{\partial W^-(0,0,0,\mu^2)}{\partial w^+} z^- + \frac{\partial W^-(0,0,0,\mu^2)}{\partial \mu^1} y.$$

Note that the coordinates z^- and y depend on the initial coordinate w^- *linearly*, so that multiplying w^- by -1 is equivalent to multiplying z^- and y simultaneously by -1. In the new coordinate system, mapping (1.28) takes the form

$$(w^+, \mu^1) \mapsto (z^-(w^+, \mu^1, \mu^2), y(w^+, \mu^1, \mu^2)) = (w^+ + A(w^+, \mu^1, \mu^2), \mu^1 + B(w^+, \mu^1, \mu^2)),$$

where $A = O_2(w^+, \mu^1)$, $B = O_2(w^+, \mu^1)$. Let also $z^+ = w^+ + A(w^+, \mu^1, \mu^2)$ be the new coordinate in $L^+ = \mathbb{R}^p$. Then involution (1.20), the family of differential equations on M determined by vector fields (1.21), and mapping (1.22) take the form

$$G : (x, z^+, z^-, y) \mapsto (-x, z^+, -z^-, -y), \tag{1.29}$$

$$\begin{aligned} \dot{x} &= \omega(\mu^2) + O(z^+, z^-, y, \mu^1), \\ \dot{z}^+ &= O(z^-, y) = a(\mu)z^- + b(\mu)y + O_2(z^+, z^-, y), \\ \dot{z}^- &= z^+ + O_2(z^-, y), \\ \dot{y} &= \mu^1 + O_2(z^+, z^-, y, \mu^1), \end{aligned} \tag{1.30}$$

and

$$\mathbb{R}^{p+s} \to \mathbb{R}^{m+p}, \quad (z^+, \mu) \mapsto (z^+, \mu^1 + O_2(z^+, \mu^1)), \tag{1.31}$$

respectively. For $\mu^1 = 0$ and each value of μ^2, torus $\{z^+ = z^- = 0, y = 0\}$ is invariant under both involution G and the flow of $X^\mu = X^{(0,\mu^2)}$ and carries parallel dynamics with frequency vector $\omega(\mu^2)$. We thus obtain an $(s-m)$-parameter analytic family of invariant n-tori with parallel dynamics for X^μ.

In this and the next sections, we follow closely our paper [61].

1.4 Summary

1.4.1 A heuristic principle

To summarize, in all the above contexts we have found l-parameter *analytic* families of invariant n-tori with parallel dynamics for some $l \in \mathbb{Z}_+$ provided that the vector fields X^μ are *integrable*. The question is what will happen to these families if one perturbs the (families of) vector fields. Attempting to answer this question, first suppose that the perturbation is *still integrable*, so that it just slightly shifts the initial family of tori. In essence, to perturb X^μ within the integrable realm means to change $\omega(\mu)$ and $\Omega(\mu)$ in (1.11), $\omega(y,\mu)$ in (1.12), $F(y,\mu)$ and $K(y,\mu)$ in (1.16), $H^\mu(y)$ in (1.19), $\omega(y,\mu)$ and $\Omega(y,\mu)$ in (1.27), or $\omega(\mu^2)$, $a(\mu)$ and $b(\mu)$ in (1.30). At first glance, this would not lead to anything interesting. However, when examining the behavior of the frequency vectors of the tori under the perturbation, three following different situations can be met:

(i) all the tori in the unperturbed family are nonresonant, and this property is preserved by perturbations;

(ii) for a generic unperturbed family of vector fields, some tori are resonant, some are not, and this property is preserved by perturbations;

(iii) by an arbitrarily small perturbation, one can make all the tori resonant.

Situation (i) always takes place for $n = 0$ and $n = 1$, since equilibria and closed trajectories treated as invariant tori with parallel dynamics of dimensions 0 and 1, respectively, are always nonresonant. This situation also occurs in the Hamiltonian coisotropic (n,p,s) context when $p = (n-1)/2$ and the straight line $U^\perp$ does not lie in any resonant hyperplane in $\mathbb{R}^n$.

Situation (iii) corresponds to $n > 1$ and $l = 0$, i.e., to the cases where $n > 1$ and the family $\{X^\mu\}_\mu$ of integrable vector fields possesses a single n-torus [this takes place in the dissipative (n,p,s) context for $s = 0$, in the volume preserving (n,p,s) context for $p > 1$, $s = 0$, in the reversible (n,m,p,s) context 1 for $m = s = 0$, and in the reversible (n,m,p,s) context 2 for $s = m$]. By an arbitrarily small perturbation, it is possible to make that single torus resonant. However, situation (iii) is also realized in the Hamiltonian coisotropic (n,p,s) context (with $l = n - 2p + s \geq 1$) when the subspace $U^\perp$ lies in one of the resonant hyperplanes. In the latter case, not only one can make all the tori resonant by arbitrarily small perturbations, but also all the tori of the unperturbed family of vector fields are always resonant.

The "most familiar" situation is (ii). It occurs in all the cases except those indicated above as pertaining to situations (i) or (iii).

Now we can formulate the following heuristic principle describing the fate of analytic families of invariant tori with parallel dynamics of the initial family of vector fields under *nonintegrable* perturbations:

A small *generic* perturbation in the respective situations:

(i) preserves the initial family of tori and leaves it analytic;

(ii) preserves the family of tori, but makes it Cantor-like;

(iii) destroys the initial family of tori completely.

Remark 1. In the first two situations, the unperturbed family of vector fields is assumed to satisfy some nondegeneracy and nonresonance conditions. Moreover, in the first two situations, all the invariant tori with parallel dynamics of the perturbed family of vector fields are nonresonant (i.e., quasi-periodic), and even Diophantine.

Remark 2. In the sequel, while discussing any property of families of vector fields, we will use the terms "generic" and "typical" in the following sense. The term *generic* means that families of vector fields with the property indicated constitute an open everywhere dense set in the space $\mathfrak{X}^\omega_s$ of all the families of vector fields, whereas the term *typical* means that families of vector fields with the property indicated constitute a set in the space $\mathfrak{X}^\omega_s$ with non-empty interior.

As we saw before in § 1.2.1, a generic dynamical system admits no resonant tori (a corollary of the Kupka–Smale theorem, cf. [260, 265]), so it is not surprising that generically perturbed families of vector fields no longer possess invariant tori with parallel dynamics in situation (iii) or l-parameter *analytic* families of invariant tori with parallel dynamics in situation (ii). Note, however, that in situation (iii) with $l = 0$, i.e., when the unperturbed family of vector fields has a single n-torus with parallel dynamics, a perturbation does not destroy this torus as an invariant submanifold: a perturbed family still has an invariant torus close to the unperturbed one [67, 115, 158, 356], but the dynamics on the perturbed torus is generically no longer parallel (cf. § 1.2.1). What is highly unexpected is that a perturbed family of vector fields does possess many quasi-periodic invariant tori in situations (i) and (ii). However, the principle above is indeed true "as the first approximation". To make the latter statement more precise, let us find out what this principle tells in each context for nontrivial torus dimensions $n > 1$:

(α) The dissipative (n, p, s) context: a *generic* family of vector fields has no invariant n-tori with parallel dynamics for $s = 0$ (i.e., when there are no parameters) while a *typical* family of vector fields possesses s-parameter Cantor-like families of Diophantine invariant n-tori for $s \geq 1$.

(β) The volume preserving (n, p, s) context with $p > 1$: a *generic* family of globally divergence-free vector fields has no invariant n-tori with parallel dynamics for $s = 0$ (i.e., when there are no parameters) while a *typical* family of globally divergence-free vector fields possesses s-parameter Cantor-like families of Diophantine invariant n-tori for $s \geq 1$.

(γ) The volume preserving (n, p, s) context with $p = 1$: a *typical* family of globally divergence-free vector fields possesses $(s+1)$-parameter Cantor-like families of Diophantine invariant n-tori.

(δ) The Hamiltonian isotropic (n, p, s) context: a *typical* family of Hamiltonian vector fields possesses $(n + s)$-parameter Cantor-like families of Diophantine isotropic invariant n-tori.

(ϵ) The Hamiltonian coisotropic (n, p, s) context: depending on the global properties of the symplectic structure on the phase space, either **1)** a *generic* family of Hamiltonian vector fields has no coisotropic invariant n-tori with parallel dynamics, or **2)** a *typical* family of Hamiltonian vector fields possesses $(n - 2p + s)$-parameter Cantor-like [for $1 \leq$

$p < (n-1)/2$] or analytic [for $p = (n-1)/2$] families of Diophantine coisotropic invariant n-tori.

(ζ) The reversible (n, m, p, s) context 1: a *generic* family of reversible vector fields has no invariant n-tori with parallel dynamics for $m = s = 0$ while a *typical* family of reversible vector fields possesses $(m+s)$-parameter Cantor-like families of Diophantine invariant n-tori for $m + s \geq 1$.

(η) The reversible (n, m, p, s) context 2: a *generic* family of reversible vector fields has no invariant n-tori with parallel dynamics for $s = m$ while a *typical* family of reversible vector fields possesses $(s-m)$-parameter Cantor-like families of Diophantine invariant n-tori for $s > m$.

Remark. Like statement (δ) for $p = 0$, one observes that statement (ϵ) *for any* $p < (n-1)/2$ implies that non-ergodicity is a typical property of Hamiltonian systems (in both cases, the union of the tori has positive Lebesgue measure).

Are these seven statements (α)–(η) actually true? We start with the least studied cases. To the best of the authors' knowledge, the reversible context 2 has not been considered yet in the literature, and nothing can be said about quasi-periodic invariant tori in this context [although statement (η) above seems very likely].

1.4.2 The Hamiltonian coisotropic context

Statement (ϵ) above concerning the Hamiltonian coisotropic context is true (however, that the space $U^{\perp}$ does not lie in any resonant hyperplane in R^n does not guarantee the presence of many quasi-periodic coisotropic invariant n-tori in small perturbations of generic integrable Hamiltonians; one should also impose some Diophantine conditions on $U^{\perp}$). More precisely, the following two theorems hold (in the formulations of these theorems, we will suppose for simplicity that $s = 0$, i.e., there are no external parameters). Let Y be some finite open domain in R^{n-2p}.

Theorem 1.5 {Parasyuk [263, 264] and Herman (see [361])} *Let $n \geq 4$, $1 \leq p \leq (n-2)/2$, the symplectic structure ω^2 with constant coefficients on $M = \{(x, y)\} = T^n \times Y \subset T^n \times R^{n-2p}$ satisfy some Diophantine conditions (the set of structures which do not meet those conditions is of measure zero), and the unperturbed integrable Hamiltonian $H(y)$ satisfy some nondegeneracy conditions. Then any Hamiltonian vector field on M with Hamilton function $H(y)+\Delta(x, y)$ sufficiently close to $H(y)$ possesses Floquet Diophantine coisotropic invariant analytic n-tori, the $2(n-p)$-dimensional Lebesgue measure of the union of these tori tending to $\mathrm{meas}_{2(n-p)} M = (2\pi)^n \mathrm{meas}_{n-2p} Y$ as the perturbation magnitude tends to 0. The Floquet $(n-2p) \times (n-2p)$ matrix of each of the tori is zero.*

Theorem 1.6 (Herman [155, 156]) *Let $n \geq 3$ be odd, $p = (n-1)/2$, the symplectic structure ω^2 with constant coefficients on $M = \{(x, y)\} = T^n \times Y \subset T^n \times R$ satisfy some Diophantine conditions (the set of structures which do not meet those conditions is of measure zero), and the unperturbed integrable Hamiltonian $H(y)$ satisfy the nondegeneracy condition $dH/dy \neq 0$. Then each energy level of any Hamilton function $H(y) + \Delta(x, y)$ sufficiently close to $H(y)$ is a Floquet Diophantine coisotropic invariant n-torus of the corresponding Hamiltonian vector field. The frequency vectors of these tori are proportional to one and the same vector ω^0 which depends on neither the torus nor the Hamiltonian* *but is determined by the symplectic structure only. The Floquet* 1×1 *matrix of each of the tori is zero.*

In the context of Theorem 1.6, we have the following amazing picture [155, 156]. The energy levels of the unperturbed Hamiltonian $H(y)$ are n-tori with Diophantine motion, all the frequency vectors being proportional to some fixed vector ω^0. Now we perturb this Hamiltonian arbitrarily. Obviously, the energy levels of a perturbed Hamiltonian will be still n-tori close to the unperturbed ones, but it turns out that the motion on those perturbed tori *will be still Diophantine* with frequency vectors proportional to the same vector ω^0! In fact, all the Hamilton functions close to $H(y)$ are integrable. This phenomenon is a direct consequence of the fact that the Hamiltonian nature of a vector field imposes very severe restrictions on the motion on invariant tori of small codimensions.

For exact formulations of Diophantine conditions for ω^2 and nondegeneracy conditions for $H(y)$, we refer the reader to the original papers [155, 156, 263, 264], see also [361] for the case $p = (n-2)/2$.

1.5 Small divisors and Whitney-smoothness in 1-bite problems

In this section we consider the so called 1-bite small divisor problem, which has a linear character and can be solved directly by the Fourier series. In accordance with most of the formulations in the sequel we restrict ourselves to the setting of vector fields. In the more complicated nonlinear small divisor problems below, such 1-bite problems occur in a (Newtonian) iteration process (compare also Section 1.2 above where we noticed this correspondence for circle mappings). First we give some examples where the problems will be formally solved. After this we deal with the convergence and the regularity of the solutions in all the variables they depend on. Here the concept of Whitney-smoothness will be met in technical detail.

Indeed, instead of the homological equation (1.8) in § 1.2.1 we shall consider a linear partial differential equation of the form

$$\left\langle \omega, \frac{\partial u}{\partial x} \right\rangle = f \tag{1.32}$$

where $f = f(x, \omega)$ is a given function in $x = (x_1, \ldots, x_n)$ modulo $2\pi \in \mathrm{T}^n$ and $\omega = (\omega_1, \ldots, \omega_n) \in \mathrm{R}^n$. Assuming the T^n-average $[f(\cdot, \omega)]$ of function f to vanish identically (with respect to ω), we try to solve this equation for $u = u(x, \omega)$.

1.5.1 Examples and formal solutions

As a motivation to the above problem we consider two examples, that are also interesting for their own sake.

Reducibility to the Floquet form for codimension 1 tori

We here return to the *reducibility problem* mentioned in § 1.1.1. On $\mathrm{T}^n \times \mathrm{R}^m$ with coordinates $x = (x_1, \ldots, x_n)$ and $y = (y_1, \ldots, y_m)$, consider a vector field X determining the system (1.3) of differential equations

$$\begin{aligned}\dot{x} &= \omega + O(y)\\ \dot{y} &= \Omega(x)y + O_2(y)\end{aligned}$$

with $\omega \in \mathrm{R}^n$ and $\Omega : \mathrm{T}^n \to gl(m, \mathrm{R})$. The vector field X has therefore $\mathrm{T}^n \times \{0\}$ as an invariant n-torus with parallel dynamics and frequency vector ω. The problem then is to find a transformation of $\mathrm{T}^n \times \mathrm{R}^m$, e.g., of the form $(x, y) \mapsto (x, z) = (x, A(x)y)$ for some $A : \mathrm{T}^n \to GL(m, \mathrm{R})$, such that in the transformed system

$$\begin{aligned}\dot{x} &= \omega + O(z)\\ \dot{z} &= \Lambda z + O_2(z)\end{aligned}$$

the matrix $\Lambda \in gl(m, \mathrm{R})$ is x-independent, i.e., this system is in the Floquet form [cf. (1.2)]. For more details here see, e.g., [5, 12, 42, 62, 64, 148, 149, 162, 167, 168, 170, 171, 184]. From, e.g., papers [64, 148, 149] it is known that for $n, m \geq 2$ reducibility does not always take place. Also it is well known, see [5, 12], that for $m = 1$ and arbitrary n reducibility does hold, provided that ω is Diophantine (cf. the discussion in § 1.1.1). The proof of this fact involves a 1-bite small divisor problem as we shall show now.

Indeed, if we apply the transformation $(x, y) \mapsto (x, z) = (x, A(x)y)$ to our vector field X, we get the equation

$$\begin{aligned}\dot{z} &= A\dot{y} + \dot{A}y = A\Omega y + \left(\sum_{j=1}^{n} \omega_j \frac{\partial A}{\partial x_j}\right) y + O_2(y)\\ &= \left\{A\Omega A^{-1} + \left\langle \omega, \frac{\partial A}{\partial x}\right\rangle A^{-1}\right\} z + O_2(z),\end{aligned}$$

and we just require the matrix in the curly braces to be constant. For $m = 1$ the matrix product commutes, and our problem is therefore to make all the nonzero Fourier coefficients of the function

$$\Omega + \left\langle \omega, \frac{\partial \log A}{\partial x}\right\rangle$$

vanish:

$$\frac{\partial}{\partial x}\left(\Omega + \left\langle \omega, \frac{\partial \log A}{\partial x}\right\rangle\right) = 0. \tag{1.33}$$

As before [cf. (1.8)], we approach this problem by expanding Ω and $\log A$ in the Fourier series:

$$\Omega(x) = \sum_{k \in Z^n} \Omega_k e^{i\langle x, k\rangle}, \quad \log A(x) = \sum_{k \in Z^n} B_k e^{i\langle x, k\rangle}$$

and comparing the nonzero Fourier coefficients:

$$\Omega_k + i\langle \omega, k\rangle B_k = 0.$$

This yields for $k \in \mathrm{Z}^n \setminus \{0\}$

$$B_k = \frac{i\Omega_k}{\langle \omega, k\rangle},$$

while B_0 is free to choose. First note that for general Ω a formal solution exists if and only if the frequency vector ω is nonresonant. However, even then the convergence of the formal series

$$\sum_{k\in\mathbb{Z}^n} B_k e^{i\langle x,k\rangle}$$

is problematic, since the denominators $\langle\omega,k\rangle$, $k\in\mathbb{Z}^n\setminus\{0\}$, have 0 as an accumulation point (if $n>1$). This is just the small divisor problem as met in § 1.2.1, but since no further iteration is involved we speak of a 1-bite problem. It will be solved below in § 1.5.2.

Quasi-periodic responses: the linear case

In the second example we return to the problem presented in § 1.2.3, regarding response solutions in a quasi-periodically forced oscillator, here restricting to the linear case. To be more precise, we consider the linear second order differential equation

$$\ddot{y}+c\dot{y}+ay=f(t,a,c)$$

with $y,\ t\in\mathbb{R}$, where a and c with $a>0$ and $c^2<4a$ serve as parameters. As before f is assumed to be quasi-periodic in t, i.e., $f(t,a,c)=F(t\omega_1,t\omega_2,\ldots,t\omega_n,a,c)$ for some function $F:\mathbb{T}^n\times\mathbb{R}^2\to\mathbb{R}$ and a fixed nonresonant frequency vector $\omega\in\mathbb{R}^n$. Again the problem is to find a response solution, which now leads to a 1-bite small divisor problem as follows.

The corresponding vector field on $\mathbb{T}^n\times\mathbb{R}^2$ in this case determines the system of differential equations

$$\begin{aligned}\dot{x}&=\omega\\ \dot{y}&=z\\ \dot{z}&=-ay-cz+F(x,a,c),\end{aligned}$$

where $x=(x_1,x_2,\ldots,x_n)$ modulo 2π. We are looking for an invariant torus which is a graph $(y,z)=(y(x,a,c),z(x,a,c))$ over $\mathbb{T}^n$ with dynamics $\dot{x}=\omega$.

For technical convenience we complexify the problem as follows:

$$\lambda=-\frac{c}{2}+i\sqrt{a-\frac{c^2}{4}}\quad\text{and}\quad\zeta=z-\bar{\lambda}y.$$

Notice that $\lambda^2+c\lambda+a=0$, $\operatorname{Im}\lambda>0$, and

$$\ddot{y}+c\dot{y}+ay=\left[\frac{d}{dt}-\lambda\right]\left[\frac{d}{dt}-\bar{\lambda}\right]y.$$

Then our differential equation gets the form

$$\begin{aligned}\dot{x}&=\omega\\ \dot{\zeta}&=\lambda\zeta+F(x,\lambda),\end{aligned}$$

where we use complex multiplication and identify $\mathbb{R}^2\cong\mathbb{C}$. The desired torus-graph $\zeta=\zeta(x,\lambda)$ with dynamics $\dot{x}=\omega$ now has to be obtained from the linear equation

$$\left\langle\omega,\frac{\partial\zeta}{\partial x}\right\rangle=\lambda\zeta+F, \tag{1.34}$$

as one easily verifies. Again we expand F and ζ in the Fourier series

$$F(x,\lambda)=\sum_{k\in Z^n} F_k(\lambda)e^{i\langle x,k\rangle}, \quad \zeta(x,\lambda)=\sum_{k\in Z^n}\zeta_k(\lambda)e^{i\langle x,k\rangle}$$

and compare the coefficients. It then appears that for all $k\in Z^n$ we get

$$\zeta_k(\lambda)=\frac{F_k(\lambda)}{i\langle\omega,k\rangle-\lambda}.$$

Here the denominators vanish at the resonance points $\lambda=i\langle\omega,k\rangle$, $k\in Z^n$, so in a dense subset of the imaginary λ-axis (if $n>1$). In general a formal solution exists if and only if λ is not in this set, and as before the convergence is problematic because of the corresponding small divisors. Nevertheless we here only have a 1-bite problem.

1.5.2 Convergence and Whitney-smoothness

We now arrive at the representative equation (1.32), inspired by (1.33) and (1.34), which is of the 1-bite small divisor type. To be precise, let $T^n\times R^n$ have coordinates $x=(x_1,\ldots,x_n)\in T^n$ and $\omega=(\omega_1,\ldots,\omega_n)\in R^n$. Then, given $f: T^n\times R^n\to R$ with the T^n-average $[f(\cdot,\omega)]\equiv 0$, we consider the linear partial differential equation (1.32)

$$\left\langle \omega,\frac{\partial u}{\partial x}\right\rangle = f$$

for a function $u: T^n\times R^n\to R$. Again we expand f in the Fourier series

$$f(x,\omega)=\sum_{k\neq 0} f_k(\omega)e^{i\langle x,k\rangle}, \tag{1.35}$$

then for u one obtains

$$u(x,\omega)=u_0(\omega)+\sum_{k\neq 0}\frac{f_k(\omega)}{i\langle\omega,k\rangle}e^{i\langle x,k\rangle}, \tag{1.36}$$

the constant term $u_0(\omega)$ being arbitrary. As before, we have to confine ourselves with nonresonant ω's. However, now go further and impose the *Diophantine* conditions (1.1) on ω: for *fixed* $\tau>n-1$ and $\gamma>0$ we require for all $k\in Z^n\setminus\{0\}$ that

$$|\langle\omega,k\rangle|\geq\gamma|k|^{-\tau}. \tag{1.37}$$

The corresponding subset of R^n denoted in the sequel by R^n_γ possesses the following properties.

(i) If $\omega\in R^n_\gamma$ and $c\geq 1$, then also $c\omega\in R^n_\gamma$; so the set R^n_γ is the union of closed half lines (rays).

(ii) Let $S^{n-1}\subset R^n$ be the unit sphere, then $S^{n-1}\cap R^n_\gamma$ is the union of a Cantor set and a countable set.

(iii) The complement of this Cantor set in S^{n-1} has Lebesgue measure of the order of γ as $\gamma\downarrow 0$.

Property (i) is obvious. Property (ii) follows from the Cantor–Bendixson theorem, see, e.g., Hausdorff [147], since $\mathbb{R}^n_\gamma$ and hence $S^{n-1} \cap \mathbb{R}^n_\gamma$ is a closed set. In fact this theorem splits every closed subset of a metric space with a countable base into a countable set and a perfect set. Here the perfect set is totally disconnected and therefore a Cantor set. Indeed, the intersections of S^{n-1} with the hyperplanes $\langle \omega, k \rangle = 0$, $k \in \mathbb{Z}^n \setminus \{0\}$, densely fill the complement of $S^{n-1} \cap \mathbb{R}^n_\gamma$ in S^{n-1}. Cantor sets are topologically small (nowhere dense), but this one by (iii) has large measure as soon as $\gamma > 0$ is small. For similar arguments concerning Cantor sets on $\mathbb{R}$, see § 1.2.1. The verification of property (iii) is straightforward, cf. [337] (see also much more general statements in § 2.5.1 and Section 6.4).

Observe that $\mathbb{R}^n_\gamma$ is a Whitney-smooth foliation of closed half lines (or more generally smooth manifolds with boundary) parametrized over a Cantor set. Colloquially we will call such a foliation a "Cantor set".

We shall consider the solution $u = u(x, \omega)$ of our 1-bite problem for $(x, \omega) \in \mathbb{T}^n \times \mathbb{R}^n_\gamma$. The regularity in the ω-direction is problematic since $\mathbb{R}^n_\gamma$ does not contain any interior points. For a function on such a closed set, however, the concept of *Whitney-smoothness* is appropriate [160, 278, 338, 355] (see also § 6.1.4 below). In this case it means that the function u can be extended as an ordinary smooth function to a neighborhood of the set $\mathbb{T}^n \times \mathbb{R}^n_\gamma$ in $\mathbb{T}^n \times \mathbb{R}^n$.

Theorem 1.7 (Whitney-smooth solutions of the 1-bite problem) *Let the function (1.35) $f : \mathbb{T}^n \times \mathbb{R}^n \to \mathbb{R}$ be analytic in all its arguments. Then the function (1.36) $u = u(x, \omega)$ is analytic in x and of class Whitney-C^∞ in ω, where (x, ω) varies over $\mathbb{T}^n \times \mathbb{R}^n_\gamma$.*

Proof. The rest of this section is devoted to the proof of Theorem 1.7. At the end we include a remark on the case where f is less regular. At this moment we emphasize that the Whitney extension of u to a neighborhood of $\mathbb{T}^n \times \mathbb{R}^n_\gamma$ in $\mathbb{T}^n \times \mathbb{R}^n$ is not unique and that outside $\mathbb{T}^n \times \mathbb{R}^n_\gamma$ in general it *will no longer be a solution* of the 1-bite problem in question. The proof uses approximability properties of Whitney-smooth functions by analytic ones in the same way as was done by Zehnder [364] and Pöschel [278].

In the course of the proof, we will denote by $|\cdot|$ the maximum (l_∞) norm of independent variables ranging in $\mathbb{R}^n$ and $\mathbb{C}^n$ (so that $|x| = \max_{j=1}^n |x_j|$, $|\omega| = \max_{j=1}^n |\omega_j|$), but *not* in $\mathbb{Z}^n$, where $|\cdot|$ will still denote the l_1-norm: $|k| = \sum_{j=1}^n |k_j|$.

The Paley–Wiener lemma and a set-up of the proof. Let $\Gamma \subset \mathbb{R}^n$ be a compact domain and put $\Gamma_\gamma = \Gamma \cap \mathbb{R}^n_\gamma$. By analyticity f has a holomorphic extension to a neighborhood of $\mathbb{T}^n \times \Gamma$ in $(\mathbb{C}/2\pi\mathbb{Z})^n \times \mathbb{C}^n$. For some constants $\kappa, \rho \in]0,1[$ this neighborhood contains the set $\mathrm{cl}\left((\mathbb{T}^n + \kappa) \times (\Gamma + \rho)\right)$ where "cl" is for the closure and

$$\begin{aligned} \Gamma + \rho &:= \bigcup_{\omega \in \Gamma} \{\omega' \in \mathbb{C}^n \, : \, |\omega' - \omega| < \rho\}, \\ \mathbb{T}^n + \kappa &:= \bigcup_{x \in \mathbb{T}^n} \{x' \in (\mathbb{C}/2\pi\mathbb{Z})^n \, : \, |x' - x| < \kappa\}. \end{aligned}$$

Let M be the supremum of f over the set $\mathrm{cl}\left((\mathbb{T}^n + \kappa) \times (\Gamma + \rho)\right)$, then one has

Lemma 1.8 (Paley–Wiener) ([259], see also [12, 173]) *Let $f = f(x,\omega)$ be analytic as above, with Fourier series*

$$f(x,\omega) = \sum_{k\in \mathbb{Z}^n} f_k(\omega)e^{i\langle x,k\rangle}$$

[cf. (1.35)]. Then, for all $k \in \mathbb{Z}^n$ and all $\omega \in \mathrm{cl}\,(\Gamma + \rho)$,

$$|f_k(\omega)| \leq Me^{-\kappa|k|}.$$

Proof. By definition

$$f_k(\omega) = (2\pi)^{-n} \oint_{\mathbb{T}^n} f(x,\omega)e^{-i\langle x,k\rangle}\, dx.$$

Firstly, assume that for $k = (k_1,\ldots,k_n)$ all the entries $k_j \neq 0$. Then take $z = (z_1,\ldots,z_n)$ with

$$z_j = x_j - i\kappa\,\mathrm{sign}\,(k_j), \quad 1 \leq j \leq n,$$

and use the Cauchy theorem to obtain

$$f_k(\omega) = (2\pi)^{-n} \oint f(z,\omega)e^{-i\langle z,k\rangle}\, dz = (2\pi)^{-n} \oint_{\mathbb{T}^n} f(z,\omega)e^{-i\langle x,k\rangle-\kappa|k|}\, dx,$$

whence $|f_k(\omega)| \leq Me^{-\kappa|k|}$, as desired.

Secondly, if for some j $(1 \leq j \leq n)$ we have $k_j = 0$, we just do not shift the integration torus in that direction. □

Remark. In fact, this is the so called "easy" Paley–Wiener lemma. Its inverse is also true [12, 173, 259]: a function on $\mathbb{T}^n$ whose Fourier coefficients decay exponentially is analytic and can be extended holomorphically to the complex domain $(\mathbb{C}/2\pi\mathbb{Z})^n$ by a distance determined by the decay rate (i.e., κ in our notations).

Next, for $j \in \mathbb{Z}_+$ we consider the truncation

$$u_j(x,\omega) = \sum_{0<|k|\leq 2^j} \frac{f_k(\omega)}{i\langle\omega,k\rangle}e^{i\langle x,k\rangle} \tag{1.38}$$

of u, cf. (1.36). From the Paley–Wiener lemma (Lemma 1.8) it follows directly that for each fixed $\omega \in \mathbb{R}^n_\gamma$, the sequence $\{u_j(\cdot,\omega)\}_{j\geq 1}$ converges to a holomorphic function. In order to study the ω-dependence of this limit we consider the geometric sequence

$$r_j = \gamma 2^{-(j\tau+j+1)}, \quad j \in \mathbb{N}$$

and the corresponding complex domains $\Gamma_\gamma + r_j$ $(j \in \mathbb{N})$. Notice that for j sufficiently large $\Gamma_\gamma + r_j \subset \Gamma + \rho$. The next result ensures that, for j sufficiently large, the truncation u_j is a well defined holomorphic function on $(\mathbb{T}^n + \kappa) \times (\Gamma_\gamma + r_j)$.

Proposition 1.9 *For all $\omega \in \Gamma_\gamma + r_j$ and all $k \in \mathbb{Z}^n$ with $0 < |k| \leq 2^j$, one has*

$$|\langle\omega,k\rangle| \geq \tfrac{1}{2}\gamma|k|^{-\tau}.$$

Proof. For $\omega \in \Gamma_\gamma + r_j$ there exists $\omega^* \in \Gamma_\gamma$ such that $|\omega - \omega^*| < r_j$. It then follows for all k with $0 < |k| \leq 2^j$ that

$$\begin{aligned} |\langle \omega, k\rangle| &\geq |\langle \omega^*, k\rangle| - |\omega - \omega^*|\,|k| \\ &\geq \gamma|k|^{-\tau} - r_j 2^j = \gamma|k|^{-\tau} - \tfrac{1}{2}\gamma(2^j)^{-\tau} \geq \tfrac{1}{2}\gamma|k|^{-\tau}. \end{aligned}$$

□

It is now our aim to show the following. On the sequence of domains

$$\{(\mathrm{T}^n + \tfrac{1}{2}\kappa) \times (\Gamma_\gamma + r_j)\}_{j\geq 1},$$

which shrinks geometrically to $(\mathrm{T}^n + \frac{1}{2}\kappa) \times \Gamma_\gamma$, the corresponding sequence $\{u_j\}_{j\geq 1}$ satisfies the conditions of the Inverse Approximation Lemma 6.14 (see § 6.1.4). To be precise, for arbitrary $\alpha \in \mathrm{R}_+ \setminus \mathrm{Z}_+$, there exists a constant $N_\alpha > 0$ such that for all $j \in \mathrm{N}$ and all $(x, \omega) \in (\mathrm{T}^n + \frac{1}{2}\kappa) \times (\Gamma_\gamma + r_j)$, one has

$$|(u_j - u_{j-1})(x, \omega)| \leq N_\alpha r_j^\alpha. \tag{1.39}$$

This property yields that the pointwise limit $u(x, \omega)$ of u_j is Whitney-C^α. Since $\alpha \notin \mathrm{N}$ is arbitrary, this implies that $u(x, \omega)$ is Whitney-C^∞. Notice again that u is analytic in x by the uniformity of the convergence.

The estimates. It now remains to carry out the estimates sufficient for (1.39). We will show that on the relevant domain, $|u_j - u_{j-1}|$ decays superexponentially as $j \to +\infty$. Indeed, for $(x, \omega) \in (\mathrm{T}^n + \frac{1}{2}\kappa) \times (\Gamma_\gamma + r_j)$ one has

$$\begin{aligned} |(u_j - u_{j-1})(x, \omega)| &\leq \sum_{2^{j-1} < |k| \leq 2^j} \left|\frac{f_k(\omega)}{i\langle \omega, k\rangle}\right| |e^{i\langle x, k\rangle}| \\ &\leq \frac{2M}{\gamma} \sum_{|k| > 2^{j-1}} |k|^\tau e^{(|\operatorname{Im} x| - \kappa)|k|} \leq \frac{2M}{\gamma} \sum_{|k| > 2^{j-1}} |k|^\tau e^{-\kappa|k|/2}, \end{aligned}$$

where we have used Lemma 1.8 and Proposition 1.9. In order to estimate the last sum first realize that for $l \in \mathrm{N}$ the number

$$\#\{k \in \mathrm{Z}^n \,:\, |k| = l\} \leq 2^n l^{n-1} \tag{1.40}$$

(see [4, 5, 12, 278]). Consequently,

$$\sum_{|k| > 2^{j-1}} |k|^\tau e^{-\kappa|k|/2} \leq 2^n \sum_{l > 2^{j-1}} l^{\tau+n-1} e^{-\kappa l/2}. \tag{1.41}$$

Finally set $m = 2^{j-1}$ and notice that for j sufficiently large $m\kappa \geq 2(\tau + n)$, whence

$$\begin{aligned} \sum_{l=m+1}^{\infty} l^{\tau+n-1} e^{-\kappa l/2} &\leq \int_m^\infty s^{\tau+n-1} e^{-\kappa s/2} ds \\ &= m^{\tau+n-1} e^{-\kappa m/2} \int_m^\infty \left(\frac{s}{m}\right)^{\tau+n-1} e^{-\kappa(s-m)/2} ds \\ &\leq m^{\tau+n-1} e^{-\kappa m/2} \int_m^\infty \exp\left[\left(\frac{\tau+n-1}{m} - \frac{\kappa}{2}\right)(s-m)\right] ds \\ &\leq m^{\tau+n-1} e^{-\kappa m/2} \int_m^\infty e^{-(s-m)/m} ds = m^{\tau+n} e^{-\kappa m/2}, \end{aligned}$$

which ensures superexponential decay of $|u_j - u_{j-1}|$ in j. Here we have used the fact that $y \le e^{y-1}$ for all y ($y = s/m$ in our calculation) and once more that $m\kappa \ge 2(\tau + n)$. This completes the proof of Theorem 1.7. □

Remark 1. Application of the Whitney Extension Theorem 6.15, see § 6.1.4 below and [160, 278, 338, 355] as well, now gives a C^∞ function on $\mathrm{T}^n \times \mathrm{int}(\Gamma)$ ("int" is for the interior), since the periodicity in the x-variables can be preserved in this extension. The same is possible for analyticity in the x-direction.

Remark 2. Similar results hold when f is only (in-)finitely differentiable. In that case f is approximated by analytic functions, e.g., using the Approximation Lemma, cf. Zehnder [364], for a summary also see [62, 162] and § 6.1.4. The above method is complicated for nonanalytic functions f by a diagonal procedure. In the case of finite differentiability u is less regular than f, the losses of differentiability being not equal for the x- and the ω-directions. For a detailed analysis of this so called *anisotropic differentiability*, see Pöschel [278].

Remark 3. The Diophantine condition (1.1), (1.37) is suitable to control the small divisors in view of the Paley–Wiener lemma (Lemma 1.8). However, one can also consider more general Diophantine conditions of the form

$$|\langle \omega, k\rangle| \ge \frac{\gamma}{\Delta(|k|)} \quad \forall k \in \mathrm{Z}^n \setminus \{0\} \tag{1.42}$$

where Δ is an *approximation function* [279, 305], i.e, an arbitrary monotonically increasing continuous function $\Delta : [1, +\infty[\to \mathrm{R}$ such that $\Delta(1) > 0$ and

$$\int_1^\infty \frac{\log \Delta(u)\, du}{u^2} < +\infty.$$

The Diophantine condition (1.1), (1.37) corresponds to $\Delta(u) = u^\tau$ and is used almost always but other types of approximation functions are also sometimes exploited [279, 302, 303, 305]. Similarly, one can consider the conditions

$$|\langle \omega, k\rangle + \langle \omega^N, \ell\rangle| \ge \frac{\gamma}{\Delta(|k|)} \quad \forall k \in \mathrm{Z}^n \setminus \{0\} \;\; \forall \ell \in \mathrm{Z}^r,\; |\ell| \le 2 \tag{1.43}$$

instead of (1.4).

Remark 4. Theorem 1.7 directly applies to the Floquet problem. In the example of the quasi-periodic response, we adapt this approach slightly. Indeed, for a fixed nonresonant ω we study the formal solution

$$\zeta(x, \lambda) = \sum_{k \in Z^n} \frac{F_k(\lambda)}{i\langle \omega, k\rangle - \lambda} e^{i\langle x, k\rangle}.$$

The Diophantine condition here is that for given $\tau > n-1$ and $\gamma > 0$ and for all $k \in \mathrm{Z}^n \setminus \{0\}$

$$|\lambda - i\langle \omega, k\rangle| \ge \gamma |k|^{-\tau}.$$

This condition excludes a countable number of open discs from the upper-half λ-plane, indexed by k. This "bead string" has a measure of the order of γ^2 as $\gamma \downarrow 0$. The complement of the "bead string" in the imaginary λ-axis is a Cantor set. Here we can apply the above kind of argument and obtain the Whitney-smoothness of the solution in λ.

It may be of interest to study the latter problem for its own sake. A question is what happens as $\gamma \downarrow 0$? The intersection of the "bead strings" over all $\gamma > 0$ in the imaginary λ-axis leaves out a residual set of measure zero containing the resonance points $i\langle\omega, k\rangle$, $k \in \mathbb{Z}^n \setminus \{0\}$. In the interior of the complement of this intersection our solution is analytic, both in x and in λ. For $\operatorname{Re}\lambda \neq 0$ this also follows by writing $\zeta = T(F)$, where

$$(T(F))(x,\lambda) = \int_0^\infty e^{\lambda s} F(x - s\,\omega, \lambda)\, ds$$

(for simplicity we restricted to the case $\operatorname{Re}\lambda < 0$). Notice that in the operator (supremum) norm we have $\|T\| = |\operatorname{Re}\lambda|^{-1}$, which explodes on the imaginary axis. For more details see Braaksma & Broer [47]. What happens to our solution in the residual set mentioned above? And what can be said of the losses of differentiability in the finitely differentiable case?

Chapter 2

The conjugacy theory

In this chapter, we present precise formulations of the quasi-periodicity persistence theorems in the dissipative context, volume preserving context, Hamiltonian isotropic context and reversible context 1 (see Sections 1.3–1.4). The parameter μ labeling the vector fields is always assumed to vary in an open domain $P \subset \mathbb{R}^s$, $s \geq 0$. All the quantities δ_j, ϵ_j, α_j, β_j are supposed to be real. The symbols $\mathcal{O}_d(a)$ denote a neighborhood of a point a in the Euclidean space $\mathbb{R}^d$ [we write $\mathcal{O}(a)$ instead of $\mathcal{O}_1(a)$]. The letter Y denotes an open domain in $\mathbb{R}^d$. Also, $\mathbb{R}P^d$ is the d-dimensional real projective space ($d \in \mathbb{Z}_+$) and $\Pi : \mathbb{R}^d \setminus \{0\} \to \mathbb{R}P^{d-1}$ for $d \in \mathbb{N}$ denotes the natural projection.

2.1 Preliminary considerations

The aim of this section is to elucidate the meaning of more advanced Diophantine conditions (1.4) involving *both* the internal and normal frequencies of Floquet quasi-periodic invariant tori. We examine the dissipative (n, p, s) context in more detail. Consider an s-parameter family $\tilde{X} = \{\tilde{X}^\mu\}_\mu$ of vector fields

$$\tilde{X}^\mu = [\omega(\mu) + f(x, z, \mu)]\frac{\partial}{\partial x} + [\Omega(\mu)z + h(x, z, \mu)]\frac{\partial}{\partial z} \tag{2.1}$$

where $x \in \mathbb{T}^n$, $z \in \mathcal{O}_p(0)$, $\mu \in P \subset \mathbb{R}^s$, $\omega : P \to \mathbb{R}^n$, $\Omega : P \to gl(p, \mathbb{R})$, and functions f and h are treated as small perturbations (of the order of $\epsilon > 0$), cf. (1.11). As a naïve attempt to prove the existence of Floquet invariant tori with parallel dynamics in the family (2.1), one tries to find a family of mappings

$$\Phi_\mu : (\chi, \zeta) \mapsto (\chi + U(\chi, \mu),\ \zeta + A(\chi, \mu) + B(\chi, \mu)\zeta)$$

and a parameter shift $\mu \mapsto \mu + \Lambda(\mu)$ such that the mapping Φ_μ^{-1} transforms the vector field $\tilde{X}^{\mu+\Lambda(\mu)}$ into the field

$$\tilde{X}_0^\mu = [\omega(\mu) + a(\chi, \zeta, \mu)]\frac{\partial}{\partial \chi} + [\Omega(\mu)\zeta + b(\chi, \zeta, \mu)]\frac{\partial}{\partial \zeta} \tag{2.2}$$

with $a = O(\zeta)$ and $b = O_2(\zeta)$. Here $\chi \in \mathbb{T}^n$ and $\zeta \in \mathcal{O}_p(0)$. The set $\{\zeta = 0\}$ is a Floquet invariant n-torus of field (2.2) with parallel dynamics, frequency vector $\omega(\mu)$ and Floquet matrix $\Omega(\mu)$. Consequently, $\{\Phi_\mu(\chi, 0) : \chi \in \mathbb{T}^n\}$ is the desired Floquet invariant n-torus of field $\tilde{X}^{\mu+\Lambda(\mu)}$ with parallel dynamics, frequency vector $\omega(\mu)$ and Floquet matrix $\Omega(\mu)$ [provided that Φ_μ does exist].

The equality $(\Phi_\mu)_*\tilde{X}_0^\mu = \tilde{X}^{\mu+\Lambda(\mu)}$ means that

$$\omega + a + \frac{\partial U}{\partial \chi}(\omega + a) = \omega(\mu + \Lambda) + f(\chi + U,\ \zeta + A + B\zeta,\ \mu + \Lambda),$$

$$\Omega\zeta + b + \frac{\partial A}{\partial \chi}(\omega + a) + \frac{\partial B}{\partial \chi}(\omega + a)\zeta + B(\Omega\zeta + b) =$$
$$\Omega(\mu + \Lambda)(\zeta + A + B\zeta) + h(\chi + U,\ \zeta + A + B\zeta,\ \mu + \Lambda) \tag{2.3}$$

[cf. (1.7)]. Here the argument of the functions ω, Ω, and Λ is μ (if not stated otherwise), the arguments of the functions U, A, and B are (χ, μ), while the arguments of the functions a and b are (χ, ζ, μ).

Recall that the perturbations f and h are of the order of ϵ. It is natural to suppose that the functions U, A, B, a, b, and Λ are also of the order of ϵ. Then, neglecting terms of the order of ϵ^2 in the system (2.3), we obtain

$$a + \frac{\partial U}{\partial \chi}\omega = \frac{\partial \omega}{\partial \mu}\Lambda + f, \tag{2.4}$$

$$b + \frac{\partial A}{\partial \chi}\omega + \frac{\partial B}{\partial \chi}\omega\zeta + B\Omega\zeta = \frac{\partial \Omega}{\partial \mu}\Lambda\zeta + \Omega A + \Omega B\zeta + h. \tag{2.5}$$

Here the argument of the functions ω, Ω, and Λ is μ, the arguments of the functions U, A, and B are (χ, μ), while the arguments of the functions f, h, a, and b are (χ, ζ, μ).

Let

$$f(\chi, \zeta, \mu) = u(\chi, \mu) + O(\zeta),$$
$$h(\chi, \zeta, \mu) = v(\chi, \mu) + w(\chi, \mu)\zeta + O_2(\zeta).$$

Omitting in (2.4) all the terms of the form $O(\zeta)$, we get

$$\frac{\partial U}{\partial \chi}\omega = \frac{\partial \omega}{\partial \mu}\Lambda + u. \tag{2.6}$$

Similarly, omitting in (2.5) all the terms of the form $O_2(\zeta)$ and equaling separately the ζ-independent terms and the terms linear in ζ, we get

$$\frac{\partial A}{\partial \chi}\omega = \Omega A + v, \tag{2.7}$$

$$\frac{\partial B}{\partial \chi}\omega + B\Omega = \frac{\partial \Omega}{\partial \mu}\Lambda + \Omega B + w. \tag{2.8}$$

The system of equations (2.6)–(2.8) with respect to functions $U(\chi, \mu)$, $A(\chi, \mu)$, $B(\chi, \mu)$, $\Lambda(\mu)$ is a "linearization" of the system (2.3) in the same sense as the homological equation (1.8) is a "linearization" of the equation (1.7). However, even system (2.6)–(2.8) cannot, generally speaking, be solved in any *open* domain of the parameter space $P \subset \mathbb{R}^s$, cf. the examples in § 1.5.1. We fix therefore a value $\mu_\star \in P$ of parameter μ and try to answer the following question:

Under what conditions on $\omega(\mu_\star)$, $\partial\omega(\mu_\star)/\partial\mu$, $\Omega(\mu_\star)$, and $\partial\Omega(\mu_\star)/\partial\mu$, can system (2.6)–(2.8) with respect to functions $U(\chi, \mu_\star)$, $A(\chi, \mu_\star)$, $B(\chi, \mu_\star)$, and vector $\Lambda(\mu_\star)$ be solved for any functions $u(\chi, \mu_\star)$, $v(\chi, \mu_\star)$, and $w(\chi, \mu_\star)$?

These conditions are prerequisites for the existence of a Floquet invariant n-torus with parallel dynamics, frequency vector $\omega(\mu_*)$ and Floquet matrix $\Omega(\mu_*)$ in the family $\widetilde{X}$ of vector fields (2.1).

Introduce the notations

$$\omega(\mu_*) = \omega^*, \quad \frac{\partial\omega(\mu_*)}{\partial\mu} = (D\omega)^*, \quad \Omega(\mu_*) = \Omega^*, \quad \frac{\partial\Omega(\mu_*)}{\partial\mu} = (D\Omega)^*, \quad \Lambda(\mu_*) = \Lambda^*.$$

Let u_k, v_k, w_k, U_k, A_k, B_k be the Fourier coefficients of the functions $u(\chi,\mu_*)$, $v(\chi,\mu_*)$, $w(\chi,\mu_*)$, $U(\chi,\mu_*)$, $A(\chi,\mu_*)$, $B(\chi,\mu_*)$, respectively ($k \in \mathbb{Z}^n$). System (2.6)–(2.8) at $\mu = \mu_*$ can be rewritten as

$$(D\omega)^*\Lambda^* = -u_0, \tag{2.9}$$
$$i\langle\omega^*, k\rangle U_k = u_k \qquad (k \neq 0), \tag{2.10}$$
$$\Omega^* A_0 = -v_0, \tag{2.11}$$
$$i\langle\omega^*, k\rangle A_k - \Omega^* A_k = v_k \qquad (k \neq 0), \tag{2.12}$$
$$(D\Omega)^*\Lambda^* + \Omega^* B_0 - B_0\Omega^* = -w_0, \tag{2.13}$$
$$i\langle\omega^*, k\rangle B_k + B_k\Omega^* - \Omega^* B_k = w_k \qquad (k \neq 0). \tag{2.14}$$

Note that U_0 does not enter these equations and can therefore be chosen in an arbitrary way. For instance, one can set $U_0 = 0$.

Find first the conditions for the formal solvability of the system of equations (2.9)–(2.14). Let $\lambda_1(\mu), \ldots, \lambda_p(\mu)$ be the eigenvalues of the matrix $\Omega(\mu)$ and $\lambda_j^* = \lambda_j(\mu_*)$, $1 \le j \le p$. The eigenvalues of the linear operator

$$\mathbb{C}^p \to \mathbb{C}^p, \qquad A \mapsto i\langle\omega^*, k\rangle A - \Omega^* A$$

are $i\langle\omega^*, k\rangle - \lambda_j^*$, $1 \le j \le p$. The eigenvalues of the linear operator

$$gl(p, \mathbb{C}) \to gl(p, \mathbb{C}), \qquad B \mapsto i\langle\omega^*, k\rangle B + B\Omega^* - \Omega^* B$$

are $i\langle\omega^*, k\rangle + \lambda_j^* - \lambda_l^*$, $1 \le j \le p$, $1 \le l \le p$ [196].[1] Therefore, the equations (2.10), (2.11), (2.12), (2.14) can be solved with respect to U_k, A_0, A_k, B_k for any u_k, v_0, v_k, w_k ($k \neq 0$) if and only if

$$\langle\omega^*, k\rangle \neq 0, \tag{2.15}$$
$$\lambda_j^* \neq 0 \qquad (1 \le j \le p), \tag{2.16}$$
$$i\langle\omega^*, k\rangle \neq \lambda_j^* \qquad (1 \le j \le p), \tag{2.17}$$
$$i\langle\omega^*, k\rangle \neq \lambda_j^* - \lambda_l^* \qquad (1 \le j \le p,\ 1 \le l \le p) \tag{2.18}$$

[of course, (2.15) is contained in (2.18)].

Suppose that among the numbers $\lambda_1^*, \ldots, \lambda_p^*$, there are r pairs of complex conjugate numbers $\alpha_1^* \pm i\beta_1^*, \ldots, \alpha_r^* \pm i\beta_r^*$ ($\beta_1^* > 0, \ldots, \beta_r^* > 0$) and $p - 2r$ real numbers ($0 \le r \le p/2$), so that $\beta_1^*, \ldots, \beta_r^*$ are the normal frequencies of the torus sought for. One very easily verifies that the system of inequalities (2.15)–(2.18) is implied by the system

$$\det\Omega^* \neq 0, \tag{2.19}$$
$$\langle\omega^*, k\rangle \neq \langle\beta^*, \ell\rangle \quad \forall \ell \in \mathbb{Z}^r,\ |\ell| \le 2. \tag{2.20}$$

[1] For the determination of the Jordan normal form of the linear operator $\mathrm{ad}_\Omega : gl(p, \mathbb{C}) \to gl(p, \mathbb{C})$, $\mathrm{ad}_\Omega B = B\Omega - \Omega B$ from that of an arbitrary linear operator $\Omega \in gl(p, \mathbb{C})$ see, e.g., Golubchikov [135].

Now it remains to examine the equations (2.9) and (2.13). We will assume that all the p numbers $\lambda_1^*, \ldots, \lambda_p^*$ are pairwise distinct.

Lemma 2.1 *The system of equations (2.9) and (2.13) is solvable with respect to* Λ^*, B_0 *for any* u_0, w_0 *if and only if the mapping* $\mu \mapsto (\omega(\mu), \lambda(\mu))$ *is submersive at* $\mu = \mu_*$.

Proof. Introduce the notation $\partial\lambda_j(\mu_*)/\partial\mu = (D\lambda_j)^*$, $1 \le j \le p$. Besides, we will denote by $[L]_{jl}$ the entries of any matrix $L \in gl(p, \mathbf{C})$, $1 \le j \le p$, $1 \le l \le p$. Let a matrix $M \in GL(p, \mathbf{C})$ diagonalize Ω^*, i.e., $M\Omega^*M^{-1} = diag(\lambda_1^*, \ldots, \lambda_p^*)$. Rewrite (2.13) as

$$M(D\Omega)^*\Lambda^*M^{-1} + (M\Omega^*M^{-1})(MB_0M^{-1}) - (MB_0M^{-1})(M\Omega^*M^{-1}) = -Mw_0M^{-1},$$

i.e.,

$$\frac{\partial[M\Omega(\mu_*)M^{-1}]_{jl}}{\partial\mu}\Lambda^* + (\lambda_j^* - \lambda_l^*)[MB_0M^{-1}]_{jl} = -[Mw_0M^{-1}]_{jl}, \tag{2.21}$$

$1 \le j \le p$, $1 \le l \le p$. On the other hand, the eigenvalues of matrix $M\Omega(\mu)M^{-1}$ with diagonal entries equal to $\lambda_j^* + \sigma_j(\mu - \mu_*) + O_2(\mu - \mu_*)$, $1 \le j \le p$, and off-diagonal entries equal to $O(\mu - \mu_*)$ are $\lambda_j^* + \sigma_j(\mu - \mu_*) + O_2(\mu - \mu_*)$ [here σ_j are certain row-vectors in $\mathbf{C}^s$]. This means that

$$\frac{\partial[M\Omega(\mu_*)M^{-1}]_{jj}}{\partial\mu} = (D\lambda_j)^*.$$

Consequently, (2.21) translates to

$$(D\lambda_j)^*\Lambda^* = -[Mw_0M^{-1}]_{jj}, \tag{2.22}$$

$$(\lambda_j^* - \lambda_l^*)[MB_0M^{-1}]_{jl} = -[Mw_0M^{-1}]_{jl} - \frac{\partial[M\Omega(\mu_*)M^{-1}]_{jl}}{\partial\mu}\Lambda^*, \quad j \ne l. \tag{2.23}$$

The statement of Lemma 2.1 follows straightforwardly from (2.9), (2.22), and (2.23). Note that B_0 is not determined uniquely by (2.13), because $[MB_0M^{-1}]_{jj}$ can be chosen in an arbitrary way. For instance, one can set $[MB_0M^{-1}]_{jj} = 0$, $1 \le j \le p$. □

Thus, the *formal* solvability conditions for system (2.6)–(2.8) at $\mu = \mu_*$ are (2.19), (2.20) together with the submersivity of the mapping $\mu \mapsto (\omega(\mu), \lambda(\mu))$ at $\mu = \mu_*$ (provided that all the eigenvalues of matrix Ω^* are simple). To ensure convergence of the Fourier series with coefficients U_k, A_k, B_k, one has to replace (2.20) with some Diophantine inequality (cf. the discussion in § 1.5.2), say,

$$|\langle\omega^*, k\rangle + \langle\beta^*, \ell\rangle| \ge \gamma|k|^{-\tau} \quad \forall \ell \in \mathbf{Z}^r,\ |\ell| \le 2, \tag{2.24}$$

γ and τ being positive constants independent of $k \in \mathbf{Z}^n \setminus \{0\}$. This coincides with (1.4) for $\omega^N = \beta$.

2.2 Whitney-smooth families of tori: a definition

Now we are ready to give a precise definition of "a Whitney-smooth l-parameter family of Floquet Diophantine invariant n-tori" of an s-parameter family of vector fields (cf. Sevryuk [326, 328] and Broer, Huitema & Sevryuk [61]). Recall that the expression "a Whitney-smooth function on a closed set $\Xi \subset \mathbf{R}^l$" means a function on Ξ that can be extended to a smooth function defined on a neighborhood of Ξ or the whole space $\mathbf{R}^l$ [160, 278, 338, 355], compare a detailed discussion in §§ 1.5.2 and 6.1.4.

Definition 2.2 *Let $X = \{X^\mu\}_\mu$ be an analytic s-parameter family of analytic vector fields on a K-dimensional manifold M, parameter μ varying in an open domain $P \subset \mathbb{R}^s$. A Whitney-smooth l-parameter family of Floquet Diophantine invariant analytic n-tori of X is the image*

$$\mathcal{F}(\mathbb{T}^n \times \{0\} \times \Xi)$$

of the set $\mathbb{T}^n \times \{0\} \times \Xi$ ($\Xi \subset \mathbb{R}^l$) under a mapping

$$\mathcal{F} : \mathbb{T}^n \times \mathcal{O} \times \mathbb{R}^l \to M \times P, \quad 0 \in \mathcal{O} \subset \mathbb{R}^{K-n}$$

possessing the following properties (below x, w, and ξ are the coordinates in $\mathbb{T}^n$, $\mathbb{R}^{K-n}$, and $\mathbb{R}^l$, respectively).

a) *$\mathcal{O} = \mathcal{O}_{K-n}(0)$ is a neighborhood of the origin in $\mathbb{R}^{K-n}$ while Ξ is a closed subset of $\mathbb{R}^l$ of positive Lebesgue measure.*

b) *The mapping $\mathcal{F}$ is analytic in $x \in \mathbb{T}^n$ and $w \in \mathcal{O}$ and is of class C^∞ in $\xi \in \mathbb{R}^l$. The restriction of $\mathcal{F}$ to $\mathbb{T}^n \times \{0\} \times \Xi$ is injective and the inverse mapping*

$$\mathcal{F}^{-1} : \mathcal{F}(\mathbb{T}^n \times \{0\} \times \Xi) \to \mathbb{T}^n \times \mathbb{R}^{K-n} \times \mathbb{R}^l$$

can be extended to a C^∞ map defined on $M \times P$.

c) *For any $\xi \in \Xi$, the set $\mathcal{F}(\mathbb{T}^n \times \mathcal{O} \times \{\xi\})$ lies in one of the "fibers" $M \times \{\Lambda(\xi)\}$ (the function $\Lambda : \xi \mapsto \Lambda(\xi) \in P$ is defined for $\xi \in \Xi$ only). Thus, for $\xi \in \Xi$ the restriction $\mathcal{F}^\xi$,*

$$\mathcal{F}^\xi : \mathbb{T}^n \times \mathcal{O} \to M, \quad \mathcal{F}(x, w, \xi) = (\mathcal{F}^\xi(x, w), \Lambda(\xi)),$$

is well defined.

d) *For any $\xi \in \Xi$, $\mathcal{F}^\xi$ is a diffeomorphism of $\mathbb{T}^n \times \mathcal{O}$ onto its image, and the vector field $(\mathcal{F}^\xi)_*^{-1} X^{\Lambda(\xi)}$ has the form*

$$[\omega(\xi) + O(w)]\frac{\partial}{\partial x} + [\Omega(\xi)w + O_2(w)]\frac{\partial}{\partial w} \tag{2.25}$$

where $\omega(\xi)$ is some constant vector in $\mathbb{R}^n$ and $\Omega(\xi)$ is some constant matrix in $gl(K-n, \mathbb{R})$.

e) *Moreover, for all $\xi \in \Xi$, the vectors $\omega(\xi)$ are uniformly Diophantine, i.e., there exist constants $\tau > 0$ and $\gamma > 0$ independent of ξ and such that*

$$|\langle \omega(\xi), k\rangle| \geq \gamma |k|^{-\tau} \quad \forall k \in \mathbb{Z}^n \setminus \{0\}.$$

Condition d) implies that each set

$$T_\xi = \mathcal{F}^\xi(\mathbb{T}^n \times \{0\}) \subset M$$

($\xi \in \Xi$) is an invariant n-torus of field $X^{\Lambda(\xi)}$ carrying parallel dynamics, and, moreover, this torus is Floquet. Condition e) implies that the parallel dynamics on each torus T_ξ is in fact Diophantine. Since $meas_l\, \Xi > 0$ [condition a)] and the inverse of the restriction of $\mathcal{F}$ to $\mathbb{T}^n \times \{0\} \times \Xi$ is Whitney-smooth [condition b)], the $(n+l)$-dimensional Hausdorff measure of the union of all the tori T_ξ is positive. However, a Whitney-smooth l-parameter family of Floquet Diophantine invariant n-tori is not just a collection of Floquet Diophantine invariant n-tori such that the $(n+l)$-dimensional measure of the union of the tori is positive. We also require that the tori depend on the labeling l-dimensional parameter ξ in a Whitney-smooth way and that the normalizing coordinate system around each torus [in which the corresponding vector field takes form (2.25), cf. (1.2)] can be also chosen to depend on ξ Whitney-smoothly.

If one deals with Hamiltonian or reversible vector fields, the following additional conditions are imposed on Whitney-smooth families of invariant tori.

(i) *In the Hamiltonian isotropic context, we also require each torus T_ξ to be isotropic.*

(ii) *In the Hamiltonian coisotropic context, we also require each torus T_ξ to be coisotropic.*

(iii) *In the reversible context (where all the vector fields X^μ are reversible with respect to involution $G : M \to M$ of the phase space), we also require that for each $\xi \in \Xi$, the involution $(\mathcal{F}^\xi)^{-1} G \mathcal{F}^\xi$ has the form*

$$(x, w) \mapsto (-x, Rw) \tag{2.26}$$

with ξ-independent involutive matrix R. In particular, each torus T_ξ is invariant not only under the flow of field $X^{\Lambda(\xi)}$, but also under the reversing involution G.

Recall that whereas the components of the vector $\omega(\xi)$ in (2.25) are called *internal frequencies* of torus T_ξ, the positive imaginary parts of the eigenvalues of the matrix $\Omega(\xi)$ are called *normal frequencies* of this torus [60, 62, 162], see § 1.1.1. They constitute the *normal frequency vector* that will be denoted in the sequel by $\omega^N(\xi)$. Thus, if, e.g., the eigenvalues of $\Omega(\xi)$ are $\delta_1, \ldots, \delta_{N_1}, \alpha_1 \pm i\beta_1, \ldots, \alpha_{N_2} \pm i\beta_{N_2}$ where $N_1 + 2N_2 = K - n$ and $\delta \in \mathbb{R}^{N_1}$, $\alpha \in \mathbb{R}^{N_2}$, $\beta \in \mathbb{R}^{N_2}$, $\beta_j > 0$, all the numbers $\beta_1, \ldots, \beta_{N_2}$ being distinct, then $\omega^N(\xi) = \beta$. *The collections $(\omega(\xi), \omega^N(\xi))$ of internal and normal frequencies of the tori T_ξ are usually required to satisfy further uniform Diophantine conditions of the form (1.4).*

The differentiability of Cantor-like families of invariant tori in dynamical systems was first established by Lazutkin [197–200] for mappings of the two-dimensional cylinder (or annulus) $\{(x, y) \ : \ x \in \mathbb{S}^1, y \in \mathbb{R}\}$ possessing the so called *intersection property*: each circle $y = \varphi(x)$ [or, more generally, each circle homological to $\{y = const\}$] intersects its image under the mapping (such mappings are slight generalizations of exact symplectic diffeomorphisms, see § 5.1.1). Lazutkin's results were carried over to higher dimensions by Svanidze [340]. Analogous theorems for Hamiltonian vector fields were first proven by Pöschel [278] (see also [277]) and Chierchia & Gallavotti [93], and those for reversible vector fields, by Pöschel [278]. For a very recent and detailed exposition of these results concerning the case where external parameters are absent, the reader is referred to Lazutkin's book [201]. Whitney-smooth families of Floquet Diophantine invariant tori in dynamical systems depending on external parameters were obtained in a general set-up by Broer, Huitema & Takens [60, 62, 162]. Dissipative, globally divergence-free, and Hamiltonian vector fields were considered in [62, 162] while reversible vector fields were examined in [162] {the reversible $(n, m, 0, s)$ context 1} and [60] {the reversible (n, m, p, s) context 1 for arbitrary $p \geq 0$}. Certain Whitney-smooth families of Diophantine invariant $(n-1)$-tori in Hamiltonian systems with n degrees of freedom were constructed recently by Rudnev & Wiggins [295] (see § 4.2.2).

Remark 1. Sometimes, Whitney-smooth l-parameter families of Diophantine invariant analytic n-tori which are *not necessarily Floquet* are considered. The definition of such families is obtained from Definition 2.2 by omitting the factor $\mathcal{O} \subset \mathbb{R}^{K-n}$ and the variable $w \in \mathbb{R}^{K-n}$ throughout; condition d) having to be replaced with the following condition:

For any $\xi \in \Xi$, the set $\mathcal{F}^\xi(\mathrm{T}^n) \subset M$ is invariant under the flow of the field $X^{\Lambda(\xi)}$, and the vector field on T^n induced by $X^{\Lambda(\xi)}$ via $\mathcal{F}^\xi$ has the form $\omega(\xi)\partial/\partial x$.

This situation is often encountered in the infinite dimensional KAM theory, in particular, while constructing quasi-periodic solutions for partial differential equations (see [185, 186, 279–281] and references therein), where in many problems, one does not control the normal directions (and, moreover, all of them are elliptic, i.e., the whole phase space is the center manifold).

Also, sometimes the map $\mathcal{F}$ is only finitely differentiable in $\xi \in \mathbb{R}^l$, in which case one speaks of a *finitely* Whitney-smooth family, or Whitney-C^k family for $k < \infty$. For instance, Chierchia & Gallavotti [94] considered finitely Whitney-smooth families of Diophantine invariant $(n-1)$-tori in (analytic) Hamiltonian systems with n degrees of freedom (see also [90]).

Remark 2. In [61], we required the restriction of $\mathcal{F}$ to $\mathrm{T}^n \times \mathcal{O} \times \Xi$ to be injective in condition b). That was not quite correct, because in many situations this restriction is not injective (e.g., the dimension $K + l$ of $\mathrm{T}^n \times \mathcal{O} \times \mathbb{R}^l$ can well be greater than the dimension $K + s$ of $M \times P$). One is able only to require the restriction of $\mathcal{F}$ to $\mathrm{T}^n \times \{0\} \times \Xi$ to be injective.

2.3 Quasi-periodic stability

The main idea of the multiparameter KAM theory as invented by Moser [244, 246] and developed further by Broer, Huitema & Takens [60, 62, 162] is to consider integrable vector fields depending on a parameter of dimension *large enough* to guarantee the existence of a Floquet invariant torus with parallel dynamics possessing any collection (ω, Ω) of internal and normal data (cf. Section 2.1). From the practical viewpoint, this requirement means *the submersivity* of the mapping

$$\text{the parameter labeling the tori} \mapsto (\omega,\ \text{the spectrum of } \Omega)$$

[cf. the mappings (2.28), (2.36)–(2.38), (2.40), (2.51), (2.62) below]. In a sufficiently small nonintegrable perturbation of this multiparameter family of integrable vector fields, one then looks for Floquet invariant tori with parallel dynamics whose internal ω and normal ω^N frequencies satisfy certain Diophantine conditions, and it turns out to be possible to find tori with *all* the collections (ω, Ω) of internal and normal data meeting those conditions (the so called quasi-periodic stability [62, 162], cf. §§ 1.2.1 and 6.1.1). Following Moser, the external parameters needed to achieve stability are sometimes called "modifying terms". Papers [62, 162] contain a detailed comparison of the "unfolding theory" presented therein and Moser's original approach [246], see also a discussion in Sevryuk [326]. The multiparameter KAM theorems were also obtained by Bogolyubov, Mitropol'skiĭ & Samoĭlenko [42] and Bibikov [39] and refined by Matveev [227].

In this section, we give precise formulations of our "main" quasi-periodic stability theorems for all the five contexts under consideration. The exposition follows closely our paper [61]. In all the theorems below (except for Theorem 2.5 pertaining to the volume preserving context with $p = 1$), the unperturbed vector fields X^μ are not, strictly speaking, integrable [equivariant with respect to the free action of T^n]. The family X of vector fields is just assumed to possess an *analytic* family of Floquet invariant analytic n-tori with parallel dynamics.

2.3.1 The dissipative context

Consider an analytic family of analytic vector fields on $T^n \times R^p$ ($n \geq 1, p \geq 0$):

$$X = \{X^\mu\}_\mu = [\omega(\mu) + f(x, z, \mu)]\frac{\partial}{\partial x} + [\Omega(\mu)z + h(x, z, \mu)]\frac{\partial}{\partial z} \tag{2.27}$$

[cf. (1.11)], where $x \in T^n$, $z \in \mathcal{O}_p(0)$, $\mu \in P \subset R^s$, $\omega : P \to R^n$, $\Omega : P \to gl(p, R)$, $f = O(z)$, $h = O_2(z)$. Let for $\mu \in \Gamma \subset P$ [Γ being diffeomorphic to a closed s-dimensional ball]

1) all the eigenvalues

$$\delta_1, \ldots, \delta_{N_1}, \quad \alpha_1 \pm i\beta_1, \ldots, \alpha_{N_2} \pm i\beta_{N_2}$$

($\beta_j > 0$) of matrix $\Omega(\mu)$ are simple and other than zero ($N_1 + 2N_2 = p$);

2) the mapping

$$R^s \ni \mu \mapsto (\omega, \delta, \alpha, \beta) \in R^{n+p} \tag{2.28}$$

is submersive (so that $s \geq n + p$).

Fix $\tau > n - 1$. Set $\omega^N = \beta \in R^r$, where $r = N_2$. By Γ_γ, where $\gamma > 0$, denote the set [which is "Cantor" for $n \geq 2$]

$$\Gamma_\gamma := \{\mu \in \Gamma : \forall k \in Z^n \setminus \{0\}\ \forall \ell \in Z^r, |\ell| \leq 2,\ |\langle\omega, k\rangle + \langle\omega^N, \ell\rangle| \geq \gamma|k|^{-\tau}\}. \tag{2.29}$$

Since mapping (2.28) is submersive, this set is a Whitney-smooth foliation (cf. § 1.5.2) of $(s - n - r + 1)$-dimensional analytic (maybe, with boundary) surfaces Γ_γ^ι, each surface being a part of the preimage of some point $\iota \in RP^{n+r-1}$ under the mapping

$$R^s \ni \mu \mapsto \Pi(\omega, \omega^N) \in RP^{n+r-1}. \tag{2.30}$$

Note that $\operatorname{meas}_s \Gamma_\gamma / \operatorname{meas}_s \Gamma \to 1$ as $\gamma \downarrow 0$.

Theorem 2.3 (Main theorem in the dissipative context) [62, 162] *Let X be an analytic family of analytic vector fields (2.27) on $T^n \times R^p$ satisfying the conditions above. Then for any $\gamma > 0$ and any neighborhood $\mathfrak{O}$ of zero in the space of all C^∞-mappings*

$$T^n \times P \to R^n \times R^p \times gl(p, R) \times R^s, \quad (x, \mu) \mapsto (\chi, \zeta_1, \zeta_2, \lambda), \tag{2.31}$$

analytic in x and such that λ does not depend on x, there exists a neighborhood $\mathcal{X}$ of the family X in the space of all analytic families of analytic vector fields

$$\begin{aligned}\widetilde{X} = \{\widetilde{X}^\mu\}_\mu \ &= \ \left[\omega(\mu) + f(x, z, \mu) + \widetilde{f}(x, z, \mu)\right]\frac{\partial}{\partial x} + \\ &\quad \left[\Omega(\mu)z + h(x, z, \mu) + \widetilde{h}(x, z, \mu)\right]\frac{\partial}{\partial z}\end{aligned} \tag{2.32}$$

such that for any $\widetilde{X} \in \mathcal{X}$ there is a mapping in $\mathfrak{D}$

$$(x, \mu) \mapsto (\chi(x, \mu), \zeta_1(x, \mu), \zeta_2(x, \mu), \lambda(\mu)) \tag{2.33}$$

with the following property: for each $\mu_0 \in \Gamma_\gamma$ the vector field $(\Phi_{\mu_0})_*^{-1} \widetilde{X}^{\mu_0+\lambda(\mu_0)}$, where

$$\Phi_{\mu_0} : (\bar{x}, \bar{z}) \mapsto (\bar{x} + \chi(\bar{x}, \mu_0), \bar{z} + \zeta_1(\bar{x}, \mu_0) + \zeta_2(\bar{x}, \mu_0)\bar{z}), \tag{2.34}$$

has the form

$$[\omega(\mu_0) + O(\bar{z})]\frac{\partial}{\partial \bar{x}} + [\Omega(\mu_0)\bar{z} + O_2(\bar{z})]\frac{\partial}{\partial \bar{z}}. \tag{2.35}$$

Moreover, the mapping Φ_{μ_0} depends on μ_0 analytically when μ_0 varies on each of the surfaces Γ_γ^ι.

In a "less topological" language, this theorem runs as follows: for any $\gamma > 0$, $N^* \in \mathbb{N}$ and $\epsilon^* > 0$, there exists $\delta^* > 0$ such that the following holds. For any analytic family of analytic vector fields (2.32), where $|\tilde{f}(x, z, \mu)|$ and $|\tilde{h}(x, z, \mu)|$ are less than δ^* in a fixed (independent of γ, N^*, ϵ^*) complex neighborhood of $\mathbb{T}^n \times \{0\} \times \Gamma \subset \mathbb{T}^n \times \mathbb{R}^p \times \mathbb{R}^s$, there exists a C^∞ mapping

$$\mathbb{T}^n \times P \to \mathbb{R}^n \times \mathbb{R}^p \times gl(p, \mathbb{R}) \times \mathbb{R}^s, \quad (x, \mu) \mapsto (\chi(x, \mu), \zeta_1(x, \mu), \zeta_2(x, \mu), \lambda(\mu))$$

analytic in x and possessing the following properties: **(1)** all the partial derivatives of the functions χ, ζ_1, ζ_2, λ of orders 0, 1, ..., N^* are less than ϵ^* in $\mathbb{T}^n \times P$ and **(2)** for each $\mu_0 \in \Gamma_\gamma$ the vector field $(\Phi_{\mu_0})_*^{-1} \widetilde{X}^{\mu_0+\lambda(\mu_0)}$ [Φ_{μ_0} being defined by (2.34)] has the form (2.35). Moreover, the mapping Φ_{μ_0} depends on μ_0 analytically when μ_0 varies on each of the surfaces Γ_γ^ι.

For the proof of Theorem 2.3 in the simplest case $p = 0$, see Section 6.1.

2.3.2 The volume preserving context ($p \geq 2$)

Consider an analytic family of analytic vector fields (2.27) on $\mathbb{T}^n \times \mathbb{R}^p$ ($n \geq 1$, $p \geq 2$) globally divergence-free with respect to the volume element $dx \wedge dz$, where $x \in \mathbb{T}^n$, $z \in O_p(0)$, $\mu \in P \subset \mathbb{R}^s$, $\omega : P \to \mathbb{R}^n$, $\Omega : P \to sl(p, \mathbb{R})$, $f = O(z)$, $h = O_2(z)$. Here one should distinguish the cases $p = 2$ and $p > 2$. The reason is that the presence of purely imaginary eigenvalues of a 2×2 real matrix with trace zero is a typical possibility whereas for $p > 2$, all the eigenvalues of a generic $p \times p$ real matrix with trace zero have nonzero real parts. Assume that for $\mu \in \Gamma \subset P$ [Γ being diffeomorphic to a closed s-dimensional ball] the following holds.

For $p = 2$ (hyperbolic case): 1) the eigenvalues

$$\delta_1, \delta_2, \quad \delta_1 + \delta_2 = 0$$

of matrix $\Omega(\mu)$ are simple;

2) the mapping

$$\mathbb{R}^s \ni \mu \mapsto (\omega, \delta) \in \mathbb{R}^n \times L \cong \mathbb{R}^{n+1} \tag{2.36}$$

is submersive, where

$$L = \left\{\delta \in \mathbb{R}^2 \ : \ \delta_1 + \delta_2 = 0\right\} \cong \mathbb{R}$$

(so that $s \geq n+1$).
Set $r = 0$, $\omega^N = 0 \in \mathbb{R}^0 = \{0\}$.

For $p = 2$ (elliptic case): 1) the eigenvalues

$$\pm i\varepsilon$$

of matrix $\Omega(\mu)$ are simple (and $\varepsilon > 0$);
2) the mapping

$$\mathbb{R}^s \ni \mu \mapsto (\omega, \varepsilon) \in \mathbb{R}^{n+1} \tag{2.37}$$

is submersive (so that $s \geq n+1$).
Set $r = 1$, $\omega^N = \varepsilon \in \mathbb{R}$.

For $p > 2$: 1) all the eigenvalues

$$\delta_1, \ldots, \delta_{N_1}, \quad \alpha_1 \pm i\beta_1, \ldots, \alpha_{N_2} \pm i\beta_{N_2}, \quad \delta_1 + \cdots + \delta_{N_1} + 2(\alpha_1 + \cdots + \alpha_{N_2}) = 0$$

($\beta_j > 0$) of matrix $\Omega(\mu)$ with trace zero are simple and other than zero ($N_1 + 2N_2 = p$);
2) the mapping

$$\mathbb{R}^s \ni \mu \mapsto (\omega, \delta, \alpha, \beta) \in \mathbb{R}^n \times L \times \mathbb{R}^{N_2} \cong \mathbb{R}^{n+p-1} \tag{2.38}$$

is submersive, where

$$L = \left\{(\delta, \alpha) \in \mathbb{R}^{N_1+N_2} \ : \ \delta_1 + \cdots + \delta_{N_1} + 2(\alpha_1 + \cdots + \alpha_{N_2}) = 0\right\} \cong \mathbb{R}^{N_1+N_2-1}$$

(so that $s \geq n+p-1$).
Set $r = N_2$, $\omega^N = \beta \in \mathbb{R}^r$.

Fix $\tau > n-1$. By Γ_γ, where $\gamma > 0$, denote the set (2.29) [which is "Cantor" for $n \geq 2$]. As in the dissipative context, this set is a Whitney-smooth foliation of $(s-n-r+1)$-dimensional analytic (maybe, with boundary) surfaces Γ^ι_γ, each surface being a part of the preimage of some point $\iota \in \mathbb{R}P^{n+r-1}$ under mapping (2.30). Also, $meas_s \Gamma_\gamma / meas_s \Gamma \to 1$ as $\gamma \downarrow 0$.

Theorem 2.4 (Main theorem in the volume preserving context with $p \geq 2$) [62, 162] *Let X be an analytic family of analytic globally divergence-free vector fields (2.27) on $\mathbb{T}^n \times \mathbb{R}^p$ satisfying the conditions above. Then for any $\gamma > 0$ and any neighborhood $\mathfrak{D}$ of zero in the space of all C^∞-mappings (2.31) analytic in x and such that λ does not depend on x, there exists a neighborhood $\mathcal{X}$ of the family X in the space of all analytic families of analytic globally divergence-free vector fields (2.32) such that for any $\widetilde{X} \in \mathcal{X}$ there is a mapping (2.33) in $\mathfrak{D}$ with the following property: for each $\mu_0 \in \Gamma_\gamma$ the vector field $(\Phi_{\mu_0})_*^{-1} \widetilde{X}^{\mu_0+\lambda(\mu_0)}$ [Φ_{μ_0} being defined by (2.34)] has the form (2.35). Moreover, the mapping Φ_{μ_0} depends on μ_0 analytically when μ_0 varies on each of the surfaces Γ^ι_γ. Finally, all the mappings Φ_μ are volume preserving.*

2.3.3 The volume preserving context ($p=1$)

Consider an analytic family of analytic vector fields on $T^n \times R$ ($n \geq 1$) globally divergence-free with respect to the volume element $dx \wedge dy$:

$$X = \{X^\mu\}_\mu = \omega(y,\mu)\frac{\partial}{\partial x} \tag{2.39}$$

[cf. (1.12)], where $x \in T^n$, $y \in Y \subset R$, $\mu \in P \subset R^s$, $\omega : Y \times P \to R^n$. Let for $(y,\mu) \in \Gamma \subset Y \times P$ [Γ being diffeomorphic to a closed $(s+1)$-dimensional ball] the mapping

$$R^{s+1} \ni (y,\mu) \mapsto \omega \in R^n \tag{2.40}$$

is submersive (so that $s \geq n-1$).

Fix $\tau > n-1$. By Γ_γ, where $\gamma > 0$, denote the set [which is "Cantor" for $n \geq 2$]

$$\Gamma_\gamma := \{(y,\mu) \in \Gamma \, : \, \forall k \in Z^n \setminus \{0\} \ |\langle \omega, k\rangle| \geq \gamma |k|^{-\tau}\}. \tag{2.41}$$

Since mapping (2.40) is submersive, this set is a Whitney-smooth foliation of $(s-n+2)$-dimensional analytic (maybe, with boundary) surfaces Γ_γ^ι, each surface being a part of the preimage of some point $\iota \in RP^{n-1}$ under the mapping

$$R^{s+1} \ni (y,\mu) \mapsto \Pi\omega \in RP^{n-1}. \tag{2.42}$$

Note also that $meas_{s+1}\Gamma_\gamma / \, meas_{s+1}\Gamma \to 1$ as $\gamma \downarrow 0$.

Theorem 2.5 (Main theorem in the volume preserving context with $p=1$) [62, 162] *Let X be an analytic family of analytic globally divergence-free vector fields (2.39) on $T^n \times R$ satisfying the conditions above. Then for any $\gamma > 0$ and any neighborhood $\mathfrak{D}$ of zero in the space of all C^∞-mappings*

$$T^n \times Y \times P \to R^n \times R \times R \times R^s, \quad (x,y,\mu) \mapsto (\chi, \eta_1, \eta_2, \lambda), \tag{2.43}$$

analytic in x and such that λ does not depend on x, there exists a neighborhood $\mathcal{X}$ of the family X in the space of all analytic families of analytic globally divergence-free vector fields

$$\widetilde{X} = \{\widetilde{X}^\mu\}_\mu = \left[\omega(y,\mu) + \widetilde{f}(x,y,\mu)\right]\frac{\partial}{\partial x} + \widetilde{g}(x,y,\mu)\frac{\partial}{\partial y} \tag{2.44}$$

such that for any $\widetilde{X} \in \mathcal{X}$ there is a mapping in $\mathfrak{D}$

$$(x,y,\mu) \mapsto (\chi(x,y,\mu), \eta_1(x,y,\mu), \eta_2(x,y,\mu), \lambda(y,\mu)) \tag{2.45}$$

with the following property: for each $(y_0,\mu_0) \in \Gamma_\gamma$ the vector field $(\Phi_{y_0\mu_0})_*^{-1}\widetilde{X}^{\mu_0+\lambda(y_0,\mu_0)}$, where

$$\Phi_{y_0\mu_0} : (\bar{x},\bar{y}) \mapsto (\bar{x} + \chi(\bar{x},y_0,\mu_0), \bar{y} + \eta_1(\bar{x},y_0,\mu_0) + \eta_2(\bar{x},y_0,\mu_0)(\bar{y}-y_0)), \tag{2.46}$$

has the form

$$[\omega(y_0,\mu_0) + O(\bar{y}-y_0)]\frac{\partial}{\partial \bar{x}} + O_2(\bar{y}-y_0)\frac{\partial}{\partial \bar{y}}. \tag{2.47}$$

Moreover, the mapping $\Phi_{y_0\mu_0}$ depends on (y_0,μ_0) analytically when the point (y_0,μ_0) varies on each of the surfaces Γ_γ^ι.

2.3.4 The Hamiltonian isotropic context

Consider an analytic family of vector fields on $T^n \times R^{n+2p}$ ($n \geq 1$, $p \geq 0$) which are Hamiltonian with respect to the symplectic structure (1.13) with the analytic Hamilton function

$$H^\mu(x,y,z) = F(y,\mu) + \tfrac{1}{2}\langle z, K(y,\mu)z\rangle + \Delta(x,y,z,\mu) \tag{2.48}$$

[cf. (1.16)], where $x \in T^n$, $y \in Y \subset R^n$, $z \in O_{2p}(0)$, $\mu \in P \subset R^s$, $F : Y \times P \to R$, $K : Y \times P \to gl(2p, R)$, $K(y,\mu)$ is symmetric for all values of y and μ, $\Delta = O_3(z)$. This family has the form

$$X = \{X^\mu\}_\mu = [\omega(y,\mu) + O_2(z)]\frac{\partial}{\partial x} + O_3(z)\frac{\partial}{\partial y} + [\Omega(y,\mu)z + O_2(z)]\frac{\partial}{\partial z} \tag{2.49}$$

[cf. (1.17)], where $\omega = \partial F/\partial y$ and $\Omega = JK$, the $2p \times 2p$ matrix J being defined by (1.14). Let for $(y,\mu) \in \Gamma \subset Y \times P$ [Γ being diffeomorphic to a closed $(n+s)$-dimensional ball]

1) all the eigenvalues

$$\pm\, \delta_1, \ldots, \pm\delta_{N_1}, \qquad \pm i\varepsilon_1, \ldots, \pm i\varepsilon_{N_2}, \qquad \pm\alpha_1 \pm i\beta_1, \ldots, \pm\alpha_{N_3} \pm i\beta_{N_3} \tag{2.50}$$

($\delta_j > 0$, $\varepsilon_j > 0$, $\alpha_j > 0$, $\beta_j > 0$) of Hamiltonian matrix $\Omega(y,\mu)$ [cf. (1.18)] are simple ($N_1 + N_2 + 2N_3 = p$);

2) the mapping

$$R^{n+s} \ni (y,\mu) \mapsto (\omega, \delta, \varepsilon, \alpha, \beta) \in R^{n+p} \tag{2.51}$$

is submersive (so that $s \geq p$).

Fix $\tau > n - 1$. Set $\omega^N = (\varepsilon, \beta) \in R^r$, where $r = N_2 + N_3$. By Γ_γ, where $\gamma > 0$, denote the set [which is "Cantor" for $n \geq 2$]

$$\Gamma_\gamma := \{(y,\mu) \in \Gamma \;:\; \forall k \in Z^n \setminus \{0\}\ \forall \ell \in Z^r, |\ell| \leq 2,\ |\langle\omega, k\rangle + \langle\omega^N, \ell\rangle| \geq \gamma|k|^{-\tau}\}. \tag{2.52}$$

Since mapping (2.51) is submersive, this set is a Whitney-smooth foliation of $(s - r + 1)$-dimensional analytic (maybe, with boundary) surfaces Γ_γ^ι, each surface being a part of the preimage of some point $\iota \in RP^{n+r-1}$ under the mapping

$$R^{n+s} \ni (y,\mu) \mapsto \Pi(\omega, \omega^N) \in RP^{n+r-1}. \tag{2.53}$$

Note also that $meas_{n+s}\Gamma_\gamma / meas_{n+s}\Gamma \to 1$ as $\gamma \downarrow 0$.

Theorem 2.6 (Main theorem in the Hamiltonian isotropic context) [62, 162] *Let X be an analytic family of analytic Hamiltonian vector fields (2.49) on $T^n \times R^{n+2p}$ satisfying the conditions above. Then for any $\gamma > 0$ and any neighborhood $\mathfrak{O}$ of zero in the space of all C^∞-mappings*

$$T^n \times Y \times O_n(0) \times O_{2p}(0) \times P \to R^n \times R^n \times R^{2p} \times R^s, \quad (x, y, \hat{y}, z, \mu) \mapsto (\chi, \eta, \zeta, \lambda), \tag{2.54}$$

affine in $\hat{y}$ and z, analytic in x and such that χ does not depend on $\hat{y}$ and z while λ does not depend on x, $\hat{y}$ and z, there exists a neighborhood $\mathcal{X}$ of the Hamilton function H^μ in the space of all analytic and analytically μ-dependent Hamiltonians

$$\widetilde{H}^\mu(x,y,z) = F(y,\mu) + \tfrac{1}{2}\langle z, K(y,\mu)z\rangle + \Delta(x,y,z,\mu) + \widetilde{\Delta}(x,y,z,\mu) \tag{2.55}$$

(Hamilton function $\widetilde{H}^\mu$ determines the family $\{\widetilde{X}^\mu\}_\mu$ of vector fields) such that for any $\widetilde{H}^\mu \in \mathcal{X}$ there is a mapping in $\mathfrak{D}$

$$(x, y, \hat{y}, z, \mu) \mapsto (\chi(x, y, \mu), \eta(x, y, \hat{y}, z, \mu), \zeta(x, y, \hat{y}, z, \mu), \lambda(y, \mu)) \tag{2.56}$$

with the following property: for each $(y_0, \mu_0) \in \Gamma_\gamma$ the vector field $(\Phi_{y_0\mu_0})_*^{-1} \widetilde{X}^{\mu_0+\lambda(y_0,\mu_0)}$, where

$$\Phi_{y_0\mu_0} : (\bar{x}, \bar{y}, \bar{z}) \mapsto (\bar{x}+\chi(\bar{x}, y_0, \mu_0), \bar{y}+\eta(\bar{x}, y_0, \bar{y}-y_0, \bar{z}, \mu_0), \bar{z}+\zeta(\bar{x}, y_0, \bar{y}-y_0, \bar{z}, \mu_0)), \tag{2.57}$$

has the form

$$[\omega(y_0, \mu_0) + O(|\bar{y} - y_0| + |\bar{z}|)]\frac{\partial}{\partial \bar{x}} + O_2(|\bar{y} - y_0| + |\bar{z}|)\frac{\partial}{\partial \bar{y}} + \\ [\Omega(y_0, \mu_0)\bar{z} + O_2(|\bar{y} - y_0| + |\bar{z}|)]\frac{\partial}{\partial \bar{z}}. \tag{2.58}$$

Moreover, the mapping $\Phi_{y_0\mu_0}$ depends on (y_0, μ_0) analytically when the point (y_0, μ_0) varies on each of the surfaces Γ^c_γ. Finally, the invariant n-torus

$$\{\Phi_{y_0\mu_0}(\bar{x}, y_0, 0) \;:\; \bar{x} \in \mathrm{T}^n\} \tag{2.59}$$

of the field $\widetilde{X}^{\mu_0+\lambda(y_0,\mu_0)}$ is isotropic.

Remark. The literature devoted to the Hamiltonian isotropic KAM theory is now immense. For the "classical" isotropic $(n, 0, 0)$ context, see, e.g., Bost [43] for a review and a large bibliography. The *lower-dimensional* invariant tori[2] in Hamiltonian vector fields [mainly the Hamiltonian isotropic (n, p, s) context with $p \geq 1$ in our terminology] are studied in [36, 37, 39, 61, 62, 72, 87, 90, 92, 94, 100, 109, 113, 122, 125, 126, 137, 138, 142, 162, 169, 172, 184, 203, 231, 232, 246, 248–250, 279, 295, 330, 331, 342, 364]. Note that these lower-dimensional tori are said to be *elliptic* if $N_1 = N_3 = 0$ in (2.50) [109, 279], and said to be *hyperbolic* if $N_2 = 0$ [37, 137, 342, 364]. The existence of lower-dimensional invariant tori in integrable Hamiltonian systems is sometimes referred to as the *limit degeneracy* (see Arnol'd [5] for details). One also speaks of the so called *properly degenerate* (or *intrinsically degenerate*, or *superintegrable*) framework, where the unperturbed Hamiltonian does not depend on some action variables y_j ([5, 20], for recent works see, e.g., Hanßmann [144–146] and Broer [57]).

2.3.5 The reversible context 1

Consider an analytic family of analytic G-reversible vector fields on $\mathrm{T}^n \times \mathrm{R}^{m+2p}$ ($n \geq 1$, $m \geq 0$, $p \geq 0$):

$$X = \{X^\mu\}_\mu = [\omega(y, \mu) + f(x, y, z, \mu)]\frac{\partial}{\partial x} + g(x, y, z, \mu)\frac{\partial}{\partial y} + \\ [\Omega(y, \mu)z + h(x, y, z, \mu)]\frac{\partial}{\partial z} \tag{2.60}$$

[cf. (1.27)],

$$G : (x, y, z) \mapsto (-x, y, Rz) \tag{2.61}$$

[2] whose dimensions are less than the number of degrees of freedom

[cf. (1.26)], where $x \in \mathrm{T}^n$, $y \in Y \subset \mathrm{R}^m$, $z \in \mathcal{O}_{2p}(0)$, $\mu \in P \subset \mathrm{R}^s$, R is an involutive $2p \times 2p$ real matrix whose 1- and (-1)-eigenspaces are p-dimensional, $\omega : Y \times P \to \mathrm{R}^n$, $\Omega : Y \times P \to \{L \in sl(2p, \mathrm{R}) : LR + RL = 0\}$, $f = O(z)$, $g = O_2(z)$, $h = O_2(z)$. The reversibility with respect to G imposes the following conditions on the terms f, g, and h:

$$f(-x, y, Rz, \mu) \equiv f(x, y, z, \mu), \qquad g(-x, y, Rz, \mu) \equiv -g(x, y, z, \mu),$$
$$h(-x, y, Rz, \mu) \equiv -Rh(x, y, z, \mu).$$

Let for $(y, \mu) \in \Gamma \subset Y \times P$ [Γ being diffeomorphic to a closed $(m+s)$-dimensional ball]

1) all the eigenvalues (2.50) of infinitesimally R-reversible matrix $\Omega(y, \mu)$ are simple;

2) the mapping

$$\mathrm{R}^{m+s} \ni (y, \mu) \mapsto (\omega, \delta, \varepsilon, \alpha, \beta) \in \mathrm{R}^{n+p} \tag{2.62}$$

is submersive (so that $s \geq n - m + p$).

Fix $\tau > n - 1$. Set $\omega^N = (\varepsilon, \beta) \in \mathrm{R}^r$, where $r = N_2 + N_3$. By Γ_γ, where $\gamma > 0$, denote the set (2.52) [which is "Cantor" for $n \geq 2$]. Since mapping (2.62) is submersive, this set is a Whitney-smooth foliation of $(s+m-n-r+1)$-dimensional analytic (maybe, with boundary) surfaces Γ_γ^ι, each surface being a part of the preimage of some point $\iota \in \mathrm{RP}^{n+r-1}$ under the mapping

$$\mathrm{R}^{m+s} \ni (y, \mu) \mapsto \Pi(\omega, \omega^N) \in \mathrm{RP}^{n+r-1}. \tag{2.63}$$

Also, $\mathrm{meas}_{m+s}\, \Gamma_\gamma / \mathrm{meas}_{m+s}\, \Gamma \to 1$ as $\gamma \downarrow 0$.

Theorem 2.7 (Main theorem in the reversible context 1) [60] *Let X be an analytic family of analytic G-reversible vector fields (2.60) on $\mathrm{T}^n \times \mathrm{R}^{m+2p}$ satisfying the conditions above. Then for any $\gamma > 0$ and any neighborhood $\mathfrak{O}$ of zero in the space of all C^∞-mappings*

$$\mathrm{T}^n \times Y \times \mathcal{O}_m(0) \times \mathcal{O}_{2p}(0) \times P \to \mathrm{R}^n \times \mathrm{R}^m \times \mathrm{R}^{2p} \times \mathrm{R}^s, \quad (x, y, \hat{y}, z, \mu) \mapsto (\chi, \eta, \zeta, \lambda), \tag{2.64}$$

affine in $\hat{y}$ and z, analytic in x and such that χ does not depend on $\hat{y}$ and z while λ does not depend on x, $\hat{y}$ and z, there exists a neighborhood $\mathcal{X}$ of the family X in the space of all analytic families of analytic G-reversible vector fields

$$\begin{aligned} \widetilde{X} = \{\widetilde{X}^\mu\}_\mu \;=\; & \left[\omega(y, \mu) + f(x, y, z, \mu) + \tilde{f}(x, y, z, \mu)\right] \frac{\partial}{\partial x} + \\ & \left[g(x, y, z, \mu) + \tilde{g}(x, y, z, \mu)\right] \frac{\partial}{\partial y} + \\ & \left[\Omega(y, \mu) z + h(x, y, z, \mu) + \tilde{h}(x, y, z, \mu)\right] \frac{\partial}{\partial z} \end{aligned} \tag{2.65}$$

such that for any $\widetilde{X} \in \mathcal{X}$ there is a mapping (2.56) in $\mathfrak{O}$ with the following properties:

$$\begin{aligned} &\chi(-x, y, \mu) \equiv -\chi(x, y, \mu), \qquad \eta(-x, y, \hat{y}, Rz, \mu) \equiv \eta(x, y, \hat{y}, z, \mu), \\ &\zeta(-x, y, \hat{y}, Rz, \mu) \equiv R\zeta(x, y, \hat{y}, z, \mu) \end{aligned} \tag{2.66}$$

and for each $(y_0, \mu_0) \in \Gamma_\gamma$ the vector field $(\Phi_{y_0\mu_0})_*^{-1} \widetilde{X}^{\mu_0 + \lambda(y_0, \mu_0)}$ [$\Phi_{y_0\mu_0}$ being defined by (2.57)] has the form (2.58). Moreover, the mapping $\Phi_{y_0\mu_0}$ depends on (y_0, μ_0) analytically when the point (y_0, μ_0) varies on each of the surfaces Γ_γ^ι.

Remark 1. If $g = O(z)$, then the statement of Theorem 2.7 is still true with the term $O(|\tilde{y} - y_0| + |\tilde{z}|)\partial/\partial\tilde{y}$ instead of $O_2(|\tilde{y} - y_0| + |\tilde{z}|)\partial/\partial\tilde{y}$ in (2.58). Nevertheless, the invariant n-torus (2.59) of the field $\widetilde{X}^{\mu_0+\lambda(y_0,\mu_0)}$ is still Floquet and its Floquet matrix is similar to $\mathbf{0}_m \oplus \Omega(y_0, \mu_0)$ where $\mathbf{0}_m$ is the $m \times m$ zero matrix.

Remark 2. For an extensive bibliography on the reversible KAM theory, see [287, 320, 326]. Here we confine ourselves by some references on the *lower-dimensional* invariant tori[3] in reversible vector fields [mainly the reversible (n, m, p, s) context 1 with $p \geq 1$], namely, [249, 250, 273, 310–312, 316, 318, 320] (for $m = n$), [323, 325] (for $m \geq n$), and [39, 60, 61, 326] (for arbitrary n and m). As in the Hamiltonian isotropic context, these lower-dimensional tori are said to be *elliptic* if $N_1 = N_3 = 0$ in (2.50), and said to be *hyperbolic* if $N_2 = 0$.

All the five Theorems 2.3–2.7 can be proven by the KAM technique of an infinite sequence of coordinate transformations within a unified Lie-algebraic approach (originally invented by Moser [246, 248]), see Broer, Huitema & Takens [60, 62, 162]. Theorem 2.3 for $p = 0$ is proven in Section 6.1.

2.4 The preliminary parameter reduction

The mappings Φ_{μ_0} or $\Phi_{y_0\mu_0}$ defined by (2.34), (2.46), (2.57) in Theorems 2.3–2.7 preserve all the internal and normal data of the unperturbed tori satisfying the appropriate Diophantine conditions. However, the latter are formulated in terms of the internal ω and normal ω^N frequencies only [to be more precise, these conditions consist in that the point μ_0 or (y_0, μ_0) labeling the unperturbed tori belongs to the set Γ_γ defined by (2.29), (2.41), (2.52)]. Moreover, if some pair (ω^0, ω^{N0}) satisfies these conditions, so does any pair $(c\omega^0, c\omega^{N0})$ with $c \geq 1$ [cf. § 1.5.2]. Consequently, it turns out to be possible to reduce the required number of parameters in Theorems 2.3–2.7, thereby weakening the correspondence between the unperturbed tori and perturbed ones, namely, obtaining the preservation of the internal and normal frequencies only, and up to proportionality (cf. [62, 162]). The precise formulations of these "relaxed" theorems follow.

2.4.1 The dissipative context

Let all the conditions of Theorem 2.3 be met except for that just the mapping

$$\mathbf{R}^{s+1} \ni (c, \mu) \mapsto (c\omega(\mu), c\omega_-^N(\mu)) \in \mathbf{R}^{n+r}, \quad c \in \mathcal{O}(1) \subset \mathbf{R} \tag{2.67}$$

is submersive for $\mu \in \Gamma$, or, equivalently, the mapping (2.30) is submersive for $\mu \in \Gamma$ (so that $s \geq n + r - 1$).

Theorem 2.8 (Relaxed theorem in the dissipative context) *Let X be an analytic family of analytic vector fields (2.27) on $\mathbf{T}^n \times \mathbf{R}^p$ satisfying the conditions above. Then for any $\gamma > 0$ and any neighborhood $\mathfrak{O}$ of zero in the space of all C^∞-mappings*

$$\mathbf{T}^n \times P \to \mathbf{R}^n \times \mathbf{R}^p \times gl(p, \mathbf{R}) \times \mathbf{R}^s \times \mathbf{R}, \quad (x, \mu) \mapsto (\chi, \zeta_1, \zeta_2, \lambda, \phi - 1), \tag{2.68}$$

[3] whose dimensions are less than the phase space codimension of the fixed point manifold of the reversing involution

analytic in x and such that λ and ϕ do not depend on x, there exists a neighborhood $\mathcal{X}$ of the family X in the space of all analytic families of analytic vector fields (2.32) such that for any $\widetilde{X} \in \mathcal{X}$ there is a mapping in $\mathfrak{O}$

$$(x,\mu) \mapsto (\chi(x,\mu), \zeta_1(x,\mu), \zeta_2(x,\mu), \lambda(\mu), \phi(\mu) - 1) \tag{2.69}$$

with the following property: for each $\mu_0 \in \Gamma_\gamma$ the vector field $(\Phi_{\mu_0})_*^{-1} \widetilde{X}^{\mu_0+\lambda(\mu_0)}$ [Φ_{μ_0} being defined by (2.34)] has the form

$$[\phi(\mu_0)\omega(\mu_0) + O(\tilde{z})]\frac{\partial}{\partial \tilde{x}} + [\Omega' \tilde{z} + O_2(\tilde{z})]\frac{\partial}{\partial \tilde{z}}, \tag{2.70}$$

where the eigenvalues of matrix Ω' are

$$\delta'_1, \ldots, \delta'_{N_1}, \quad \alpha'_1 \pm i\phi(\mu_0)\beta_1(\mu_0), \ldots, \alpha'_{N_2} \pm i\phi(\mu_0)\beta_{N_2}(\mu_0).$$

Moreover, the mapping Φ_{μ_0} depends on μ_0 analytically when μ_0 varies on each of the surfaces Γ^ι_γ.

2.4.2 The volume preserving context ($p \geq 2$)

Let all the conditions of Theorem 2.4 be met except for that just the mapping (2.67) is submersive for $\mu \in \Gamma$, or, equivalently, the mapping (2.30) is submersive for $\mu \in \Gamma$ (so that $s \geq n + r - 1$).

Theorem 2.9 (Relaxed theorem in the volume preserving context with $p \geq 2$) *Let X be an analytic family of analytic globally divergence-free vector fields (2.27) on $T^n \times R^p$ satisfying the conditions above. Then for any $\gamma > 0$ and any neighborhood $\mathfrak{O}$ of zero in the space of all C^∞-mappings (2.68) analytic in x and such that λ and ϕ do not depend on x, there exists a neighborhood $\mathcal{X}$ of the family X in the space of all analytic families of analytic globally divergence-free vector fields (2.32) such that for any $\widetilde{X} \in \mathcal{X}$ there is a mapping (2.69) in $\mathfrak{O}$ with the following property: for each $\mu_0 \in \Gamma_\gamma$ the vector field $(\Phi_{\mu_0})_*^{-1} \widetilde{X}^{\mu_0+\lambda(\mu_0)}$ [Φ_{μ_0} being defined by (2.34)] has the form (2.70), where the eigenvalues of matrix Ω' are*

$$\delta'_1, \ldots, \delta'_{N_1}, \quad \alpha'_1 \pm i\phi(\mu_0)\beta_1(\mu_0), \ldots, \alpha'_{N_2} \pm i\phi(\mu_0)\beta_{N_2}(\mu_0),$$

$$\delta'_1 + \cdots + \delta'_{N_1} + 2(\alpha'_1 + \cdots + \alpha'_{N_2}) = 0$$

[for $p = 2$ (hyperbolic case) and $p > 2$] or

$$\pm i\phi(\mu_0)\varepsilon(\mu_0)$$

[for $p = 2$ (elliptic case)]. Moreover, the mapping Φ_{μ_0} depends on μ_0 analytically when μ_0 varies on each of the surfaces Γ^ι_γ. Finally, all the mappings Φ_μ are volume preserving.

2.4.3 The volume preserving context ($p = 1$)

Let all the conditions of Theorem 2.5 be met except for that just the mapping

$$R^{s+2} \ni (c, y, \mu) \mapsto c\omega(y,\mu) \in R^n, \quad c \in \mathcal{O}(1) \subset R \tag{2.71}$$

is submersive for $(y,\mu) \in \Gamma$, or, equivalently, the mapping (2.42) is submersive for $(y,\mu) \in \Gamma$ (so that $s \geq n - 2$).

Theorem 2.10 (Relaxed theorem in the volume preserving context with $p = 1$) *Let X be an analytic family of analytic globally divergence-free vector fields (2.39) on $T^n \times R$ satisfying the conditions above. Then for any $\gamma > 0$ and any neighborhood $\mathfrak{O}$ of zero in the space of all C^∞-mappings*

$$T^n \times Y \times P \to R^n \times R \times R \times R^s \times R, \quad (x, y, \mu) \mapsto (\chi, \eta_1, \eta_2, \lambda, \phi - 1), \tag{2.72}$$

analytic in x and such that λ and ϕ do not depend on x, there exists a neighborhood $\mathcal{X}$ of the family X in the space of all analytic families of analytic globally divergence-free vector fields (2.44) such that for any $\widetilde{X} \in \mathcal{X}$ there is a mapping in $\mathfrak{O}$

$$(x, y, \mu) \mapsto (\chi(x, y, \mu), \eta_1(x, y, \mu), \eta_2(x, y, \mu), \lambda(y, \mu), \phi(y, \mu) - 1) \tag{2.73}$$

with the following property: for each $(y_0, \mu_0) \in \Gamma_\gamma$ the vector field $(\Phi_{y_0\mu_0})_*^{-1} \widetilde{X}^{\mu_0+\lambda(y_0,\mu_0)}$ [$\Phi_{y_0\mu_0}$ being defined by (2.46)] has the form

$$[\phi(y_0, \mu_0)\omega(y_0, \mu_0) + O(\bar{y} - y_0)]\frac{\partial}{\partial \bar{x}} + O_2(\bar{y} - y_0)\frac{\partial}{\partial \bar{y}}. \tag{2.74}$$

Moreover, the mapping $\Phi_{y_0\mu_0}$ depends on (y_0, μ_0) analytically when the point (y_0, μ_0) varies on each of the surfaces Γ_γ^t.

2.4.4 The Hamiltonian isotropic context

Let all the conditions of Theorem 2.6 be met except for that just the mapping

$$R^{n+s+1} \ni (c, y, \mu) \mapsto (c\omega(y, \mu), c\omega^N(y, \mu)) \in R^{n+r}, \quad c \in \mathcal{O}(1) \subset R \tag{2.75}$$

is submersive for $(y, \mu) \in \Gamma$, or, equivalently, the mapping (2.53) is submersive for $(y, \mu) \in \Gamma$ (so that $s \geq r - 1$).

Theorem 2.11 (Relaxed theorem in the Hamiltonian isotropic context) *Let X be an analytic family of analytic Hamiltonian vector fields (2.49) on $T^n \times R^{n+2p}$ satisfying the conditions above. Then for any $\gamma > 0$ and any neighborhood $\mathfrak{O}$ of zero in the space of all C^∞-mappings*

$$\begin{aligned} &T^n \times Y \times \mathcal{O}_n(0) \times \mathcal{O}_{2p}(0) \times P \to R^n \times R^n \times R^{2p} \times R^s \times R, \\ &(x, y, \hat{y}, z, \mu) \mapsto (\chi, \eta, \zeta, \lambda, \phi - 1), \end{aligned} \tag{2.76}$$

affine in $\hat{y}$ and z, analytic in x and such that χ does not depend on $\hat{y}$ and z while λ and ϕ do not depend on x, $\hat{y}$ and z, there exists a neighborhood $\mathcal{X}$ of the Hamilton function H^μ in the space of all analytic and analytically μ-dependent Hamiltonians (2.55) (Hamilton function $\widetilde{H}^\mu$ determines the family $\{\widetilde{X}^\mu\}_\mu$ of vector fields) such that for any $\widetilde{H}^\mu \in \mathcal{X}$ there is a mapping in $\mathfrak{O}$

$$(x, y, \hat{y}, z, \mu) \mapsto (\chi(x, y, \mu), \eta(x, y, \hat{y}, z, \mu), \zeta(x, y, \hat{y}, z, \mu), \lambda(y, \mu), \phi(y, \mu) - 1) \tag{2.77}$$

with the following property: for each $(y_0, \mu_0) \in \Gamma_\gamma$ the vector field $(\Phi_{y_0\mu_0})_*^{-1} \widetilde{X}^{\mu_0+\lambda(y_0,\mu_0)}$ [$\Phi_{y_0\mu_0}$ being defined by (2.57)] has the form

$$\begin{aligned} &[\phi(y_0, \mu_0)\omega(y_0, \mu_0) + O(|\bar{y} - y_0| + |\bar{z}|)]\frac{\partial}{\partial \bar{x}} + O_2(|\bar{y} - y_0| + |\bar{z}|)\frac{\partial}{\partial \bar{y}} + \\ &[\Omega' \bar{z} + O_2(|\bar{y} - y_0| + |\bar{z}|)]\frac{\partial}{\partial \bar{z}}, \end{aligned} \tag{2.78}$$

where the eigenvalues of matrix Ω' *are*

$$\begin{aligned}&\pm\delta'_1,\ldots,\pm\delta'_{N_1}, \quad \pm i\phi(y_0,\mu_0)\varepsilon_1(y_0,\mu_0),\ldots,\pm i\phi(y_0,\mu_0)\varepsilon_{N_2}(y_0,\mu_0),\\ &\pm\alpha'_1 \pm i\phi(y_0,\mu_0)\beta_1(y_0,\mu_0),\ldots,\pm\alpha'_{N_3} \pm i\phi(y_0,\mu_0)\beta_{N_3}(y_0,\mu_0).\end{aligned} \tag{2.79}$$

Moreover, the mapping $\Phi_{y_0\mu_0}$ *depends on* (y_0,μ_0) *analytically when the point* (y_0,μ_0) *varies on each of the surfaces* Γ^ι_γ. *Finally, the invariant n-torus (2.59) of the field* $\widetilde{X}^{\mu_0+\lambda(y_0,\mu_0)}$ *is isotropic.*

2.4.5 The reversible context 1

Let all the conditions of Theorem 2.7 be met except for that just the mapping

$$R^{m+s+1} \ni (c,y,\mu) \mapsto (c\omega(y,\mu), c\omega^N(y,\mu)) \in R^{n+r}, \quad c \in \mathcal{O}(1) \subset R \tag{2.80}$$

is submersive for $(y,\mu)\in\Gamma$, or, equivalently, the mapping (2.62) is submersive for $(y,\mu)\in\Gamma$ (so that $s \geq n-m+r-1$).

Theorem 2.12 (Relaxed theorem in the reversible context 1) *Let X be an analytic family of analytic G-reversible vector fields (2.60) on* $T^n\times R^{m+2p}$ *satisfying the conditions above. Then for any* $\gamma>0$ *and any neighborhood* $\mathfrak{D}$ *of zero in the space of all* C^∞*-mappings*

$$\begin{aligned}&T^n\times Y\times\mathcal{O}_m(0)\times\mathcal{O}_{2p}(0)\times P \to R^n\times R^m\times R^{2p}\times R^s\times R,\\ &(x,y,\hat{y},z,\mu)\mapsto(\chi,\eta,\zeta,\lambda,\phi-1),\end{aligned} \tag{2.81}$$

affine in $\hat{y}$ *and* z, *analytic in* x *and such that* χ *does not depend on* $\hat{y}$ *and* z *while* λ *and* ϕ *do not depend on* x, $\hat{y}$ *and* z, *there exists a neighborhood* $\mathcal{X}$ *of the family X in the space of all analytic families of analytic G-reversible vector fields (2.65) such that for any* $\widetilde{X}\in\mathcal{X}$ *there is a mapping (2.77) in* $\mathfrak{D}$ *satisfying identities (2.66) and possessing the following property: for each* $(y_0,\mu_0)\in\Gamma_\gamma$ *the vector field* $(\Phi_{y_0\mu_0})_*^{-1}\widetilde{X}^{\mu_0+\lambda(y_0,\mu_0)}$ *[*$\Phi_{y_0\mu_0}$ *being defined by (2.57)] has the form (2.78), the eigenvalues of matrix* Ω' *being of the form (2.79). Moreover, the mapping* $\Phi_{y_0\mu_0}$ *depends on* (y_0,μ_0) *analytically when the point* (y_0,μ_0) *varies on each of the surfaces* Γ^ι_γ.

Remark. If $g=O(z)$, then the statement of Theorem 2.12 is still true with the term $O(|\bar{y}-y_0|+|\bar{z}|)\partial/\partial\bar{y}$ instead of $O_2(|\bar{y}-y_0|+|\bar{z}|)\partial/\partial\bar{y}$ in (2.78). Nevertheless, the invariant n-torus (2.59) of the field $\widetilde{X}^{\mu_0+\lambda(y_0,\mu_0)}$ is still Floquet and its Floquet matrix is similar to $0_m\oplus\Omega'$.

2.4.6 The proof in the Hamiltonian isotropic context

The "relaxed" Theorems 2.8–2.12 can be easily obtained from their "main" counterparts 2.3–2.7 by the Implicit Function Theorem. Below we demonstrate this reduction technique in the Hamiltonian isotropic context. The proofs for the other contexts are quite similar.

Proof of Theorem 2.11. Let $\nu\in\mathcal{O}_t(0)$ be an additional t-dimensional parameter and let analytic mappings

$$F^{\text{new}}: Y\times P\times\mathcal{O}_t(0)\to R, \quad K^{\text{new}}: Y\times P\times\mathcal{O}_t(0)\to gl(2p,R)$$

[$K^{\text{new}}(y,\mu,\nu)$ is symmetric for all values of y, μ and ν] possess the following properties:

i) $F^{\text{new}}(y,\mu,0) \equiv F(y,\mu)$, $K^{\text{new}}(y,\mu,0) \equiv K(y,\mu)$;

ii) the mapping

$$\mathbb{R}^{n+s+t} \ni (y,\mu,\nu) \mapsto (\omega^{\text{new}}, \delta^{\text{new}}, \varepsilon^{\text{new}}, \alpha^{\text{new}}, \beta^{\text{new}}) \in \mathbb{R}^{n+p}$$

is submersive for $(y,\mu) \in \Gamma$, $\nu = 0$, where $\omega^{\text{new}} = \partial F^{\text{new}}/\partial y$ and δ^{new}, ε^{new}, α^{new}, β^{new} pertain to $\Omega^{\text{new}} = JK^{\text{new}}$ (so that $t \geq p - s$).

The possibility of extending $K(y,\mu)$ in this way follows from the theory of versal unfoldings for Hamiltonian matrices [120, 161, 178].

One can apply Theorem 2.6 to the Hamilton function $\widetilde{H}^{\mu,\nu}_{\text{new}}$ having been obtained from the function $\widetilde{H}^{\mu}$ (2.55) by replacing $F(y,\mu)$ and $K(y,\mu)$ with $F^{\text{new}}(y,\mu,\nu)$ and $K^{\text{new}}(y,\mu,\nu)$, respectively. Let Hamiltonian $\widetilde{H}^{\mu,\nu}_{\text{new}}$ determine the family $\left\{\widetilde{X}^{\mu,\nu}_{\text{new}}\right\}_{\mu,\nu}$ of vector fields. One has $\widetilde{H}^{\mu,0}_{\text{new}} = \widetilde{H}^{\mu}$ and $\widetilde{X}^{\mu,0}_{\text{new}} = \widetilde{X}^{\mu}$. By $\Gamma^{\text{new}}_{\gamma}$, where $\gamma > 0$, denote the set

$$\begin{aligned}\Gamma^{\text{new}}_{\gamma} := \{(y,\mu,\nu) \in \Gamma \times O_t(0) \ : \ & \forall k \in \mathbb{Z}^n \setminus \{0\}\ \forall \ell \in \mathbb{Z}^r, |\ell| \leq 2, \\ & |\langle \omega^{\text{new}}, k\rangle + \langle (\omega^N)^{\text{new}}, \ell\rangle| \geq \gamma |k|^{-\tau}\}\end{aligned} \tag{2.82}$$

where $(\omega^N)^{\text{new}} = (\varepsilon^{\text{new}}, \beta^{\text{new}})$. Now for any $\gamma > 0$ and any $(y_0,\mu_0,\nu_0) \in \Gamma^{\text{new}}_{\gamma/2}$, Theorem 2.6 provides a mapping

$$\begin{aligned}\Phi^{\text{new}}_{y_0\mu_0\nu_0} : (\bar{x},\bar{y},\bar{z}) \ \mapsto \ & (\bar{x} + \chi^{\text{new}}(\bar{x}, y_0, \mu_0, \nu_0), \\ & \bar{y} + \eta^{\text{new}}(\bar{x}, y_0, \bar{y} - y_0, \bar{z}, \mu_0, \nu_0), \\ & \bar{z} + \zeta^{\text{new}}(\bar{x}, y_0, \bar{y} - y_0, \bar{z}, \mu_0, \nu_0))\end{aligned}$$

such that the vector field

$$\left(\Phi^{\text{new}}_{y_0\mu_0\nu_0}\right)^{-1}_{*} \widetilde{X}^{\mu_0+\lambda^{\text{new}}(y_0,\mu_0,\nu_0),\nu_0+\theta(y_0,\mu_0,\nu_0)}_{\text{new}}$$

has the form

$$\begin{aligned}&[\omega^{\text{new}}(y_0,\mu_0,\nu_0) + O(|\bar{y} - y_0| + |\bar{z}|)]\frac{\partial}{\partial \bar{x}} + O_2(|\bar{y} - y_0| + |\bar{z}|)\frac{\partial}{\partial \bar{y}} + \\ &[\Omega^{\text{new}}(y_0,\mu_0,\nu_0)\bar{z} + O_2(|\bar{y} - y_0| + |\bar{z}|)]\frac{\partial}{\partial \bar{z}}.\end{aligned}$$

One can solve the equation $\nu + \theta(y,\mu,\nu) = 0$ with respect to ν and obtain $\nu = \xi(y,\mu)$ where the function ξ is C^∞-small. Now consider the system of equations

$$\omega^{\text{new}}(y,\mu,\xi(y,\mu)) = c\,\omega(y_0,\mu_0), \quad (\omega^N)^{\text{new}}(y,\mu,\xi(y,\mu)) = c\,\omega^N(y_0,\mu_0) \tag{2.83}$$

with respect to (c,y,μ) for $(y_0,\mu_0) \in \Gamma$. As the mapping (2.75) is submersive, system (2.83) can be solved by the Implicit Function Theorem (nonuniquely for $s > r - 1$):

$$c = \phi(y_0,\mu_0), \quad y = \Theta(y_0,\mu_0), \quad \mu = M(y_0,\mu_0)$$

where the functions $\phi(y,\mu)-1$, $\Theta(y,\mu)-y$, and $M(y,\mu)-\mu$ are C^∞-small. Let $\kappa(y,\mu)=\xi(\Theta(y,\mu),M(y,\mu))$. Obviously, $(\Theta(y_0,\mu_0),M(y_0,\mu_0),\kappa(y_0,\mu_0))\in\Gamma^{\text{new}}_{\gamma/2}$ whenever $(y_0,\mu_0)\in\Gamma_\gamma$. It remains to set

$$\begin{aligned}
&\chi(x,y,\mu)=\chi^{\text{new}}(x,\Theta(y,\mu),M(y,\mu),\kappa(y,\mu)),\\
&\eta(x,y,\hat{y},z,\mu)=\eta^{\text{new}}(x,\Theta(y,\mu),\hat{y},z,M(y,\mu),\kappa(y,\mu))+\Theta(y,\mu)-y,\\
&\zeta(x,y,\hat{y},z,\mu)=\zeta^{\text{new}}(x,\Theta(y,\mu),\hat{y},z,M(y,\mu),\kappa(y,\mu)),\\
&\lambda(y,\mu)=\lambda^{\text{new}}(\Theta(y,\mu),M(y,\mu),\kappa(y,\mu))+M(y,\mu)-\mu.
\end{aligned}\tag{2.84}$$

Indeed, let $(y_0,\mu_0)\in\Gamma_\gamma$. We will write $\Theta(y_0,\mu_0)=\Theta_0$, $M(y_0,\mu_0)=M_0$, $\xi(\Theta_0,M_0)=\kappa(y_0,\mu_0)=\kappa_0$. Taking into account that $\kappa_0+\theta(\Theta_0,M_0,\kappa_0)=0$, we have for the mapping $\Phi_{y_0\mu_0}$ defined by (2.57) and (2.84):

$$(\Phi_{y_0\mu_0})_*^{-1}\,\widetilde{X}^{\mu_0+\lambda(y_0,\mu_0)}=\left(\Phi^{\text{new}}_{\Theta_0M_0\kappa_0}\mathfrak{Y}_{\Theta_0-y_0}\right)_*^{-1}\widetilde{X}^{M_0+\lambda^{\text{new}}(\Theta_0,M_0,\kappa_0),0}_{\text{new}}=\tag{2.85}$$

$$\begin{aligned}
&[\omega^{\text{new}}(\Theta_0,M_0,\kappa_0)+O(|\bar{y}+(\Theta_0-y_0)-\Theta_0|+|\bar{z}|)]\frac{\partial}{\partial\bar{x}}+\\
&O_2(|\bar{y}+(\Theta_0-y_0)-\Theta_0|+|\bar{z}|)\frac{\partial}{\partial\bar{y}}+\\
&[\Omega^{\text{new}}(\Theta_0,M_0,\kappa_0)\bar{z}+O_2(|\bar{y}+(\Theta_0-y_0)-\Theta_0|+|\bar{z}|)]\frac{\partial}{\partial\bar{z}}=\\
&[\omega^{\text{new}}(\Theta_0,M_0,\kappa_0)+O(|\bar{y}-y_0|+|\bar{z}|)]\frac{\partial}{\partial\bar{x}}+O_2(|\bar{y}-y_0|+|\bar{z}|)\frac{\partial}{\partial\bar{y}}+\\
&[\Omega^{\text{new}}(\Theta_0,M_0,\kappa_0)\bar{z}+O_2(|\bar{y}-y_0|+|\bar{z}|)]\frac{\partial}{\partial\bar{z}}.
\end{aligned}$$

Here $\mathfrak{Y}_a:(\bar{x},\bar{y},\bar{z})\mapsto(\bar{x},\bar{y}+a,\bar{z})$ for $a\in R^n$, the factor $\mathfrak{Y}_{\Theta_0-y_0}$ in (2.85) comes from the term $\Theta(y,\mu)-y$ in definition (2.84) of $\eta(x,y,\hat{y},z,\mu)$. Now it suffices to note that

$$\omega^{\text{new}}(\Theta_0,M_0,\kappa_0)=\phi(y_0,\mu_0)\omega(y_0,\mu_0),\quad (\omega^N)^{\text{new}}(\Theta_0,M_0,\kappa_0)=\phi(y_0,\mu_0)\omega^N(y_0,\mu_0)$$

which completes the proof. □

2.5 The final parameter reduction

The number s of external parameters required in Theorems 2.8–2.12 is still too large compared to the heuristic predictions of § 1.4.1. The reason is that in those theorems, although already "relaxed" with respect to the information on the perturbed tori (we no longer control the real parts of the eigenvalues of the Floquet matrices of the tori), there is still a correspondence between the perturbed tori and unperturbed ones: to each perturbed torus, there corresponds an unperturbed torus with the same collection of the internal and normal frequencies (up to proportionality). If we get rid of any intention to connect the perturbed tori and unperturbed ones and wish to prove just that a perturbed system possesses a Whitney-smooth family of Floquet Diophantine invariant n-tori (and that the relative measure of the union of these tori tends to unit as the perturbation size tends to zero), then we would be able to reduce the number of parameters to the heuristic values

[$s = 1$ for the dissipative context and volume preserving context with $p \geq 2$, $s = 0$ for the volume preserving context with $p = 1$ and Hamiltonian isotropic context, and, finally, $s = \max\{1 - m, 0\}$ for the reversible context 1]. However, this final parameter reduction requires more advanced mathematical tools than the Implicit Function Theorem we used for the preliminary reduction, namely, the Diophantine approximations on submanifolds of the Euclidean space (*Diophantine approximations of dependent quantities* in Sprindžuk's terminology [337]).

The exposition of the Diophantine approximation theory and the "miniparameter" KAM theorems in this section follows closely our paper [61].

2.5.1 Diophantine approximations on submanifolds

In the theory of Diophantine approximations on submanifolds, one looks for Diophantine points (i.e., those satisfying suitable arithmetical conditions) that lie on a given submanifold $\mathfrak{W}$ of the Euclidean space R^n. A surface $\mathfrak{W} \subset \mathrm{R}^n$ may contain many Diophantine points (in the sense of the Hausdorff measure) due to *algebraic* reasons (if this surface is algebraic and meets certain conditions of the algebraic nature) or *geometric* reasons (e.g., if surface $\mathfrak{W}$ is sufficiently "bent") [337]. It is the latter case that is of importance for the KAM theory, the surface $\mathfrak{W}$ being in fact the image of the frequency map $\omega : P \to \mathrm{R}^n$ or $\omega : Y \times P \to \mathrm{R}^n$ (or the image of a map close to ω).

The main observation of the "miniparameter" KAM theory developed in this section is that the unperturbed mappings ω and ω^N need not satisfy any submersivity conditions (involving only the first derivatives of these mappings) to ensure the existence of many invariant tori in perturbed systems. Instead, those mappings have to meet just much weaker conditions involving partial derivatives of arbitrary orders.

We proceed now to precise statements. Let $W \subset \mathrm{R}^d$ be an open domain ($d \in \mathrm{N}$) and $\Gamma \subset W$ a subset of W diffeomorphic to a closed d-dimensional ball. Let $\mathfrak{F} : W \to \mathrm{R}^n$ be a mapping of class C^Q, $Q \in \mathrm{N}$. We will write

$$D^q\mathfrak{F}(w) = \frac{\partial^{|q|}\mathfrak{F}(w)}{\partial w^q} \in \mathrm{R}^n \quad \text{for} \quad q \in \mathrm{Z}_+^d,\ 0 \leq |q| \leq Q,$$

where $\partial w^q = \partial w_1^{q_1} \cdots \partial w_d^{q_d}$. Choose an arbitrary vector $e \in \mathrm{R}^n \setminus \{0\}$. It is obvious that if for some $w^0 \in W$ and $j \in \mathrm{Z}_+$, $0 \leq j \leq Q$, the equality

$$\sum_{|q|=j} \langle D^q\mathfrak{F}(w^0), e\rangle u^q = 0$$

holds for all $u \in \mathrm{R}^d$ (here $u^q = u_1^{q_1} \cdots u_d^{q_d}$), then all the $(d + j - 1)!/j!(d - 1)!$ vectors $D^q\mathfrak{F}(w^0)$, $|q| = j$, are orthogonal to e. Consequently, if for some $w \in W$ the collection of $(d + Q)!/d!Q!$ vectors

$$D^q\mathfrak{F}(w), \quad q \in \mathrm{Z}_+^d,\ 0 \leq |q| \leq Q \tag{2.86}$$

span R^n, then the quantity

$$\rho_{\mathfrak{F}}^Q(w) := \min_{\|e\|=1} \max_{j=0}^{Q} \max_{\|u\|=1} \left| \sum_{|q|=j} \langle D^q\mathfrak{F}(w), e\rangle u^q \right|$$

is positive ($e \in \mathrm{R}^n$, $u \in \mathrm{R}^d$), here $\|e\|^2 := \sum_{i=1}^n e_i^2$ and similarly for $\|u\|$.

Let also $\mathfrak{G} : W \to \mathrm{R}$ be a function of class C^Q. For any $w \in W$ we will write

$$\Xi^Q_{\mathfrak{G}}(w) := \max_{j=0}^{Q} \max_{\|u\|=1} \left| \sum_{|q|=j} D^q\mathfrak{G}(w)u^q \right|$$

(here again $u \in \mathrm{R}^d$).

Lemma 2.13 (Diophantine approximations on submanifolds) [326] *Suppose that for each $w \in \Gamma$ the collection of $(d+Q)!/d!Q!$ vectors (2.86) span R^n. Assume also that*

$$\langle k, \mathfrak{F}(w)\rangle \neq \mathfrak{G}(w) \tag{2.87}$$

for all $w \in \Gamma$ and $k \in \mathrm{Z}^n$, $0 < \|k\| \le \Xi^Q_{\mathfrak{G}}(w) \big/ \rho^Q_{\mathfrak{F}}(w)$.

Then there is $\delta^* > 0$ such that for any C^Q mappings $\widetilde{\mathfrak{F}} : W \to \mathrm{R}^n$ and $\widetilde{\mathfrak{G}} : W \to \mathrm{R}$ subject to inequalities

$$\sup_{w\in W} \max_{|q|\le Q} \left| D^q\widetilde{\mathfrak{F}}(w) - D^q\mathfrak{F}(w)\right| < \delta^*, \quad \sup_{w\in W} \max_{|q|\le Q} \left| D^q\widetilde{\mathfrak{G}}(w) - D^q\mathfrak{G}(w)\right| < \delta^* \tag{2.88}$$

($q \in \mathrm{Z}^d_+$) the following holds. For any $\tau > nQ-1$ and $\epsilon^* > 0$ there exists $\gamma = \gamma(\tau, \epsilon^*) > 0$ such that the Lebesgue measure of the set of those points $w \in \Gamma$ for which

$$\left| \left\langle k, \widetilde{\mathfrak{F}}(w)\right\rangle - \widetilde{\mathfrak{G}}(w)\right| \ge \gamma |k|^{-\tau} \quad \textit{for all} \quad k \in \mathrm{Z}^n \setminus \{0\}$$

is greater than $(1-\epsilon^*)\,\mathrm{meas}_d\,\Gamma$.

Note that $\Xi^Q_{\mathfrak{G}}(w) \big/ \rho^Q_{\mathfrak{F}}(w)$ is bounded from above on Γ, and condition (2.87) involves therefore only finitely many resonances to be avoided.

The "homogeneous nonperturbative" analogue of Lemma 2.13 (where the function $\mathfrak{G}$ is absent and one has to estimate $\langle k, \mathfrak{F}(w)\rangle$ only) was obtained by Bakhtin [23, 25]. In turn, Bakhtin's theorem generalizes an earlier result by Pyartli ([286], see also [358, 361]) pertaining to the case $d = 1$. Xia [359] and Cheng & Sun [89] gave another sharpening of Pyartli's lemma. A statement very close to Bakhtin's theorem was also proven by Xu, You & Qiu [360]. In the latter paper, estimates on $\left\langle k, \widetilde{\mathfrak{F}}(w)\right\rangle$ were obtained as well. "Nonhomogeneous" Diophantine approximations of dependent quantities (with the function $\mathfrak{G}$ included) were first studied by Sevryuk [326], where a proof for Lemma 2.13 was presented. This proof is reproduced in Section 6.4.

Lemma 2.14 [302, 303] *For an analytic mapping $\mathfrak{F} : W \to \mathrm{R}^n$, the following two statements are equivalent:*

(1) *there is a number $Q \in \mathrm{N}$ such that the collection of $(d+Q)!/d!Q!$ vectors (2.86) span R^n for each point $w \in \Gamma$;*

(2) *the image of the mapping $\mathfrak{F} : \Gamma \to \mathrm{R}^n$ does not lie in any linear hyperplane in R^n passing through the origin.*

Proof. If the image $\mathfrak{F}(\Gamma)$ of set Γ lies in some hyperplane in R^n passing through the origin, then for any $w \in \Gamma$ and $q \in \mathrm{Z}^d_+$ the vector $D^q\mathfrak{F}(w)$ lies in this hyperplane. Thus, (1) $\Longrightarrow$ (2) (here we have not used the analyticity of $\mathfrak{F}$). On the other hand, suppose that $\mathfrak{F}(\Gamma)$ does not lie in any hyperplane in $\mathbb{R}^n$ passing through the origin. If for some point $w^0 \in \Gamma$ all the vectors $D^q\mathfrak{F}(w^0)$, $q \in \mathbb{Z}_+^d$, belong to some hyperplane in $\mathbb{R}^n$, i.e., all of them are orthogonal to some vector $e \in \mathbb{R}^n \setminus \{0\}$, then $\langle \mathfrak{F}(w), e\rangle \equiv 0$ for $w \in \Gamma$ due to the analyticity of $\mathfrak{F}$. Thus, for each point $w^0 \in \Gamma$, there exists a number $Q(w^0) \in \mathbb{N}$ such that the vectors $D^q\mathfrak{F}(w^0)$, $q \in \mathbb{Z}_+^d$, $|q| \leq Q(w^0)$, span $\mathbb{R}^n$. Now observe that the vectors $D^q\mathfrak{F}(w)$, $|q| \leq Q(w^0)$, will span $\mathbb{R}^n$ for each $w \in W$ sufficiently close to w^0. As Γ is compact, there is $Q \in \mathbb{N}$ such that the vectors $D^q\mathfrak{F}(w)$, $|q| \leq Q$, span $\mathbb{R}^n$ for each $w \in \Gamma$. Thus, (2) $\Longrightarrow$ (1). □

In fact, the conditions on $\mathfrak{F}$ pointed out in Lemma 2.14 are very weak, cf. Example 4.7 in Section 4.2.

That the collection of $(d+Q)!/d!Q!$ vectors (2.86) span $\mathbb{R}^n$ is a far generalization of the fact that the mapping $\mathfrak{F}$ or the mapping

$$(c, w) \mapsto c\mathfrak{F}(w), \qquad c \in \mathcal{O}(1) \subset \mathbb{R} \tag{2.89}$$

is submersive. Indeed, the submersivity of $\mathfrak{F}$ means that the collection of d vectors $D^q\mathfrak{F}(w)$, $q \in \mathbb{Z}_+^d$, $|q| = 1$ span $\mathbb{R}^n$, while the submersivity of mapping (2.89) means that the collection of $d+1$ vectors $D^q\mathfrak{F}(w)$, $q \in \mathbb{Z}_+^d$, $0 \leq |q| \leq 1$ span $\mathbb{R}^n$. Both these collections of vectors are subcollections of collection (2.86). Whereas the mapping $\mathfrak{F}$ can be submersive for $d \geq n$ only and the mapping (2.89) can be submersive for $d \geq n-1$ only, vectors (2.86) can span $\mathbb{R}^n$ for any d and n provided that Q is sufficiently large [if $Q \geq n-1$ then $(d+Q)!/d!Q! \geq n$ for any $d \geq 1$].

For C^∞ mappings $\mathfrak{F} : W \to \mathbb{R}^n$, condition (1) in Lemma 2.14 is much stronger than condition (2). In fact, there are C^∞ mappings $\mathfrak{F} : W \to \mathbb{R}^n$ such that for all the points w in some subset $\Xi \subset \Gamma$ of positive measure all the derivatives $D^q\mathfrak{F}(w)$, $q \in \mathbb{Z}_+^d$, vanish, but each point $w^0 \in W$ possesses a neighborhood $\mathcal{O}_d(w^0) \subset W$ whose image $\mathfrak{F}(\mathcal{O}_d(w^0))$ does not lie in any hyperplane in $\mathbb{R}^n$ passing through the origin.

Example 2.15 Let $d = 1$ and Ξ be a perfect nowhere dense (i.e., Cantor) subset of segment Γ of positive measure. Let $\mathcal{F} : W \to \mathbb{R}$ be a C^∞ function such that $\{w \in W : \mathcal{F}(w) = 0\} = \Xi$ (see, e.g., [157, § 2.2]). Then $\mathcal{F}^{(l)}(w) = 0$ for each point $w \in \Xi$ and integer $l \in \mathbb{Z}_+$, since Ξ has no isolated points. We can further require that the set of points $w \in W$ for which $d\mathcal{F}(w)/dw = 0$ be nowhere dense. Now set $\mathfrak{F} = (\mathcal{F}, \mathcal{F}^2, \ldots, \mathcal{F}^n)$. This C^∞ mapping possesses the desired properties.

To the best of the authors' knowledge, the first paper to exploit Diophantine approximations of dependent quantities in the KAM theory was Parasyuk's note [262] devoted to the reversible $(n, m, 0, 0)$ context 1 for $n > m$. Unfortunately, this paper has not been translated into English.

2.5.2 The dissipative context

Let $s \geq 1$. Introduce the following notations:

$$\rho^Q(\mu) := \min_{\|e\|=1} \max_{j=0}^{Q} \max_{\|u\|=1} \left| \sum_{|q|=j} \langle D^q\omega(\mu), e\rangle u^q \right| \tag{2.90}$$

($q \in \mathbb{Z}_+^s$, $e \in \mathbb{R}^n$, $u \in \mathbb{R}^s$) where $Q \in \mathbb{N}$,

$$\Xi_\ell^Q(\mu) := \max_{j=0}^{Q} \max_{\|u\|=1} \left| \sum_{|q|=j} \langle D^q \omega^N(\mu), \ell \rangle u^q \right| \tag{2.91}$$

($q \in \mathbb{Z}_+^s$, $u \in \mathbb{R}^s$) where $Q \in \mathbb{N}$ and $\ell \in \mathbb{Z}^r$.

Let all the conditions of Theorem 2.3 be met except for that the mapping $\mu \mapsto (\omega, \delta, \alpha, \beta)$ is assumed to possess the following properties instead of submersivity:

a) there exists $Q \in \mathbb{N}$ such that for any $\mu \in \Gamma$ the collection of $(s+Q)!/s!Q!$ vectors

$$D^q \omega(\mu) \in \mathbb{R}^n, \quad q \in \mathbb{Z}_+^s, \ 0 \le |q| \le Q \tag{2.92}$$

span $\mathbb{R}^n$ [in particular, $(s+Q)!/s!Q! \ge n$, whence $s \ge n-1$ for $Q=1$ but any value of $s \ge 1$ is allowed for $Q \ge n-1$], or, equivalently, the image of the map $\omega : \Gamma \to \mathbb{R}^n$ does not lie in any linear hyperplane passing through the origin (see Lemma 2.14); this condition ensures that $\rho^Q(\mu) > 0$ for any $\mu \in \Gamma$,

b) for each $\mu \in \Gamma$, $\ell \in \mathbb{Z}^r$, $1 \le |\ell| \le 2$, and $k \in \mathbb{Z}^n$, $0 < \|k\| \le \Xi_\ell^Q(\mu)/\rho^Q(\mu)$, the following inequality holds:

$$\langle \omega(\mu), k \rangle \ne \langle \omega^N(\mu), \ell \rangle. \tag{2.93}$$

Theorem 2.16 (Miniparameter theorem in the dissipative context) *Let X be an analytic family of analytic vector fields (2.27) on $\mathbb{T}^n \times \mathbb{R}^p$ satisfying the conditions above. Then for any $\tau > nQ-1$ fixed, any $\gamma > 0$ and any neighborhood $\mathfrak{D}$ of zero in the space of all C^∞-mappings*

$$\mathbb{T}^n \times P \to \mathbb{R}^n \times \mathbb{R}^p \times gl(p, \mathbb{R}), \quad (x, \mu) \mapsto (\chi, \zeta_1, \zeta_2), \tag{2.94}$$

analytic in x, there exists a neighborhood $\mathcal{X}$ of the family X in the space of all analytic families of analytic vector fields (2.32) such that for any $\tilde{X} \in \mathcal{X}$ there are a set $\mathcal{G} \subset \Gamma$ and a mapping in $\mathfrak{D}$

$$(x, \mu) \mapsto (\chi(x, \mu), \zeta_1(x, \mu), \zeta_2(x, \mu)) \tag{2.95}$$

with the following properties: $\operatorname{meas}_s \mathcal{G} \ge (1-\gamma) \operatorname{meas}_s \Gamma$ and for each $\mu_0 \in \mathcal{G}$ the vector field $(\Phi_{\mu_0})_*^{-1} \tilde{X}^{\mu_0}$ [Φ_{μ_0} being defined by (2.34)] has the form

$$[\omega' + O(\bar{z})] \frac{\partial}{\partial \bar{x}} + [\Omega' \bar{z} + O_2(\bar{z})] \frac{\partial}{\partial \bar{z}}, \tag{2.96}$$

where the eigenvalues of matrix Ω' are

$$\delta'_1, \ldots, \delta'_{N_1}, \quad \alpha'_1 \pm i\beta'_1, \ldots, \alpha'_{N_2} \pm i\beta'_{N_2}$$

and

$$\forall k \in \mathbb{Z}^n \setminus \{0\} \ \forall \ell \in \mathbb{Z}^r, |\ell| \le 2, \quad |\langle \omega', k \rangle + \langle \omega'^N, \ell \rangle| \ge \gamma |k|^{-\tau} \tag{2.97}$$

with $\omega'^N = \beta'$.

Here there is no correspondence between the perturbed tori and unperturbed ones, and we do not need to introduce the parameter shift $\lambda(\mu_0)$ which was present in Theorems 2.3 and 2.8.

2.5.3 The volume preserving context ($p \geq 2$)

Let $s \geq 1$ and all the conditions of Theorem 2.4 be met except for that the mapping $\mu \mapsto (\omega, \delta, \alpha, \beta)$ [for $p = 2$ (hyperbolic case) and $p > 2$] or $\mu \mapsto (\omega, \varepsilon)$ [for $p = 2$ (elliptic case)] is assumed to possess the following properties instead of submersivity:

a) there exists $Q \in \mathbb{N}$ such that for any $\mu \in \Gamma$ the collection of $(s+Q)!/s!Q!$ vectors (2.92) span $\mathbb{R}^n$ [in particular, $(s+Q)!/s!Q! \geq n$, whence $s \geq n-1$ for $Q = 1$ but any value of $s \geq 1$ is allowed for $Q \geq n-1$], or, equivalently, the image of the map $\omega : \Gamma \to \mathbb{R}^n$ does not lie in any linear hyperplane passing through the origin (see Lemma 2.14); this condition ensures that $\rho^Q(\mu) > 0$ for any $\mu \in \Gamma$, where the quantities $\rho^Q(\mu)$ and $\Xi_\ell^Q(\mu)$ are defined by (2.90) and (2.91), respectively,

b) for each $\mu \in \Gamma$, $\ell \in \mathbb{Z}^r$, $1 \leq |\ell| \leq 2$, and $k \in \mathbb{Z}^n$, $0 < \|k\| \leq \Xi_\ell^Q(\mu)/\rho^Q(\mu)$, inequality (2.93) holds.

Theorem 2.17 (Miniparameter theorem in the volume preserving context with $p \geq 2$) *Let X be an analytic family of analytic globally divergence-free vector fields (2.27) on $\mathbb{T}^n \times \mathbb{R}^p$ satisfying the conditions above. Then for any $\tau > nQ - 1$ fixed, any $\gamma > 0$ and any neighborhood $\mathfrak{O}$ of zero in the space of all C^∞-mappings (2.94) analytic in x, there exists a neighborhood $\mathcal{X}$ of the family X in the space of all analytic families of analytic globally divergence-free vector fields (2.32) such that for any $\tilde{X} \in \mathcal{X}$ there are a set $\mathcal{G} \subset \Gamma$ and a mapping (2.95) in $\mathfrak{O}$ with the following properties: $\mathrm{meas}_s\, \mathcal{G} \geq (1-\gamma)\, \mathrm{meas}_s\, \Gamma$ and for each $\mu_0 \in \mathcal{G}$ the vector field $(\Phi_{\mu_0})_*^{-1} \tilde{X}^{\mu_0}$ [Φ_{μ_0} being defined by (2.34)] has the form (2.96), where the eigenvalues of matrix Ω' are*

$$\delta_1', \ldots, \delta_{N_1}', \quad \alpha_1' \pm i\beta_1', \ldots, \alpha_{N_2}' \pm i\beta_{N_2}', \quad \delta_1' + \cdots + \delta_{N_1}' + 2(\alpha_1' + \cdots + \alpha_{N_2}') = 0$$

[for $p = 2$ (hyperbolic case) and $p > 2$; set $\omega'^N = \beta'$] or

$$\pm i\varepsilon'$$

[for $p = 2$ (elliptic case); set $\omega'^N = \varepsilon'$] and inequalities (2.97) hold. Moreover, all the mappings Φ_μ are volume preserving.

2.5.4 The volume preserving context ($p = 1$)

Let all the conditions of Theorem 2.5 be met except for that the mapping $(y, \mu) \mapsto \omega$ is assumed to possess the following property instead of submersivity:

there exists $Q \in \mathbb{N}$ such that for any $(y, \mu) \in \Gamma$ the collection of $(s+Q+1)!/(s+1)!Q!$ vectors

$$D^q \omega(y, \mu) \in \mathbb{R}^n, \quad q \in \mathbb{Z}_+^{s+1},\ 0 \leq |q| \leq Q \tag{2.98}$$

span $\mathbb{R}^n$ [in particular, $(s+Q+1)!/(s+1)!Q! \geq n$, whence $s \geq n-2$ for $Q = 1$ but any value of $s \geq 0$ is allowed for $Q \geq n-1$], or, equivalently, the image of the map $\omega : \Gamma \to \mathbb{R}^n$ does not lie in any linear hyperplane passing through the origin (see Lemma 2.14).

Theorem 2.18 (Miniparameter theorem in the volume preserving context with $p = 1$) *Let X be an analytic family of analytic globally divergence-free vector fields (2.39) on $\mathbb{T}^n \times \mathbb{R}$ satisfying the conditions above. Then for any $\tau > nQ - 1$ fixed, any $\gamma > 0$ and any neighborhood $\mathfrak{O}$ of zero in the space of all C^∞-mappings*

$$\mathbb{T}^n \times Y \times P \to \mathbb{R}^n \times \mathbb{R} \times \mathbb{R}, \quad (x, y, \mu) \mapsto (\chi, \eta_1, \eta_2), \tag{2.99}$$

analytic in x, there exists a neighborhood $\mathcal{X}$ of the family X in the space of all analytic families of analytic globally divergence-free vector fields (2.44) such that for any $\tilde{X} \in \mathcal{X}$ there are a set $\mathcal{G} \subset \Gamma$ and a mapping in $\mathfrak{D}$

$$(x, y, \mu) \mapsto (\chi(x, y, \mu), \eta_1(x, y, \mu), \eta_2(x, y, \mu)) \tag{2.100}$$

with the following properties: $\text{meas}_{s+1} \mathcal{G} \geq (1-\gamma) \text{meas}_{s+1} \Gamma$ and for each $(y_0, \mu_0) \in \mathcal{G}$ the vector field $(\Phi_{y_0\mu_0})_*^{-1} \tilde{X}^{\mu_0}$ [$\Phi_{y_0\mu_0}$ being defined by (2.46)] has the form

$$[\omega' + O(\bar{y} - y_0)]\frac{\partial}{\partial \bar{x}} + O_2(\bar{y} - y_0)\frac{\partial}{\partial \bar{y}}, \tag{2.101}$$

and

$$\forall k \in \mathbb{Z}^n \setminus \{0\} \quad |\langle \omega', k \rangle| \geq \gamma |k|^{-\tau}. \tag{2.102}$$

2.5.5 The Hamiltonian isotropic context

Introduce the following notations:

$$\rho^Q(y, \mu) := \min_{\|e\|=1} \max_{j=0}^{Q} \max_{\|u\|=1} \left| \sum_{|q|=j} \langle D^q \omega(y, \mu), e \rangle u^q \right| \tag{2.103}$$

($q \in \mathbb{Z}_+^{n+s}$, $e \in \mathbb{R}^n$, $u \in \mathbb{R}^{n+s}$) where $Q \in \mathbb{N}$,

$$\Xi_\ell^Q(y, \mu) := \max_{j=0}^{Q} \max_{\|u\|=1} \left| \sum_{|q|=j} \langle D^q \omega^N(y, \mu), \ell \rangle u^q \right| \tag{2.104}$$

($q \in \mathbb{Z}_+^{n+s}$, $u \in \mathbb{R}^{n+s}$) where $Q \in \mathbb{N}$ and $\ell \in \mathbb{Z}^r$.

Let all the conditions of Theorem 2.6 be met except for that the mapping $(y, \mu) \mapsto (\omega, \delta, \varepsilon, \alpha, \beta)$ is assumed to possess the following properties instead of submersivity:

a) there exists $Q \in \mathbb{N}$ such that for any $(y, \mu) \in \Gamma$ the collection of $(n+s+Q)!/(n+s)!Q!$ vectors

$$D^q \omega(y, \mu) \in \mathbb{R}^n, \quad q \in \mathbb{Z}_+^{n+s}, \ 0 \leq |q| \leq Q \tag{2.105}$$

span $\mathbb{R}^n$, or, equivalently, the image of the map $\omega : \Gamma \to \mathbb{R}^n$ does not lie in any linear hyperplane passing through the origin (see Lemma 2.14); this condition ensures that $\rho^Q(y, \mu) > 0$ for any $(y, \mu) \in \Gamma$,

b) for each $(y, \mu) \in \Gamma$, $\ell \in \mathbb{Z}^r$, $1 \leq |\ell| \leq 2$, and $k \in \mathbb{Z}^n$, $0 < \|k\| \leq \Xi_\ell^Q(y, \mu)/\rho^Q(y, \mu)$, the following inequality holds:

$$\langle \omega(y, \mu), k \rangle \neq \langle \omega^N(y, \mu), \ell \rangle. \tag{2.106}$$

Remark. The hypotheses of this kind were first introduced by Rüssmann [302–304]. However, these papers contained no proofs.

Theorem 2.19 (Miniparameter theorem in the Hamiltonian isotropic context) *Let X be an analytic family of analytic Hamiltonian vector fields (2.49) on $\mathbb{T}^n \times \mathbb{R}^{n+2p}$ satisfying the conditions above. Then for any $\tau > nQ - 1$ fixed, any $\gamma > 0$ and any neighborhood $\mathfrak{D}$ of zero in the space of all C^∞-mappings*

$$\mathbb{T}^n \times Y \times \mathcal{O}_n(0) \times \mathcal{O}_{2p}(0) \times P \to \mathbb{R}^n \times \mathbb{R}^n \times \mathbb{R}^{2p}, \quad (x, y, \hat{y}, z, \mu) \mapsto (\chi, \eta, \zeta), \tag{2.107}$$

affine in $\hat{y}$ and z, analytic in x and such that χ does not depend on $\hat{y}$ and z, there exists a neighborhood $\mathcal{X}$ of the Hamilton function H^μ in the space of all analytic and analytically μ-dependent Hamiltonians (2.55) (Hamilton function $\widetilde{H}^\mu$ determines the family $\{\widetilde{X}^\mu\}_\mu$ of vector fields) such that for any $\widetilde{H}^\mu \in \mathcal{X}$ there are a set $\mathcal{G} \subset \Gamma$ and a mapping in $\mathfrak{O}$

$$(x, y, \hat{y}, z, \mu) \mapsto (\chi(x, y, \mu), \eta(x, y, \hat{y}, z, \mu), \zeta(x, y, \hat{y}, z, \mu)) \tag{2.108}$$

with the following properties: $\mathrm{meas}_{n+s}\,\mathcal{G} \geq (1-\gamma)\,\mathrm{meas}_{n+s}\,\Gamma$ and for each $(y_0, \mu_0) \in \mathcal{G}$ the vector field $(\Phi_{y_0\mu_0})_*^{-1}\,\widetilde{X}^{\mu_0}$ [$\Phi_{y_0\mu_0}$ being defined by (2.57)] has the form

$$[\omega' + O(|\bar{y} - y_0| + |\bar{z}|)]\frac{\partial}{\partial \bar{x}} + O_2(|\bar{y} - y_0| + |\bar{z}|)\frac{\partial}{\partial \bar{y}} + [\Omega' \bar{z} + O_2(|\bar{y} - y_0| + |\bar{z}|)]\frac{\partial}{\partial \bar{z}}, \tag{2.109}$$

where the eigenvalues of matrix Ω' are

$$\pm\,\delta_1', \ldots, \pm\delta_{N_1}', \quad \pm i\varepsilon_1', \ldots, \pm i\varepsilon_{N_2}', \quad \pm\alpha_1' \pm i\beta_1', \ldots, \pm\alpha_{N_3}' \pm i\beta_{N_3}' \tag{2.110}$$

and inequalities (2.97) hold with $\omega'^N = (\varepsilon', \beta')$. Moreover, the invariant n-torus (2.59) of the field $\widetilde{X}^{\mu_0}$ is isotropic.

Remark. For the case where there are neither "normal coordinates" ($p = 0$) nor external parameters ($s = 0$), a somewhat weakened version of Theorem 2.19 was obtained in [89, 153, 327, 329, 360]. The case $s = 0$ for arbitrary $p \geq 0$ was considered in [331].

2.5.6 The reversible context 1

Let $m + s \geq 1$ and all the conditions of Theorem 2.7 be met except for that the mapping $(y, \mu) \mapsto (\omega, \delta, \varepsilon, \alpha, \beta)$ is assumed to possess the following properties instead of submersivity:

a) there exists $Q \in \mathbb{N}$ such that for any $(y, \mu) \in \Gamma$ the collection of $(m+s+Q)!/(m+s)!Q!$ vectors

$$D^q\omega(y, \mu) \in \mathbb{R}^n, \quad q \in \mathbb{Z}_+^{m+s},\ 0 \leq |q| \leq Q \tag{2.111}$$

span $\mathbb{R}^n$ [in particular, $(m+s+Q)!/(m+s)!Q! \geq n$, whence $s \geq n - m - 1$ for $Q = 1$ but any non-negative value of $s \geq 1 - m$ is allowed for $Q \geq n - 1$], or, equivalently, the image of the map $\omega : \Gamma \to \mathbb{R}^n$ does not lie in any linear hyperplane passing through the origin (see Lemma 2.14); this condition ensures that $\rho^Q(y, \mu) > 0$ for any $(y, \mu) \in \Gamma$, where the quantities $\rho^Q(y, \mu)$ and $\Xi_\ell^Q(y, \mu)$ are defined by (2.103) and (2.104), respectively (with $u \in \mathbb{R}^{m+s}$ instead of $u \in \mathbb{R}^{n+s}$ and $q \in \mathbb{Z}_+^{m+s}$ instead of $q \in \mathbb{Z}_+^{n+s}$),

b) for each $(y, \mu) \in \Gamma$, $\ell \in \mathbb{Z}^r$, $1 \leq |\ell| \leq 2$, and $k \in \mathbb{Z}^n$, $0 < \|k\| \leq \Xi_\ell^Q(y, \mu)/\rho^Q(y, \mu)$, inequality (2.106) holds.

Theorem 2.20 (Miniparameter theorem in the reversible context 1) *Let X be an analytic family of analytic G-reversible vector fields (2.60) on $\mathbb{T}^n \times \mathbb{R}^{m+2p}$ satisfying the conditions above. Then for any $\tau > nQ - 1$ fixed, any $\gamma > 0$ and any neighborhood $\mathfrak{O}$ of zero in the space of all C^∞-mappings*

$$\mathbb{T}^n \times Y \times \mathcal{O}_m(0) \times \mathcal{O}_{2p}(0) \times P \to \mathbb{R}^n \times \mathbb{R}^m \times \mathbb{R}^{2p}, \quad (x, y, \hat{y}, z, \mu) \mapsto (\chi, \eta, \zeta), \tag{2.112}$$

affine in $\hat{y}$ and z, analytic in x and such that χ does not depend on $\hat{y}$ and z, there exists a neighborhood $\mathcal{X}$ of the family X in the space of all analytic families of analytic

G-reversible vector fields (2.65) such that for any $\widetilde{X} \in \mathcal{X}$ there are a set $\mathcal{G} \subset \Gamma$ and a mapping (2.108) in $\mathfrak{D}$ satisfying identities (2.66) and possessing the following properties: $\operatorname{meas}_{m+s} \mathcal{G} \geq (1-\gamma) \operatorname{meas}_{m+s} \Gamma$ and for each $(y_0, \mu_0) \in \mathcal{G}$ the vector field $(\Phi_{y_0\mu_0})_*^{-1} \widetilde{X}^{\mu_0}$ [$\Phi_{y_0\mu_0}$ being defined by (2.57)] has the form (2.109), the eigenvalues of matrix Ω' being of the form (2.110) and inequalities (2.97) holding with $\omega'^N = (\varepsilon', \beta')$.

Remark. If $g = O(z)$, then the statement of Theorem 2.20 is still true with the term $O(|\bar{y}-y_0|+|\bar{z}|)\partial/\partial\bar{y}$ instead of $O_2(|\bar{y}-y_0|+|\bar{z}|)\partial/\partial\bar{y}$ in (2.109). Nevertheless, the invariant n-torus (2.59) of the field $\widetilde{X}^{\mu_0}$ is still Floquet and its Floquet matrix is similar to $\mathbf{0}_m \oplus \Omega'$.

2.5.7 The proof in the reversible context 1

The "miniparameter" Theorems 2.16–2.20 can be obtained from the "main" Theorems 2.3–2.7 using Lemma 2.13. Below we demonstrate this reduction technique in the reversible context 1. The proofs for the other contexts are entirely similar.

Proof of Theorem 2.20 (cf. § 2.4.6). Let $\nu \in \mathcal{O}_t(0)$ be an additional t-dimensional parameter and let analytic mappings

$$\omega^{\text{new}} : Y \times P \times \mathcal{O}_t(0) \to \mathbf{R}^n, \quad \Omega^{\text{new}} : Y \times P \times \mathcal{O}_t(0) \to \{L \in sl(2p, \mathbf{R}) \, : \, LR + RL = 0\}$$

possess the following properties:

i) $\omega^{\text{new}}(y, \mu, 0) \equiv \omega(y, \mu)$, $\Omega^{\text{new}}(y, \mu, 0) \equiv \Omega(y, \mu)$;

ii) the mapping

$$\mathbf{R}^{m+s+t} \ni (y, \mu, \nu) \mapsto (\omega^{\text{new}}, \delta^{\text{new}}, \varepsilon^{\text{new}}, \alpha^{\text{new}}, \beta^{\text{new}}) \in \mathbf{R}^{n+p}$$

is submersive for $(y, \mu) \in \Gamma$, $\nu = 0$ (so that $t \geq n - m + p - s$).

The possibility of extending $\Omega(y, \mu)$ in this way follows from the theory of versal unfoldings for infinitesimally reversible matrices [161, 321, 333].

One can apply Theorem 2.7 to the family of G-reversible vector fields $\widetilde{X}_{\text{new}} = \left\{\widetilde{X}^{\mu,\nu}_{\text{new}}\right\}_{\mu,\nu}$ having been obtained from the family $\widetilde{X} = \{\widetilde{X}^\mu\}_\mu$ (2.65) by replacing $\omega(y, \mu)$ and $\Omega(y, \mu)$ with $\omega^{\text{new}}(y, \mu, \nu)$ and $\Omega^{\text{new}}(y, \mu, \nu)$, respectively. One has $\widetilde{X}^{\mu,0}_{\text{new}} = \widetilde{X}^\mu$. For any $\gamma' > 0$ and any $(y_0, \mu_1, \nu_1) \in \Gamma^{\text{new}}_{\gamma'}$, where the sets $\Gamma^{\text{new}}_\gamma$ for $\gamma > 0$ are defined by (2.82), Theorem 2.7 provides a mapping

$$\begin{aligned} \Phi^{\text{new}}_{y_0\mu_1\nu_1} : (\bar{x}, \bar{y}, \bar{z}) \quad \mapsto \quad & (\bar{x} + \chi^{\text{new}}(\bar{x}, y_0, \mu_1, \nu_1), \\ & \bar{y} + \eta^{\text{new}}(\bar{x}, y_0, \bar{y} - y_0, \bar{z}, \mu_1, \nu_1), \\ & \bar{z} + \zeta^{\text{new}}(\bar{x}, y_0, \bar{y} - y_0, \bar{z}, \mu_1, \nu_1)) \end{aligned}$$

such that the vector field

$$\left(\Phi^{\text{new}}_{y_0\mu_1\nu_1}\right)_*^{-1} \widetilde{X}^{\mu_1+\lambda^{\text{new}}(y_0,\mu_1,\nu_1),\nu_1+\theta(y_0,\mu_1,\nu_1)}_{\text{new}}$$

has the form

$$\begin{aligned} & [\omega^{\text{new}}(y_0, \mu_1, \nu_1) + O(|\bar{y} - y_0| + |\bar{z}|)]\frac{\partial}{\partial \bar{x}} + O_2(|\bar{y} - y_0| + |\bar{z}|)\frac{\partial}{\partial \bar{y}} + \\ & [\Omega^{\text{new}}(y_0, \mu_1, \nu_1)\bar{z} + O_2(|\bar{y} - y_0| + |\bar{z}|)]\frac{\partial}{\partial \bar{z}}. \end{aligned}$$

One can solve the equation $\nu + \theta(y, \mu, \nu) = 0$ with respect to ν and obtain $\nu = \xi(y, \mu)$ where the function ξ is C^∞-small. Set

$$\lambda(y, \mu) = \lambda^{\text{new}}(y, \mu, \xi(y, \mu)).$$

The equation $\mu_1 + \lambda(y_0, \mu_1) = \mu_0$ for each fixed y_0 can be solved with respect to μ_1 as $\mu_1 = \kappa(y_0, \mu_0)$ where the function $\kappa(y, \mu) - \mu$ is also C^∞-small. Define the set $\mathcal{G}$ as

$$\mathcal{G} := \left\{(y_0, \mu_0) \in \Gamma \ : \ (y_0, \kappa(y_0, \mu_0)) \in \Gamma \text{ and } (y_0, \kappa(y_0, \mu_0), \xi(y_0, \kappa(y_0, \mu_0))) \in \Gamma^{\text{new}}_{\gamma'}\right\}.$$

Note that this set depends not only on X, τ, and γ', but on $\widetilde{X}$ as well.

According to Lemma 2.13, $meas_{m+s}\,\mathcal{G}/\ meas_{m+s}\,\Gamma \to 1$ as $\gamma' \downarrow 0$, and one can choose $\gamma' \le \gamma$ such that $meas_{m+s}\,\mathcal{G} \ge (1-\gamma)\ meas_{m+s}\,\Gamma$. It remains to set

$$\begin{aligned}
\chi(x, y, \mu + \lambda(y, \mu)) &= \chi^{\text{new}}(x, y, \mu, \xi(y, \mu)),\\
\eta(x, y, \hat{y}, z, \mu + \lambda(y, \mu)) &= \eta^{\text{new}}(x, y, \hat{y}, z, \mu, \xi(y, \mu)),\\
\zeta(x, y, \hat{y}, z, \mu + \lambda(y, \mu)) &= \zeta^{\text{new}}(x, y, \hat{y}, z, \mu, \xi(y, \mu)),
\end{aligned} \tag{2.113}$$

or, to be more "rigorous",

$$\begin{aligned}
\chi(x, y, \mu) &= \chi^{\text{new}}(x, y, \kappa(y, \mu), \xi(y, \kappa(y, \mu))),\\
\eta(x, y, \hat{y}, z, \mu) &= \eta^{\text{new}}(x, y, \hat{y}, z, \kappa(y, \mu), \xi(y, \kappa(y, \mu))),\\
\zeta(x, y, \hat{y}, z, \mu) &= \zeta^{\text{new}}(x, y, \hat{y}, z, \kappa(y, \mu), \xi(y, \kappa(y, \mu))).
\end{aligned}$$

Indeed, let $(y_0, \mu_0) \in \mathcal{G}$, $\mu_1 = \kappa(y_0, \mu_0)$, and $\nu_1 = \xi(y_0, \mu_1)$. Then $(y_0, \mu_1, \xi(y_0, \mu_1)) \in \Gamma^{\text{new}}_{\gamma'}$. Taking into account that $\nu_1 + \theta(y_0, \mu_1, \nu_1) = 0$ and

$$\mu_0 = \mu_1 + \lambda(y_0, \mu_1) = \mu_1 + \lambda^{\text{new}}(y_0, \mu_1, \nu_1),$$

we have for the mapping $\Phi_{y_0\mu_0}$ defined by (2.57) and (2.113):

$$\begin{aligned}
(\Phi_{y_0\mu_0})_*^{-1} \widetilde{X}^{\mu_0} &= \left(\Phi^{\text{new}}_{y_0\mu_1\nu_1}\right)_*^{-1} \widetilde{X}^{\mu_1 + \lambda^{\text{new}}(y_0, \mu_1, \nu_1), 0}_{\text{new}} =\\
&[\omega^{\text{new}}(y_0, \mu_1, \nu_1) + O(|\bar{y} - y_0| + |\bar{z}|)]\frac{\partial}{\partial \bar{x}} + O_2(|\bar{y} - y_0| + |\bar{z}|)\frac{\partial}{\partial \bar{y}} +\\
&[\Omega^{\text{new}}(y_0, \mu_1, \nu_1)\bar{z} + O_2(|\bar{y} - y_0| + |\bar{z}|)]\frac{\partial}{\partial \bar{z}}.
\end{aligned}$$

□

This parameter reduction technique was invented by Herman [153] (see also [361]) and independently by Sevryuk [326, 327, 329, 331]. Talk [153] and paper [329] consider the Hamiltonian isotropic $(n, 0, 0)$ context, paper [327] examines the Hamiltonian isotropic $(n, 0, 0)$ context and reversible $(n, n, 0, 0)$ context 1, paper [331] studies more general Hamiltonian isotropic $(n, p, 0)$ context with arbitrary n, p, and paper [326] is devoted to more general reversible $(n, m, p, 0)$ context 1 with arbitrary n, m, p. A similar technique is used in [361] for the Hamiltonian coisotropic $(n, p, 0)$ context for $n \ge 4$, $p = (n-2)/2$ and for the volume preserving $(n, 1, 0)$ context.

2.6 Converse KAM theory

The hypotheses of Theorems 2.16–2.20 which guarantee the existence of Whitney-smooth families of invariant tori in any sufficiently small perturbation $\widetilde{X}$ of the initial family X of integrable (or "almost integrable") vector fields, are in fact very close to the *necessary* conditions for the persistence of Whitney-smooth families of invariant tori. Below we prove the necessity of those hypotheses in the following three cases: the volume preserving (n,p,s) context with $p=1$, the Hamiltonian isotropic (n,p,s) context with $p=0$, and the reversible (n,m,p,s) context 1 with $p=0$.

Consider the analytic family X (2.39) of analytic vector fields on $\mathrm{T}^n\times\mathrm{R}$ which are globally divergence-free with respect to the volume element $dx\wedge dy$ and integrable.

Proposition 2.21 *If the image of the map $\omega: Y\times P\to\mathrm{R}^n$ lies in a linear hyperplane in R^n passing through the origin, then there exists an analytic family $\widetilde{X}$ of analytic globally divergence-free vector fields arbitrarily close to X in the real analytic topology which possess no invariant n-tori.*

Proof. We have $\langle\omega(y,\mu),e\rangle\equiv 0$ for some vector $e\in\mathrm{R}^n\setminus\{0\}$. Set $\widetilde{\omega}(y,\mu)=A^{-1}\omega(y,\mu)$ where $A\in GL(n,\mathrm{R})$. Then $\langle\widetilde{\omega}(y,\mu),\widetilde{e}\rangle\equiv 0$ where $\widetilde{e}=A^{t}e$ (A^{t} denoting the matrix transposed to A). One can choose A arbitrarily close to the identity $n\times n$ matrix in such a way that $\widetilde{e}$ will be proportional to an integer vector: $\langle\widetilde{\omega}(y,\mu),k\rangle\equiv 0$ with $k\in\mathrm{Z}^n\setminus\{0\}$. Now let

$$\widetilde{X}=\{\widetilde{X}^{\mu}\}_{\mu}=\widetilde{\omega}(y,\mu)\frac{\partial}{\partial x}+\epsilon\sin\langle k,x\rangle\frac{\partial}{\partial y},\qquad \epsilon\in\mathrm{R}\setminus\{0\}. \tag{2.114}$$

One can easily verify that vector fields (2.114) are globally divergence-free with respect to the volume element $dx\wedge dy$. On the other hand, for the flows determined by these vector fields, one has $d\langle k,x\rangle/dt=\langle k,\widetilde{\omega}(y,\mu)\rangle\equiv 0$. The derivative $dy/dt=\epsilon\sin\langle k,x\rangle=C$ is therefore time independent, and $C\neq 0$ provided that $\sin\langle k,x^{\mathrm{initial}}\rangle\neq 0$. Consequently, vector fields (2.114) possess no invariant n-tori. □

Now consider the analytic family of vector fields on $\mathrm{T}^n\times\mathrm{R}^n$ which are Hamiltonian with respect to the symplectic structure $dy_1\wedge dx_1+\cdots+dy_n\wedge dx_n$ with the analytic integrable Hamilton function $H^{\mu}(x,y)=F(y,\mu)$, where $x\in\mathrm{T}^n$, $y\in Y\subset\mathrm{R}^n$, $\mu\in P\subset\mathrm{R}^s$, $F:Y\times P\to\mathrm{R}$. This family has the form

$$X=\{X^{\mu}\}_{\mu}=\omega(y,\mu)\frac{\partial}{\partial x}$$

where $\omega=\partial F/\partial y$.

Proposition 2.22 (Sevryuk [327, 329]) *If the image of the map $\omega: Y\times P\to\mathrm{R}^n$ lies in a linear hyperplane in R^n passing through the origin, then there exists an analytic and analytically μ-dependent Hamiltonian $\widetilde{H}^{\mu}$ arbitrarily close to H^{μ} in the real analytic topology and such that the family $\{\widetilde{X}^{\mu}\}_{\mu}$ of vector fields determined by $\widetilde{H}^{\mu}$ possesses no invariant n-tori.*

Proof. We have $\langle\partial F(y,\mu)/\partial y,e\rangle\equiv 0$ for some vector $e\in\mathrm{R}^n\setminus\{0\}$. Set $\widetilde{F}(y,\mu)=F(A^{-1}y,\mu)$ where $A\in GL(n,\mathrm{R})$. Then $\langle\partial\widetilde{F}(y,\mu)/\partial y,\widetilde{e}\rangle\equiv 0$ where $\widetilde{e}=Ae$. One can choose A arbitrarily close to the identity $n \times n$ matrix in such a way that $\tilde{e}$ will be proportional to an integer vector: $\langle \partial\tilde{F}(y,\mu)/\partial y, k\rangle \equiv 0$ with $k \in \mathbb{Z}^n \setminus \{0\}$. Now let

$$\tilde{H}^\mu(x,y) = \tilde{F}(y,\mu) + \epsilon\cos\langle k, x\rangle, \quad \epsilon \in \mathbb{R}\setminus\{0\}.$$

This Hamilton function determines the family of vector fields

$$\tilde{X} = \{\tilde{X}^\mu\}_\mu = \tilde{\omega}(y,\mu)\frac{\partial}{\partial x} + \epsilon k\sin\langle k, x\rangle\frac{\partial}{\partial y} \tag{2.115}$$

where $\tilde{\omega} = \partial\tilde{F}/\partial y$. One has $d\langle k, x\rangle/dt = \langle k, \tilde{\omega}(y,\mu)\rangle \equiv 0$. Now in an entirely similar way as in the proof of Proposition 2.21, we conclude that vector fields (2.115) possess no invariant n-tori. □

Finally, consider the analytic family of analytic G-reversible integrable vector fields on $\mathbb{T}^n \times \mathbb{R}^m$:

$$X = \{X^\mu\}_\mu = \omega(y,\mu)\frac{\partial}{\partial x}, \quad G: (x,y) \mapsto (-x,y),$$

where $x \in \mathbb{T}^n$, $y \in Y \subset \mathbb{R}^m$, $\mu \in P \subset \mathbb{R}^s$, $\omega: Y \times P \to \mathbb{R}^n$.

Proposition 2.23 (Sevryuk [327]) *If the image of the map $\omega: Y \times P \to \mathbb{R}^n$ lies in a linear hyperplane in $\mathbb{R}^n$ passing through the origin, then there exists an analytic family $\tilde{X}$ of analytic G-reversible vector fields arbitrarily close to X in the real analytic topology which possess no invariant n-tori.*

Proof. We have $\langle \omega(y,\mu), e\rangle \equiv 0$ for some vector $e \in \mathbb{R}^n\setminus\{0\}$. Set $\tilde{\omega}(y,\mu) = A^{-1}\omega(y,\mu)$ where $A \in GL(n,\mathbb{R})$. Then $\langle \tilde{\omega}(y,\mu), \tilde{e}\rangle \equiv 0$ where $\tilde{e} = A^t e$. One can choose A arbitrarily close to the identity $n \times n$ matrix in such a way that $\tilde{e}$ will be proportional to an integer vector: $\langle \tilde{\omega}(y,\mu), k\rangle \equiv 0$ with $k \in \mathbb{Z}^n \setminus \{0\}$. Now let

$$\tilde{X} = \{\tilde{X}^\mu\}_\mu = \tilde{\omega}(y,\mu)\frac{\partial}{\partial x} + \epsilon\sin\langle k, x\rangle\frac{\partial}{\partial y}, \quad \epsilon \in \mathbb{R}^m \setminus \{0\}. \tag{2.116}$$

Obviously, vector fields (2.116) are reversible with respect to involution G. One has $d\langle k, x\rangle/dt = \langle k, \tilde{\omega}(y,\mu)\rangle \equiv 0$. Now in an entirely similar way as in the proof of Proposition 2.21, we conclude that vector fields (2.116) possess no invariant n-tori. □

All the three contexts considered above [the volume preserving $(n,1,s)$ context, the Hamiltonian isotropic $(n,0,s)$ context, and the reversible $(n,m,0,s)$ context 1] do not involve normal frequencies. Now we show that the nonresonance condition (2.106) which links the internal and normal frequencies in the Hamiltonian isotropic context and reversible context 1 is also of relevance and cannot be relaxed. Namely, we present a particular vector field $\tilde{X}$ that pertains simultaneously to the Hamiltonian isotropic $(1,1,0)$ context and reversible $(1,1,1,0)$ context 1, can be made arbitrarily close to an integrable field by varying a certain internal parameter, satisfies the nondegeneracy condition imposed on $\omega(y)$, does not satisfy the nonresonance condition (2.106), and possesses no invariant 1-tori.

Let $x \in \mathbb{S}^1$, $y \in \mathbb{R}$, $0 < a < y < b$, $z \in \mathcal{O}_2(0)$, $L \in \mathbb{N}$, and $\epsilon \in \mathbb{R}$. On the phase space $\mathbb{S}^1 \times \mathbb{R}^3$, fix the symplectic structure $dy \wedge dx + dz_1 \wedge dz_2$ and consider the Hamilton function

$$\tilde{H}_\epsilon = \tfrac{1}{8}\left[2y - L\left(z_1^2 + z_2^2\right)\right]^2 - \epsilon y^{L/2}(z_1 \sin Lx + z_2 \cos Lx). \tag{2.117}$$

This function is analytic in all its arguments and affords the vector field $\widetilde{X}_\epsilon$ determining the system of differential equations

$$\begin{aligned}
\dot{x} = \frac{\partial \widetilde{H}_\epsilon}{\partial y} &= y - \frac{L}{2}(z_1^2 + z_2^2) - \frac{\epsilon L}{2} y^{(L-2)/2}(z_1 \sin Lx + z_2 \cos Lx) \\
\dot{y} = -\frac{\partial \widetilde{H}_\epsilon}{\partial x} &= \epsilon L y^{L/2}(z_1 \cos Lx - z_2 \sin Lx) \\
\dot{z}_1 = -\frac{\partial \widetilde{H}_\epsilon}{\partial z_2} &= L z_2 \left[y - \frac{L}{2}(z_1^2 + z_2^2) \right] + \epsilon y^{L/2} \cos Lx \\
\dot{z}_2 = \frac{\partial \widetilde{H}_\epsilon}{\partial z_1} &= -L z_1 \left[y - \frac{L}{2}(z_1^2 + z_2^2) \right] - \epsilon y^{L/2} \sin Lx.
\end{aligned}$$

For $\epsilon = 0$ the Hamilton function (2.117) is of the form (2.48) with $F(y) = y^2/2$, $\frac{1}{2}\langle z, K(y)z \rangle = -\frac{1}{2}Ly(z_1^2 + z_2^2)$ [and therefore $K(y) = -\mathit{diag}(Ly, Ly)$], $\Delta(x, y, z) = \frac{1}{8}L^2(z_1^2 + z_2^2)^2$, $\omega(y) = y$, $N_1 = 0$, $N_3 = 0$, $N_2 = r = 1$, and $\omega^N(y) = Ly$. The vector field $\widetilde{X}_0$ is integrable and has the form (2.49).

Moreover, the vector field $\widetilde{X}_\epsilon$ is reversible with respect to the involution

$$G : (x, y, z_1, z_2) \mapsto (-x, y, -z_1, z_2) \tag{2.118}$$

for any ϵ. Involution (2.118) is of the form (2.61) with $R = \mathit{diag}(-1, 1)$. The vector field $\widetilde{X}_0$ has the form (2.60).

Now calculate the quantities $\rho^Q(y)$ and $\Xi_\ell^Q(y)$ given by the expressions (2.103)–(2.104). One has

$$\frac{d^q\omega(y)}{dy^q} = \begin{bmatrix} y, & \text{if} \quad q = 0 \\ 1, & \text{if} \quad q = 1 \\ 0, & \text{if} \quad q \geq 2\,, \end{bmatrix} \qquad \frac{d^q\omega^N(y)}{dy^q} = \begin{bmatrix} Ly, & \text{if} \quad q = 0 \\ L, & \text{if} \quad q = 1 \\ 0, & \text{if} \quad q \geq 2 \end{bmatrix}$$

($q \in \mathbb{Z}_+$),

$$\rho^Q(y) = \max\{y,\ 1\}, \qquad \Xi_\ell^Q(y) = L|\ell| \max\{y,\ 1\}$$

($Q \in \mathbb{N}$, $\ell \in \mathbb{Z}$), and

$$\Xi_\ell^Q(y)/\rho^Q(y) = L|\ell|.$$

Thus, the nondegeneracy condition is met for any $Q \in \mathbb{N}$. The nonresonance condition reads as follows: for each $\ell \in \{-2;\ -1;\ 1;\ 2\}$ and $k \in \mathbb{Z}$, $1 \leq |k| \leq L|\ell|$, the inequality $ky \neq \ell L y$ should hold. This condition is not satisfied, moreover, it is violated "at the limit": $ky \neq \ell Ly$ for $1 \leq |k| \leq L|\ell| - 1$, but $ky = \ell Ly$ for $k = \ell L$.

Note that the eigenvalues of the Floquet 3×3 matrix of an unperturbed $(2\pi/y)$-periodic trajectory $\{y = \mathit{const}, z = 0\}$ of $\widetilde{X}_0$ are equal to 0, $\pm Lyi$, and the Floquet multipliers (see § 1.1.1) are therefore equal to 1, 1, 1.

For $\epsilon \neq 0$, the vector field $\widetilde{X}_\epsilon$ turns out to possess *no* invariant circles. Indeed, let

$$V = y^{L/2}(z_1 \cos Lx - z_2 \sin Lx).$$

One straightforwardly verifies that

$$\frac{dV}{dt} = \tfrac{1}{2}\epsilon L^2 y^{L-1}(z_1^2 + z_2^2) + \epsilon y^L \begin{bmatrix} > 0, & \text{if} \quad \epsilon > 0 \\ < 0, & \text{if} \quad \epsilon < 0 \end{bmatrix}$$

on the whole phase space. Consequently, $\widetilde{X}_\epsilon$ has no periodic trajectories for $\epsilon \neq 0$.

Similar examples concerning Hamiltonian and reversible vector fields near equilibria are given by Sevryuk [316, 318].

Propositions 2.21–2.23 pertain to the so called *converse* KAM theory because they assert that under certain hypotheses, families of vector fields admit no invariant tori. However, as a rule, converse KAM theorems refer to vector fields or diffeomorphisms which are very far from integrable ones (see, e.g., [213, 214, 217, 218, 222, 223] and also [152, 202, 224, 229]).

In the dissipative context and the volume preserving context with $p \geq 2$, normal hyperbolicity of the unperturbed invariant tori is sufficient for the persistence of these tori as of invariant submanifolds [67, 115, 158, 356] (cf. § 1.2.1). On the other hand, the hypotheses of Theorems 2.16 and 2.17 are very close to the necessary conditions for the persistence of the unperturbed tori as of Floquet invariant tori with parallel dynamics. We will show this for the dissipative $(n, 0, s)$ context.

Consider the analytic family of linear vector fields on $\mathbb{T}^n$:

$$X = \{X^\mu\}_\mu = \omega(\mu)\frac{\partial}{\partial x},$$

where $x \in \mathbb{T}^n$, $\mu \in P \subset \mathbb{R}^s$, $\omega : P \to \mathbb{R}^n$.

Proposition 2.24 *If the image of the map $\omega : P \to \mathbb{R}^n$ lies in a linear hyperplane in $\mathbb{R}^n$ passing through the origin, then there exists an analytic family $\widetilde{X}$ of analytic vector fields arbitrarily close to X in the real analytic topology and such that $\widetilde{X}^\mu$ induces parallel dynamics on $\mathbb{T}^n$ for no $\mu \in P$.*

Proof. We have $\langle \omega(\mu), e \rangle \equiv 0$ for some vector $e \in \mathbb{R}^n \setminus \{0\}$. Set $\tilde{\omega}(\mu) = A^{-1}\omega(\mu)$ where $A \in GL(n, \mathbb{R})$. Then $\langle \tilde{\omega}(\mu), \tilde{e} \rangle \equiv 0$ where $\tilde{e} = A^t e$. One can choose A arbitrarily close to the identity $n \times n$ matrix in such a way that $\tilde{e}$ will be proportional to an integer vector: $\langle \tilde{\omega}(\mu), k \rangle \equiv 0$ with $k \in \mathbb{Z}^n \setminus \{0\}$. Now let

$$\widetilde{X} = \{\widetilde{X}^\mu\}_\mu = [\tilde{\omega}(\mu) + \epsilon \sin\langle k, x \rangle]\frac{\partial}{\partial x}, \quad \epsilon \in \mathbb{R}^n, \quad \langle \epsilon, k \rangle \neq 0. \tag{2.119}$$

One has $d\langle k, x \rangle / dt = \langle \epsilon, k \rangle \sin\langle k, x \rangle$. Therefore, each of the $(n-1)$-tori that constitute the set $\{x \in \mathbb{T}^n : \langle k, x \rangle \equiv 0 \bmod \pi\}$ is either an attractor or a repeller for $\widetilde{X}^\mu$. Consequently, the dynamics determined on $\mathbb{T}^n$ by any of $\widetilde{X}^\mu$ is not parallel. □

2.7 Whitney-smooth families of tori: results

Theorems 2.8–2.12 and 2.16–2.20 present a complete description of Whitney-smooth families of Floquet Diophantine invariant tori in typical families of vector fields in various contexts. Recall that all the vector fields and their dependence on external parameters are assumed to be analytic. Let $n \geq 2$.

Theorem 2.25 (dissipative context) *For any $p \geq 0$ and $s \geq 1$, a typical s-parameter family of vector fields on an $(n + p)$-dimensional manifold possesses a Whitney-smooth*

s-parameter family of Floquet Diophantine invariant n-tori (at most one torus per each parameter value). If the number of the pairs of complex conjugate eigenvalues of the Floquet matrices of these tori is r and $s \geq n+r-1$, then the Whitney-smooth s-parameter family of n-tori consists generically of $(s-n-r+1)$-parameter analytic subfamilies.

Theorem 2.26 (volume preserving context with $p \geq 2$) *For any $p \geq 2$ and $s \geq 1$, a typical s-parameter family of globally divergence-free vector fields on an $(n+p)$-dimensional manifold possesses a Whitney-smooth s-parameter family of Floquet Diophantine invariant n-tori (at most one torus per each parameter value). If the number of the pairs of complex conjugate eigenvalues of the Floquet matrices of these tori is r and $s \geq n+r-1$, then the Whitney-smooth s-parameter family of n-tori consists generically of $(s-n-r+1)$-parameter analytic subfamilies.*

Theorem 2.27 (volume preserving context with $p = 1$) *For any $s \geq 0$, a typical s-parameter family of globally divergence-free vector fields on an $(n+1)$-dimensional manifold possesses a Whitney-smooth $(s+1)$-parameter family of Floquet Diophantine invariant n-tori. The Floquet 1×1 matrix of each of these tori is zero. If $s \geq n-2$, then the Whitney-smooth $(s+1)$-parameter family of n-tori consists generically of $(s-n+2)$-parameter analytic subfamilies.*

Theorem 2.28 (Hamiltonian isotropic context) *For any $p \geq 0$ and $s \geq 0$, a typical s-parameter family of Hamiltonian vector fields with $n+p$ degrees of freedom possesses a Whitney-smooth $(n+s)$-parameter family of Floquet Diophantine isotropic invariant n-tori. The Floquet $(n+2p) \times (n+2p)$ matrix of each of these tori has eigenvalue 0 of multiplicity n, while the remaining $2p$ eigenvalues occur in pairs $(\lambda, -\lambda)$. If the number of distinct values of the positive imaginary parts of the eigenvalues of the Floquet matrices of these tori is r and $s \geq r-1$, then the Whitney-smooth $(n+s)$-parameter family of n-tori consists generically of $(s-r+1)$-parameter analytic subfamilies.*

Theorem 2.29 (reversible context 1) *For any $m \geq 0$, $p \geq 0$ and $s \geq \max\{1-m, 0\}$, a typical s-parameter family of vector fields reversible with respect to an involution of type $(n+p, m+p)$ on an $(n+m+2p)$-dimensional manifold possesses a Whitney-smooth $(m+s)$-parameter family of Floquet Diophantine invariant n-tori. The Floquet $(m+2p) \times (m+2p)$ matrix of each of these tori has eigenvalue 0 of multiplicity m (if $m > 0$), while the remaining $2p$ eigenvalues occur in pairs $(\lambda, -\lambda)$. If the number of distinct values of the positive imaginary parts of the eigenvalues of the Floquet matrices of these tori is r and $s \geq n-m+r-1$, then the Whitney-smooth $(m+s)$-parameter family of n-tori consists generically of $(s+m-n-r+1)$-parameter analytic subfamilies.*

Of course, all the invariant tori in Theorems 2.25–2.29 are analytic.

From our viewpoint, these theorems constitute the essence of the KAM theory. It is much more important to be aware that invariant tori occur in generic systems than to know that they occur in, e.g., nearly integrable systems. Although the KAM theorems are "often said to be the first and perhaps foremost result of modern non-linear dynamics of conservative systems" [288] and of great importance for physics and astronomy [5, 13, 20, 95, 124, 140, 200–202, 288, 362], these theorems are sometimes criticized for extremely small perturbation sizes allowed by the analytic proofs. Since the KAM theory establishes the existence of invariant tori in systems very close to integrable ones, it may seem to be not very suitable for practical purposes [268, 280], and one even speaks of its "numerical inadequacies" [268]. However, this impression stems in fact from overestimating the perturbative character of the KAM theory. Suppose we have proven that any system ϵ-close to a particular integrable system admits a Whitney-smooth family of invariant tori. If ϵ is much smaller than any value encountered in physical examples (known by now), our theorem has no *direct* physical applications. Nevertheless, this does not mean uselessness. First of all, "safe" perturbation sizes obtained by a purely analytic reasoning are usually orders of magnitude smaller than true thresholds (to be found numerically or by computer-assisted analytic arguments) above which the most of the tori break up [78, 80, 139, 268]. But the main point is as follows. However small ϵ is, our theorem implies that a Whitney-smooth family of invariant tori is a *generic phenomenon* in the class of systems under consideration (cf. a discussion in § 1.3.1). The latter information is in fact the most important result and it can already be applied to physical instances. If we have found numerically invariant tori in some system then we may make the conclusion that those tori do exist (and persist under small perturbations), and we need not look for a degeneracy responsible for the tori.

"Since the KAM theory appeared, physicists have gotten an exceedingly powerful tool for studying dynamical systems. The physical intuition (so necessary in many cases where rigorous results have not been obtained yet) have stood upon some firm ground. As a rule, one can perform a perturbative analysis for most of the physical systems, find the perturbed invariant tori and, above all, be sure of their existence" (Zaslavskiĭ & Sagdeev [362, p. 79]).

Global (i.e., nonperturbative) studies of the existence of invariant tori are exemplified by [79, 154]. However, numerical evidence, especially overwhelming for the Hamiltonian context, suggests that in systems far from integrable the fraction of the phase space occupied by invariant tori is usually small [270].

Theorems 2.27–2.29 for $s = 0$ were first formulated (in the form close to that presented here) by Quispel & Sevryuk [287, 328] and then emphasized by Broer, Huitema & Sevryuk [61].

If $n = 1$ or $n = 0$ (when no small divisors are present), then no restrictions on s are needed, and Whitney-smooth families of invariant n-tori are in fact analytic. So, if $n \le 1$, then for any $s \ge 0$ typical s-parameter families of vector fields possess analytic l-parameter families of invariant n-tori, where l is equal to s, s, $s + 1$, $n + s$, $m + s$ in the contexts of Theorems 2.25, 2.26, 2.27, 2.28, 2.29, respectively.

There are also known some results [287, 328] concerning analytic families of periodic G-invariant trajectories ($n = 1$) in vector fields reversible with respect to an involution G whose fixed point manifold $\operatorname{Fix} G$ consists of connected components of different dimensions.

Chapter 3

The continuation theory

3.1 Analytic continuation of tori

In the dissipative context, volume preserving context with $p = 2$ (hyperbolic case) or $p > 2$, Hamiltonian isotropic context, and reversible context 1, it is interesting to consider the restrictions of the unperturbed family of vector fields X and perturbed one $\tilde{X}$ to *the center manifold.* Here we have excluded the volume preserving context with $p = 1$ and $p = 2$ (elliptic case) since for these cases, the center manifold always coincides with the whole phase space. The center manifold persists under perturbations [67, 115, 158, 356] but becomes, generally speaking, only finitely differentiable [12, 347]. However, we can apply the *finitely differentiable versions* of the "relaxed" Theorems 2.8, 2.9, 2.11, 2.12 to the restrictions of X and $\tilde{X}$ to the center manifold, see [151, 243, 277, 278, 306] as well as [62, 162].

In this way, if the number s of external parameters is sufficiently large, we will obtain *finitely* Whitney-smooth (see Section 2.2) l-parameter families of Floquet (within the center manifold) Diophantine invariant *finitely differentiable* n-tori, and these finitely Whitney-smooth l-parameter families will consist of l'-parameter *finitely differentiable* subfamilies [in fact, since by the parallelity those tori in the dissipative context or volume preserving context with $p \geq 2$ are r_*-normally hyperbolic for every $r_* \in \mathbb{N}$, by uniqueness it follows that they are even of class C^∞, cf. § 1.2.1]. The number l here is the same as in Theorems 2.25, 2.26, 2.28, 2.29. However, the number l' (the *regularity* of the family of tori) will be, generally speaking, *larger* than the corresponding number in Theorems 2.25, 2.26, 2.28, 2.29 because the normal frequencies β_j coming from the eigenvalues of the Floquet matrices of the tori lying in $\mathbb{C} \setminus (\mathbb{R} \cup \mathbb{R}i)$ disappear while restrictions of X and $\tilde{X}$ to the center manifold are considered. A fewer number of resonances leads to a larger regularity of the Whitney-smooth family of tori.

To make this point clearer, consider, e.g., the Hamiltonian isotropic context. In this context, according to Theorems 2.11 and 2.19, the perturbed family $\tilde{X}$ of Hamiltonian vector fields possesses a Whitney-smooth $(n+s)$-parameter family of Floquet Diophantine isotropic invariant analytic n-tori, and if $s \geq r - 1$, then this family consists of $(s - r + 1)$-parameter analytic subfamilies. The number r here is equal to $N_2 + N_3$, where integers N_2 and N_3 refer to the nonzero eigenvalues (2.50) of the Floquet matrices of the unperturbed n-tori. Now consider the restrictions of vector fields X^μ and $\tilde{X}^\mu$ to the $2(n + N_2)$-dimensional center manifold (these restrictions are still Hamiltonian). Then the eigenvalues $\pm\delta_1, \ldots, \pm\delta_{N_1}$ and $\pm\alpha_1 \pm i\beta_1, \ldots, \pm\alpha_{N_3} \pm i\beta_{N_3}$ will disappear and we will have to take into account the purely imaginary eigenvalues $\pm i\varepsilon_1, \ldots, \pm i\varepsilon_{N_2}$ only. So, for the restriction of $\widetilde{X}$ to the center manifold, the finitely differentiable versions of Theorems 2.11 and 2.19 give a finitely Whitney-smooth $(n+s)$-parameter family of Floquet (within the center manifold) Diophantine isotropic invariant finitely differentiable n-tori, and if $s \geq N_2 - 1$, then this family consists of $(s - N_2 + 1)$-parameter finitely differentiable subfamilies. We would like to emphasize that all these tori are *a priori* Floquet *within the center manifold* only. The variational equation along them may turn out to be irreducible to a constant coefficient form in the whole phase space (cf. a discussion in [172]).

We have arrived at the conclusion that the Whitney-smooth l-parameter families of Floquet Diophantine invariant analytic n-tori organized into l'_1-parameter analytic subfamilies, which are described in Theorems 2.25, 2.26, 2.28, 2.29, are *inlaid* in finitely Whitney-smooth l-parameter families of Diophantine invariant finitely differentiable n-tori (not necessarily Floquet) organized into l'_2-parameter finitely differentiable subfamilies with $l'_2 \geq l'_1$ (if the Floquet tori have normal frequencies β_j then $l'_2 > l'_1$).

A wonderful circumstance is that most of these l'_2-parameter finitely differentiable subfamilies of finitely differentiable n-tori are *analytic* and constituted of *analytic* n-tori. The resonances involving normal frequencies β_j are never obstacles to the analyticity of Diophantine invariant tori, these resonances just prevent sometimes the tori from being Floquet. So a general picture of the structure of families of tori is as follows: if the number s of external parameters is sufficiently large, the phase space typically contains l'_2-parameter analytic families of Diophantine invariant analytic n-tori. Most of these tori are Floquet. However, if one excludes all the tori that are not, then those families will become "riddled" with infinitely many "holes" and will disintegrate into l'_1-parameter analytic families.

To prove the analyticity of l'_2-parameter finitely differentiable subfamilies of invariant tori on the center manifold, one has to invoke the Bruno continuation theory [72, Part II]. According to this theory, a Floquet Diophantine invariant analytic n-torus of an analytic family of analytic vector fields is embedded, under certain conditions, in an l''-parameter analytic family of Diophantine invariant analytic n-tori not necessarily Floquet. The number l'' given by the Bruno theory *turns out to be always equal* to the number l'_2 given by the finitely differentiable versions of our "relaxed" Theorems 2.8, 2.9, 2.11, 2.12 applied to the restrictions of X and $\widetilde{X}$ to the center manifold. The Bruno theory also provides us with the *whiskers* (separatrix manifolds) of the invariant tori, but we will not discuss this topic in the present book.

Strictly speaking, to identify l'_2-parameter finitely differentiable families of finitely differentiable tori provided by our conjugacy theory and l'_2-parameter analytic families of analytic tori provided by the Bruno continuation theory, some uniqueness theorems are needed. Namely, one has to prove that

1) in the framework of the differentiable conjugacy theory (on the center manifold), each torus in question is embedded in a unique l'_2-parameter differentiable family of Floquet (within the center manifold) quasi-periodic invariant differentiable tori whose internal and normal frequencies satisfy Diophantine conditions (1.4), and

2) the analytic invariant tori in the Bruno continuation theory are Floquet within the center manifold and their internal and normal frequencies satisfy Diophantine conditions (1.4).

To the best of the authors' knowledge, such theorems are absent in the literature.

The first one seems to be just an easy exercise, whereas the second one is expected to be rather hard. However, the identity of the two kinds of l_2'-parameter families of Diophantine invariant tori described above gives rise to no doubt.

The precise formulations of the continuation theorems in each of the contexts discussed above follow.

Theorem 3.1 (Continuation theorem in the dissipative context) *Let the mapping*

$$\mathbb{R}^{s+1} \ni (c,\mu) \mapsto c\omega(\mu) \in \mathbb{R}^n, \quad c \in \mathcal{O}(1) \subset \mathbb{R} \tag{3.1}$$

be submersive for $\mu \in \Gamma$, or, equivalently, let the mapping

$$\mathbb{R}^s \ni \mu \mapsto \Pi\omega \in \mathbb{RP}^{n-1} \tag{3.2}$$

be submersive for $\mu \in \Gamma$ (so that $s \geq n-1$). Then each Floquet Diophantine invariant analytic n-torus of $\widetilde{X}$ considered in any of the Theorems 2.3, 2.8, 2.16, for which none of the nonreal eigenvalues of the Floquet matrix is purely imaginary, is embedded in an $(s-n+1)$-parameter analytic family of Diophantine invariant analytic n-tori (not necessarily Floquet). Each of these tori possesses the $(N_1^+ + 2N_2^+ + n)$-dimensional analytic repelling whisker and $(N_1^- + 2N_2^- + n)$-dimensional analytic attracting whisker, N_1^+ [N_1^-] being the number of positive [negative] real eigenvalues of the Floquet matrix and N_2^+ [N_2^-] the number of pairs of nonreal eigenvalues of the Floquet matrix with positive [negative] real parts ($N_1^+ + N_1^- = N_1$, $N_2^+ + N_2^- = N_2$).

Theorem 3.2 (Continuation theorem in the volume preserving context with $p = 2$ [hyperbolic case] or $p > 2$) *Let the mapping (3.1) be submersive for $\mu \in \Gamma$, or, equivalently, let the mapping (3.2) be submersive for $\mu \in \Gamma$ (so that $s \geq n-1$). Then each Floquet Diophantine invariant analytic n-torus of $\widetilde{X}$ considered in any of the Theorems 2.4, 2.9, 2.17, for which none of the nonreal eigenvalues of the Floquet matrix is purely imaginary, is embedded in an $(s-n+1)$-parameter analytic family of Diophantine invariant analytic n-tori (not necessarily Floquet). Each of these tori possesses the $(N_1^+ + 2N_2^+ + n)$-dimensional analytic repelling whisker and $(N_1^- + 2N_2^- + n)$-dimensional analytic attracting whisker, N_1^+ [N_1^-] being the number of positive [negative] real eigenvalues of the Floquet matrix and N_2^+ [N_2^-] the number of pairs of nonreal eigenvalues of the Floquet matrix with positive [negative] real parts ($N_1^+ + N_1^- = N_1$, $N_2^+ + N_2^- = N_2$).*

Theorem 3.3 (Continuation theorem in the Hamiltonian isotropic context) *Let the mapping*

$$\mathbb{R}^{n+s+1} \ni (c,y,\mu) \mapsto (c\omega(y,\mu), c\varepsilon(y,\mu)) \in \mathbb{R}^{n+N_2}, \quad c \in \mathcal{O}(1) \subset \mathbb{R} \tag{3.3}$$

be submersive for $(y,\mu) \in \Gamma$, or, equivalently, let the mapping

$$\mathbb{R}^{n+s} \ni (y,\mu) \mapsto \Pi(\omega,\varepsilon) \in \mathbb{RP}^{n+N_2-1} \tag{3.4}$$

be submersive for $(y,\mu) \in \Gamma$ (so that $s \geq N_2 - 1$). Then each Floquet Diophantine isotropic invariant analytic n-torus of $\widetilde{X}$ [with frequency vector $\widetilde{\omega}$ and normal frequency vector $\widetilde{\omega}^N = (\widetilde{\varepsilon}, \widetilde{\beta})$] considered in any of the Theorems 2.6, 2.11, 2.19, for which

$$\exists \widetilde{\tau} > 0\ \exists \widetilde{\gamma} > 0 : \forall k \in \mathbb{Z}^n\ \forall \ell \in \mathbb{Z}^{N_2},\ |k| + |\ell| > 0,\ \ |\langle \widetilde{\omega}, k\rangle + \langle \widetilde{\varepsilon}, \ell\rangle| \geq \widetilde{\gamma}(|k| + |\ell|)^{-\widetilde{\tau}} \tag{3.5}$$

(almost all of the tori have this property due to the submersivity condition above), is embedded in an $(s - N_2 + 1)$-parameter analytic family of Diophantine isotropic invariant analytic n-tori (not necessarily Floquet). Each of these tori possesses the $(N_1 + 2N_3 + n)$-dimensional isotropic analytic repelling whisker and $(N_1 + 2N_3 + n)$-dimensional isotropic analytic attracting whisker.

Theorem 3.4 (Continuation theorem in the reversible context 1) *Let the mapping*

$$\mathrm{R}^{m+s+1} \ni (c, y, \mu) \mapsto (c\omega(y, \mu), c\varepsilon(y, \mu)) \in \mathrm{R}^{n+N_2}, \quad c \in \mathcal{O}(1) \subset \mathrm{R} \tag{3.6}$$

be submersive for $(y, \mu) \in \Gamma$, or, equivalently, let the mapping

$$\mathrm{R}^{m+s} \ni (y, \mu) \mapsto \Pi(\omega, \varepsilon) \in \mathrm{RP}^{n+N_2-1} \tag{3.7}$$

be submersive for $(y, \mu) \in \Gamma$ (so that $s \geq n - m + N_2 - 1$). Then each Floquet Diophantine invariant analytic n-torus of $\widetilde{X}$ [with frequency vector $\widetilde{\omega}$ and normal frequency vector $\widetilde{\omega}^N = (\widetilde{\varepsilon}, \widetilde{\beta})$] considered in any of the Theorems 2.7, 2.12, 2.20, which possesses property (3.5) (almost all of the tori have this property due to the submersivity condition above), is embedded in an $(s + m - n - N_2 + 1)$-parameter analytic family of Diophantine invariant analytic n-tori (not necessarily Floquet). Each of these tori possesses the $(N_1 + 2N_3 + n)$-dimensional analytic repelling whisker and $(N_1 + 2N_3 + n)$-dimensional analytic attracting whisker.

Of Theorems 3.1–3.4, we will prove only the first one and the last one. The proofs of the remaining two theorems are quite similar. A very concise exposition of the results of Bruno's theory we will need can be found in Section 6.3. We will freely use the terminology and notations of that section (in fact, they almost coincide with the original notations of Bruno's book [72, Part II]). Note that Theorem 3.3 (and some its generalizations) is in fact proven in Bruno's book [72, Part II, § 3].

Proof of Theorem 3.1. Let $\widetilde{\omega}$ be the frequency vector of the torus in question invariant under vector field $\widetilde{X}^{\widetilde{\mu}}$. The Bruno normal form (6.43) of $\widetilde{X}$ around this torus is $\Upsilon(\mu - \widetilde{\mu})\partial/\partial U$ where $U \in \mathrm{T}^n$ and $\Upsilon(0) = \widetilde{\omega}$. The Bruno condition $\boldsymbol{\beta}$ (6.48) is satisfied because the torus is Diophantine. The Bruno $n \times n$ matrix B (6.47) is identically zero, and therefore $\mathcal{B} = \mathcal{A}$. The set $\mathcal{A} \subset \mathrm{T}^n \times \mathrm{R}^s$ is defined by $\Pi\Upsilon(\mu - \widetilde{\mu}) = \Pi\widetilde{\omega}$ [see (6.46)]. The 1-jet (the Taylor expansion up to the first order) of Υ at 0 is close to that of ω at $\widetilde{\mu}$, and due to the submersivity property of mapping (3.2), we have $\mathcal{A} = \mathrm{T}^n \times \mathcal{A}_0$ where $\mathcal{A}_0$ is an analytic $(s - n + 1)$-dimensional surface in R^s passing through $\widetilde{\mu}$. The set $\mathcal{A}$ is therefore foliated into Diophantine invariant analytic n-tori [with the frequency vectors proportional to $\widetilde{\omega}$] possessing the whiskers (lying in the set $\mathcal{A}_W$) of dimensions indicated above, the base of this foliation being $(s - n + 1)$-dimensional. □

Proof of Theorem 3.4. Let the torus in question correspond to the value $\widetilde{\mu}$ of the parameter μ. The Bruno normal form (6.43) of the differential equations determined by $\widetilde{X}$ around this torus is

$$\begin{aligned}
\dot{U} &= \Upsilon(V, \mathcal{P}, \mu - \widetilde{\mu}) = \breve{\omega}(V, \mu - \widetilde{\mu}) + O(\mathcal{P}), \\
\dot{V} &= 0, \\
\dot{\mathcal{V}}_j &= i\mathcal{V}_j \Psi_j(V, \mathcal{P}, \mu - \widetilde{\mu}) = i\mathcal{V}_j \left[\breve{K}_j(V, \mu - \widetilde{\mu}) + O(\mathcal{P})\right], \\
\dot{\mathcal{V}}_{j+N_2} &= -i\mathcal{V}_{j+N_2} \Psi_j(V, \mathcal{P}, \mu - \widetilde{\mu}) = -i\mathcal{V}_{j+N_2} \left[\breve{K}_j(V, \mu - \widetilde{\mu}) + O(\mathcal{P})\right]
\end{aligned}$$

($1 \le j \le N_2$), where $U \in \mathrm{T}^n$, $V \in \mathrm{R}^m$, $\mathcal{V} \in \mathrm{C}^{2N_2}$ [here we denote by $(V, \mathcal{V})$ the variables that are denoted by V in Section 6.3], $\mathcal{P} \in \mathrm{C}^{N_2}$ ($\mathcal{P}_j = \mathcal{V}_j \mathcal{V}_{j+N_2}$, $1 \le j \le N_2$), the reality condition for $\mathcal{V}$ is

$$\overline{\mathcal{V}_j} = -i\mathcal{V}_{j+N_2}, \quad 1 \le j \le N_2, \tag{3.8}$$

and

$$\breve{\omega}(0,0) = \widetilde{\omega}, \quad \breve{K}(0,0) = \widetilde{\varepsilon}$$

(Υ and Ψ for purely imaginary $\mathcal{P}$ and consequently $\breve{\omega}$ and $\breve{K}$ are real-valued). As one should expect, this normal form is reversible with respect to the involution

$$\breve{G} : (U, V, \mathcal{V}_1, \dots, \mathcal{V}_{N_2}, \mathcal{V}_{N_2+1}, \dots, \mathcal{V}_{2N_2}) \mapsto (-U, V, -i\mathcal{V}_{N_2+1}, \dots, -i\mathcal{V}_{2N_2}, i\mathcal{V}_1, \dots, i\mathcal{V}_{N_2})$$

[the restriction of the initial involution (2.61) to the formal central manifold]. The reality condition (3.8) is equivalent to

$$\mathcal{V}_j = (\mathcal{Z}_j + i\mathcal{Z}_{j+N_2})/\sqrt{2}, \quad \mathcal{V}_{j+N_2} = (i\mathcal{Z}_j + \mathcal{Z}_{j+N_2})/\sqrt{2}, \quad 1 \le j \le N_2,$$

where $\mathcal{Z} \in \mathrm{R}^{2N_2}$; note that $\mathcal{P}_j = i(\mathcal{Z}_j^2 + \mathcal{Z}_{j+N_2}^2)/2$ for each $1 \le j \le N_2$. Involution $\breve{G}$ is compatible with this reality condition (as one should expect) and takes the form

$$(U, V, \mathcal{Z}_1, \dots, \mathcal{Z}_{N_2}, \mathcal{Z}_{N_2+1}, \dots, \mathcal{Z}_{2N_2}) \mapsto (-U, V, \mathcal{Z}_1, \dots, \mathcal{Z}_{N_2}, -\mathcal{Z}_{N_2+1}, \dots, -\mathcal{Z}_{2N_2})$$

in coordinates $(U, V, \mathcal{Z})$. The time derivatives of the variables $\mathcal{Z}$ are equal to

$$\dot{\mathcal{Z}}_j = -\mathcal{Z}_{j+N_2}\Psi_j, \quad \dot{\mathcal{Z}}_{j+N_2} = \mathcal{Z}_j\Psi_j, \quad 1 \le j \le N_2.$$

The Bruno condition β (6.48) is satisfied due to inequalities (3.5). Let $D_1 = diag(\mathcal{V}_1, \dots, \mathcal{V}_{N_2})$ and $D_2 = diag(\mathcal{V}_{N_2+1}, \dots, \mathcal{V}_{2N_2})$ be diagonal $N_2 \times N_2$ matrices. The Bruno matrix $B \in gl(n+m+2N_2, \mathrm{C})$ [see (6.47)] is equal to

$$B = \begin{pmatrix} 0 & \partial\Upsilon/\partial V & (\partial\Upsilon/\partial\mathcal{P})D_2 & (\partial\Upsilon/\partial\mathcal{P})D_1 \\ 0 & 0 & 0 & 0 \\ 0 & iD_1(\partial\Psi/\partial V) & iD_1(\partial\Psi/\partial\mathcal{P})D_2 & iD_1(\partial\Psi/\partial\mathcal{P})D_1 \\ 0 & -iD_2(\partial\Psi/\partial V) & -iD_2(\partial\Psi/\partial\mathcal{P})D_2 & -iD_2(\partial\Psi/\partial\mathcal{P})D_1 \end{pmatrix}.$$

The matrix

$$B_0 = \begin{pmatrix} iD_1(\partial\Psi/\partial\mathcal{P})D_2 & iD_1(\partial\Psi/\partial\mathcal{P})D_1 \\ -iD_2(\partial\Psi/\partial\mathcal{P})D_2 & -iD_2(\partial\Psi/\partial\mathcal{P})D_1 \end{pmatrix}$$

is nilpotent since $B_0^2 = 0$ (the latter equality follows from $D_1D_2 = D_2D_1$). Consequently, the matrix B is also nilpotent (namely, $B^4 = 0$), and therefore $\mathcal{B} = \mathcal{A}$. The set $\mathrm{Re}\,\mathcal{A} \cap \{\mathcal{V} = 0\} \subset \mathrm{T}^n \times \mathrm{R}^m \times \mathrm{R}^s$ is defined by $\mathcal{V} = 0$, $\Pi(\Upsilon, \Psi) = \Pi(\widetilde{\omega}, \widetilde{\varepsilon})$, i.e., by $\Pi(\breve{\omega}, \breve{K}) = \Pi(\widetilde{\omega}, \widetilde{\varepsilon})$ [see (6.46)]. The 1-jets of $\breve{\omega}$ and $\breve{K}$ at $(0,0)$ are close, respectively, to those of ω and ε at $(\widetilde{y}, \widetilde{\mu}) \in \Gamma$ for some $\widetilde{y} \in Y$. Due to the submersivity property of mapping (3.7), we have $\mathrm{Re}\,\mathcal{A} \cap \{\mathcal{V} = 0\} = \mathrm{T}^n \times \mathcal{A}_0$ where $\mathcal{A}_0$ is an analytic $(s+m-n-N_2+1)$-dimensional surface in $\mathrm{R}^m \times \mathrm{R}^s$ passing through $(0, \widetilde{\mu})$. The set $\mathrm{Re}\,\mathcal{A} \cap \{\mathcal{V} = 0\}$ is therefore foliated into Diophantine invariant analytic n-tori [with the frequency vectors proportional to $\widetilde{\omega}$] possessing the $(N_1 + 2N_3 + n)$-dimensional whiskers (lying in the set $\mathrm{Re}\,\mathcal{A}_W$), the base of this foliation being $(s+m-n-N_2+1)$-dimensional. The set $\mathrm{Re}\,\mathcal{A} \cap \{\mathcal{V} \ne 0\}$ consists of invariant analytic tori of dimensions $n+1, \dots, n+N_2$. The whiskers of the latter tori also lie in the set $\mathrm{Re}\,\mathcal{A}_W$. □

3.2 Whitney-smooth families of tori: further results

Theorems 3.1–3.4 enable one to make the following addenda to Theorems 2.25, 2.26, 2.28, 2.29.

Theorem 3.5 (addendum to Theorem 2.25: dissipative context) *If $s \geq n - 1$, then each Floquet Diophantine invariant n-torus is embedded generically in an $(s-n+1)$-parameter analytic family of Diophantine invariant analytic n-tori not necessarily Floquet, the union of all the latter tori constituting a finitely Whitney-smooth s-parameter family.*

Theorem 3.6 (addendum to Theorem 2.26: volume preserving context with $p \geq 2$) *If $s \geq n-1$ and (for $p = 2$) the eigenvalues of the Floquet matrices of the tori are not purely imaginary, then each Floquet Diophantine invariant n-torus is embedded generically in an $(s - n + 1)$-parameter analytic family of Diophantine invariant analytic n-tori not necessarily Floquet, the union of all the latter tori constituting a finitely Whitney-smooth s-parameter family.*

Theorem 3.7 (addendum to Theorem 2.28: Hamiltonian isotropic context) *If the number of the pairs of purely imaginary eigenvalues of the Floquet matrices of the tori is N_* and $s \geq N_* - 1$, then almost all these Floquet Diophantine isotropic invariant n-tori are embedded generically in $(s - N_* + 1)$-parameter analytic families of Diophantine isotropic invariant analytic n-tori not necessarily Floquet, the union of all the latter tori constituting a finitely Whitney-smooth $(n + s)$-parameter family.*

Theorem 3.8 (addendum to Theorem 2.29: reversible context 1) *If the number of the pairs of purely imaginary eigenvalues of the Floquet matrices of the tori is N_* and $s \geq n-m+N_*-1$, then almost all these Floquet Diophantine invariant n-tori are embedded generically in $(s + m - n - N_* + 1)$-parameter analytic families of Diophantine invariant analytic n-tori not necessarily Floquet, the union of all the latter tori constituting a finitely Whitney-smooth $(m + s)$-parameter family.*

The integer N_* in Theorems 3.7 and 3.8 was previously denoted by N_2.

Chapter 4

Complicated Whitney-smooth families

4.1 Excitation of normal modes

Whitney-smooth families of Floquet Diophantine invariant tori in dynamical systems can form complicated hierarchical structures. The latter usually include l-parameter families of invariant n-tori with different n and l. The interaction of families of tori of different dimensions is exemplified by the so called excitation of (elliptic) normal modes in Hamiltonian and reversible systems.

4.1.1 The concept

In this section, we will describe this phenomenon in detail. Start with reversible systems and consider again an analytic s-parameter family $X = \{X^\mu\}_\mu$ of analytic G-reversible vector fields (2.60) on $\mathrm{T}^n \times \mathrm{R}^{m+2p}$, involution G being defined by (2.61). We will regard n, m, p, and s as arbitrary non-negative integers and will assume that $g = O(z)$ [as usual, $f = O(z)$ and $h = O_2(z)$]. The motion around n-tori $\{y = const, z = 0\}$ invariant under vector fields X^μ and involution G is described (in the linear approximation) by the frequency vectors $\omega(y, \mu) \in \mathrm{R}^n$ and the $2p \times 2p$ matrices $\Omega(y, \mu)$. For each y^0 and μ^0, among the eigenvalues (2.50) of matrix $\Omega(y^0, \mu^0)$, there are $N_2 \geq 0$ pairs of purely imaginary eigenvalues $\pm i\varepsilon_j(y^0, \mu^0)$. To each pair $\pm i\varepsilon_j$, there corresponds a two-dimensional linear subspace $L_j(y^0, \mu^0)$ of R^{2p} invariant under $\Omega(y^0, \mu^0)$ and R (we assume all the positive numbers ε_j to be pairwise distinct). These subspaces are sometimes called *elliptic modes* (of vector field X^{μ^0} or invariant torus $\{y = y^0, z = 0\}$) *normal* to the surface $\{z = 0\}$.[1]

Now consider the "linearized" family of vector fields

$$X^\mu_{\mathrm{lin}} = \omega(y, \mu)\frac{\partial}{\partial x} + \Omega(y, \mu)z\frac{\partial}{\partial z} \tag{4.1}$$

which are still reversible with respect to G. Having made, if necessary, a linear coordinate transformation $z = T(y, \mu)z^{\mathrm{new}}$ depending analytically on y and μ, we can assume that

[1] Thus, the word "normal" here means "orthogonal".

matrices $\Omega(y,\mu)$ have the block-diagonal form and are the direct sums of the blocks

$$\Omega(y,\mu) = \left[\bigoplus_{j=1}^{N_2} \begin{pmatrix} 0 & -\varepsilon_j(y,\mu) \\ \varepsilon_j(y,\mu) & 0 \end{pmatrix}\right] \bigoplus \Omega^{\mathrm{hyp}}(y,\mu)$$

("hyp" from "hyperbolic"), while matrix R in (2.61) has the block-diagonal form

$$R = \left[\bigoplus_{j=1}^{N_2} \begin{pmatrix} 1 & 0 \\ 0 & -1 \end{pmatrix}\right] \bigoplus R^{\mathrm{hyp}}$$

[161, 321, 333]. Here R^{hyp} is an involutive $(2p-2N_2)\times(2p-2N_2)$ matrix whose 1- and (-1)-eigenspaces are $(p-N_2)$-dimensional, and none of the eigenvalues of $(2p-2N_2)\times(2p-2N_2)$ infinitesimally R^{hyp}-reversible matrix Ω^{hyp} is purely imaginary.

Among the solutions of the differential equation determined by $X^{\mu^0}_{\mathrm{lin}}$, there are the following ones:

$$\begin{aligned} x(t) &= \omega(y^0,\mu^0) + \phi, \\ y(t) &= y^0, \\ z_{2j-1}(t) &= C_j\cos(\varepsilon_j(y^0,\mu^0)t + \psi_j), && 1\le j\le N_2, \\ z_{2j}(t) &= C_j\sin(\varepsilon_j(y^0,\mu^0)t + \psi_j), && 1\le j\le N_2, \\ z_j(t) &= 0, && 2N_2+1\le j\le 2p, \end{aligned} \tag{4.2}$$

where $\phi\in\mathrm{T}^n$, $y^0\in Y\subset\mathrm{R}^m$, $C_j\ge 0$ $(1\le j\le N_2)$, and $\psi_j\in\mathrm{S}^1$ $(1\le j\le N_2)$ are arbitrary constants. Each solution (4.2) lies on the invariant torus $\{y=y^0,\ z_{2j-1}^2+z_{2j}^2=C_j^2$ $(1\le j\le N_2)$, $z_j=0$ $(2N_2+1\le j\le 2p)\}$ of dimension $n+\nu$ where ν is the number of indices $j\le N_2$ such that $C_j>0$. This torus carries parallel dynamics with frequencies $\omega_1(y^0,\mu^0),\dots,\omega_n(y^0,\mu^0)$, $\varepsilon_{j_1}(y^0,\mu^0),\dots,\varepsilon_{j_\nu}(y^0,\mu^0)$ where $C_{j_1}>0,\dots,C_{j_\nu}>0$, and is also Floquet and G-invariant.

Thus, for each integer ν, where $0<\nu\le N_2$, the linearized family X^μ_{lin} (4.1) of vector fields possesses $N_2!/\nu!(N_2-\nu)!$ analytic $(m+s+\nu)$-parameter families of invariant $(n+\nu)$-tori with parallel dynamics, these families adjoining the "central" $(m+s)$-parameter analytic family of invariant n-tori $\{y=const, z=0\}$. Using the physical terminology, one says that in the linearized family (4.1), the normal modes are *exciting*. The Floquet matrices of the $(n+\nu)$-tori in question have $N_2-\nu$ pairs of purely imaginary eigenvalues.

It turns out that if $m+s\ge 1$ or $n\le 1$ then under suitable nondegeneracy and nonresonance conditions, all the families of invariant $(n+\nu)$-tori just described exist also in the "full" family X^μ (2.60) of vector fields but are only Whitney-smooth rather than analytic (if $n+\nu\ge 2$). The tori constituting these families are Diophantine. Moreover, these Whitney-smooth $(m+s+\nu)$-parameter families of Floquet Diophantine invariant $(n+\nu)$-tori persist under small G-reversible perturbations of the initial family $\{X^\mu\}_\mu$. It is these properties of $\{X^\mu\}_\mu$ (the existence and persistence of invariant tori of dimensions $n+1,\dots,n+N_2$) that are called the excitation of (elliptic) normal modes in reversible systems. One sometimes also says that the n-tori in the family $\{X^\mu\}_\mu$ and its perturbations are ***dressed*** with tori of larger dimensions. One should emphasize that the Whitney-smooth families of invariant tori of dimensions $n+1,\dots,n+N_2$ which adjoin the "central" family of invariant n-tori are entirely within the framework of Theorem 2.29 and can be regarded as another illustration to this theorem.

For $1 \le \nu \le N_2 - 1$, the elliptic normal modes of the invariant $(n+\nu)$-tori are in turn exciting but this does not lead to new families of invariant tori: the $(n+\nu)$-tori are dressed with the same tori of dimensions $n+\nu+1, \dots, n+N_2$ as the initial n-tori.

Let us proceed to the exact formulation and proof of the excitation theorem. The strategy of the proof is to reduce this theorem to Theorem 2.20 (if $m+s \ge 1$) or its so called "small twist" analogue (if $m = s = 0$ and $n \le 1$). Suppose for definiteness that we are interested in the excitation of the first ν normal modes ($1 \le \nu \le N_2$).

4.1.2 The reversible context ($m + s \ge 1$)

Let first $m + s \ge 1$. Assume all the eigenvalues (2.50) of matrix $\Omega(y,\mu)$ to be simple for all y and μ. Introduce the vector-valued functions

$$\begin{aligned} \widehat{\omega} &= \widehat{\omega}(y,\mu) := (\omega_1, \dots, \omega_n, \varepsilon_1, \dots, \varepsilon_\nu), \\ \widehat{\omega}^N &= \widehat{\omega}^N(y,\mu) := (\varepsilon_{\nu+1}, \dots, \varepsilon_{N_2}, \beta_1, \dots, \beta_{N_3}) \end{aligned} \tag{4.3}$$

ranging in $\mathrm{R}^{n+\nu}$ and $\mathrm{R}^{r-\nu}$, respectively [recall that $r = N_2 + N_3$], and the notations

$$\widehat{\rho}^Q(y,\mu) := \min_{\|e\|=1} \max_{j=0}^{Q} \max_{\|u\|=1} \left| \sum_{|q|=j} \langle D^q \widehat{\omega}(y,\mu), e \rangle u^q \right| \tag{4.4}$$

($q \in \mathrm{Z}_+^{m+s}$, $e \in \mathrm{R}^{n+\nu}$, $u \in \mathrm{R}^{m+s}$) where $Q \in \mathrm{N}$ [cf. (2.103)],

$$\widehat{\Xi}_\ell^Q(y,\mu) := \max_{j=0}^{Q} \max_{\|u\|=1} \left| \sum_{|q|=j} \langle D^q \widehat{\omega}^N(y,\mu), \ell \rangle u^q \right| \tag{4.5}$$

($q \in \mathrm{Z}_+^{m+s}$, $u \in \mathrm{R}^{m+s}$) where $Q \in \mathrm{N}$ and $\ell \in \mathrm{Z}^{r-\nu}$ [cf. (2.104)].

Let the mapping $(y,\mu) \mapsto (\omega, \varepsilon, \beta)$ possess the following properties:

a) there exists $Q \in \mathrm{N}$ such that for any y and μ the collection of $(m+s+Q)!/(m+s)!Q!$ vectors

$$D^q \widehat{\omega}(y,\mu) \in \mathrm{R}^{n+\nu}, \quad q \in \mathrm{Z}_+^{m+s}, \; 0 \le |q| \le Q \tag{4.6}$$

[cf. (2.111)] span $\mathrm{R}^{n+\nu}$, or, equivalently, the image of the map $\widehat{\omega}$ does not lie in any linear hyperplane passing through the origin (see Lemma 2.14); this condition ensures that $\widehat{\rho}^Q(y,\mu) > 0$ for any y and μ,

b) for each y, μ, ℓ [$\ell \in \mathrm{Z}^{r-\nu}$, $1 \le |\ell| \le 2$], and k [$k \in \mathrm{Z}^{n+\nu}$, $0 < \|k\| \le \widehat{\Xi}_\ell^Q(y,\mu)/\widehat{\rho}^Q(y,\mu)$], the following inequality holds:

$$\langle \widehat{\omega}(y,\mu), k \rangle \ne \langle \widehat{\omega}^N(y,\mu), \ell \rangle \tag{4.7}$$

[cf. (2.106)].

Theorem 4.1 *Let X be an analytic family (2.60) of analytic G-reversible vector fields on $\mathrm{T}^n \times \mathrm{R}^{m+2p}$ satisfying the conditions above. Then for any $\sigma > 0$ there exists a neighborhood $\mathcal{X}$ of the family X in the space of all analytic families of analytic G-reversible vector fields (2.65) such that any $\widetilde{X} \in \mathcal{X}$ possesses a Whitney-smooth $(m+s+\nu)$-parameter family of Floquet Diophantine invariant $(n+\nu)$-tori lying in the σ-neighborhood of the surface $\{z=0\}$, the frequency vector of an $(n+\nu)$-torus that is invariant under the field $\widetilde{X}^{\mu^0}$ and lies in the close vicinity of the n-torus $\{y=y^0, z=0\}$ being close to $\widehat{\omega}(y^0,\mu^0)$.*

Proof. Consider a perturbed family $\widetilde{X}$ (2.65) of vector fields. There exists a linear coordinate transformation $z = T(y,\mu)z^{\rm new}$ in R^{2p} which depends analytically on y and μ and reduces the matrices $\Omega(y,\mu)$ and R to the (partially) normal forms[2]

$$T^{-1}(y,\mu)\Omega(y,\mu)T(y,\mu) = \Omega^{\rm new}(y,\mu) = \left[\bigoplus_{j=1}^{\nu}\begin{pmatrix} 0 & -\varepsilon_j(y,\mu) \\ \varepsilon_j(y,\mu) & 0 \end{pmatrix}\right]\bigoplus\widehat{\Omega}(y,\mu) \quad (4.8)$$

and

$$T^{-1}(y,\mu)RT(y,\mu) = R^{\rm new} = \left[\bigoplus_{j=1}^{\nu}\begin{pmatrix} 1 & 0 \\ 0 & -1 \end{pmatrix}\right]\bigoplus\widehat{R}, \quad (4.9)$$

respectively [161, 321, 333]. Here $\widehat{R}$ is an involutive $(2p-2\nu)\times(2p-2\nu)$ matrix whose 1- and (-1)-eigenspaces are $(p-\nu)$-dimensional, matrix $\widehat{\Omega}(y,\mu)$ being infinitesimally $\widehat{R}$-reversible. In the new coordinates $(x, y, z^{\rm new})$, the vector field $\widetilde{X}^\mu$ is reversible with respect to involution

$$(x, y, z^{\rm new}) \mapsto (-x, y, R^{\rm new}z^{\rm new}) \quad (4.10)$$

and determines the system of differential equations

$$\begin{aligned} \dot{x} &= \omega + f + \tilde{f} \\ \dot{y} &= g + \tilde{g} \\ \dot{z}^{\rm new} &= \Omega^{\rm new}z^{\rm new} + T^{-1}\left[h + \tilde{h} - \frac{\partial T}{\partial y}(g+\tilde{g})z^{\rm new}\right] \end{aligned} \quad (4.11)$$

of the same form as the system determined by $\widetilde{X}^\mu$ in the initial coordinates (x, y, z) [in (4.11), the arguments of functions ω, $\Omega^{\rm new}$, and T are (y,μ) while the arguments of functions f, $\tilde{f}$, g, $\tilde{g}$, h, and $\tilde{h}$ are $(x, y, T(y,\mu)z^{\rm new}, \mu)$]. The perturbation $T^{-1}\left[\tilde{h} - (\partial T/\partial y)\tilde{g}z^{\rm new}\right]$ is small whenever $\tilde{h}$ and $\tilde{g}$ are small. We can therefore assume that the matrices $\Omega(y,\mu)$ and R have the normal forms given respectively by (4.8) and (4.9) from the very beginning.

Now introduce new ("cylindrical") coordinates (χ, η, ζ) in R^{2p} via the formulas

$$\begin{aligned} z_{2j-1} &= \sigma\eta_j\cos\chi_j, && 1\le j\le\nu, \\ z_{2j} &= \sigma\eta_j\sin\chi_j, && 1\le j\le\nu, \\ z_{2\nu+j} &= \sigma\zeta_j, && 1\le j\le 2(p-\nu), \end{aligned} \quad (4.12)$$

or

$$z = \sigma\theta(\chi,\eta,\zeta) \qquad \text{(for short)},$$

where $\chi\in T^\nu$, η_j ranges in some fixed open interval $0 < a_j < \eta_j < b_j$ $(1\le j\le\nu)$, $\zeta\in R^{2(p-\nu)}$, and σ is a small positive parameter. In the new coordinates $(x, y, \chi, \eta, \zeta)$, the system of differential equations determined by $\widetilde{X}^\mu$ takes the form

$$\begin{aligned} \dot{x} &= \omega + f + \tilde{f} \\ \dot{\chi}_j &= \varepsilon_j + \frac{1}{\sigma\eta_j}\left[(h_{2j} + \tilde{h}_{2j})\cos\chi_j - (h_{2j-1} + \tilde{h}_{2j-1})\sin\chi_j\right] \\ \dot{y} &= g + \tilde{g} \end{aligned}$$

---

[2] Here the word "normal" means "simplest" or "distinguished".

$$\dot{\eta}_j = \frac{1}{\sigma}\left[(h_{2j-1}+\widetilde{h}_{2j-1})\cos\chi_j + (h_{2j}+\widetilde{h}_{2j})\sin\chi_j\right]$$
$$\dot{\zeta}_j = \left[\widehat{\Omega}\zeta\right]_j + \frac{1}{\sigma}(h_{2\nu+j}+\widetilde{h}_{2\nu+j}) \tag{4.13}$$

[$1 \le j \le \nu$ for $\dot{\chi}$ and $\dot{\eta}$ whereas $1 \le j \le 2(p-\nu)$ for $\dot{\zeta}$], while involution G takes the form

$$(x,\chi,y,\eta,\zeta) \mapsto (-x,-\chi,y,\eta,\widehat{R}\zeta). \tag{4.14}$$

Here the arguments of functions ω, ε, and $\widehat{\Omega}$ are (y,μ) while the arguments of functions f, $\widetilde{f}$, g, $\widetilde{g}$, h, and $\widetilde{h}$ are $(x,y,\sigma\theta(\chi,\eta,\zeta),\mu)$, the symbols $\left[\widehat{\Omega}\zeta\right]_j$ designating the jth entry of vector $\widehat{\Omega}\zeta$. The key observation now is that the functions

$$f(x,y,\sigma\theta(\chi,\eta,\zeta),\mu), \quad g(x,y,\sigma\theta(\chi,\eta,\zeta),\mu), \quad \sigma^{-1}h(x,y,\sigma\theta(\chi,\eta,\zeta),\mu)$$

are small whenever σ is small because

$$f(x,y,z,\mu)=O(z), \quad g(x,y,z,\mu)=O(z), \quad h(x,y,z,\mu)=O_2(z).$$

Hence, if the perturbations $\widetilde{f}$, $\widetilde{g}$, $\sigma^{-1}\widetilde{h}$ are also small then system (4.13) satisfies all the conditions of Theorem 2.20,
the rôle of n, m, p, r being played by $n+\nu$, $m+\nu$, $p-\nu$, $r-\nu$, respectively,
the rôle of x being played by (x,χ),
the rôle of y being played by (y,η),
the rôle of z being played by ζ,
the rôle of ω being played by $\widehat{\omega}=(\omega_1,\ldots,\omega_n,\varepsilon_1,\ldots,\varepsilon_\nu)$,
the rôle of Ω being played by $\widehat{\Omega}$,
the rôle of ω^N being played by $\widehat{\omega}^N=(\varepsilon_{\nu+1},\ldots,\varepsilon_{N_2},\beta_1,\ldots,\beta_{N_3})$,
the rôle of R being played by $\widehat{R}$,
the analogues of the functions f, g, h vanishing identically,
the rôle of the perturbations $\widetilde{f}$, $\widetilde{g}$, $\widetilde{h}$ being played by the functions

$$f+\widetilde{f}, \quad \frac{1}{\sigma\eta_j}\left[(h_{2j}+\widetilde{h}_{2j})\cos\chi_j - (h_{2j-1}+\widetilde{h}_{2j-1})\sin\chi_j\right],$$
$$g+\widetilde{g}, \quad \frac{1}{\sigma}\left[(h_{2j-1}+\widetilde{h}_{2j-1})\cos\chi_j + (h_{2j}+\widetilde{h}_{2j})\sin\chi_j\right],$$
$$\frac{1}{\sigma}(h_{2\nu+j}+\widetilde{h}_{2\nu+j})$$

[$1 \le j \le \nu$ or $1 \le j \le 2(p-\nu)$]. This completes the proof. □

We emphasize that the unperturbed internal $\widehat{\omega}$ and normal $\widehat{\omega}^N$ frequency vectors for system (4.13) do not depend on a part of the action variables (namely, η).

Remark 1. Let the matrices $\Omega(y,\mu)$ and R have the normal forms given respectively by (4.8) and (4.9). Denote by $\mathcal{U}_\sigma$ the (real) σ-neighborhood of the $(n+m+s)$-dimensional set $\mathrm{T}^n \times Y \times \{z=0\} \times P$ on the $(n+m+2\nu+s)$-dimensional surface $\{z_{2\nu+1}=\cdots=z_{2p}=0\} \subset \mathrm{T}^n \times Y \times \mathrm{R}^{2p} \times P$. Denote also by $\mathcal{W}$ the union of the images of the $(n+\nu)$-tori in question (considered as tori in the space $\mathrm{T}^n \times Y \times \mathrm{R}^{2p} \times P$) under the projection

$$\mathrm{T}^n \times \mathrm{R}^{m+2p+s} \to \mathrm{T}^n \times \mathrm{R}^{m+2\nu+s}, \quad (x,y,z,\mu) \mapsto (x,y,z_1,\ldots,z_{2\nu},\mu).$$

Let $|\tilde{f}|$, $|\tilde{g}|$, $|\tilde{h}|$ be less than δ^* in a fixed complex neighborhood of $T^n \times Y \times \{0\} \times P \subset T^n \times R^m \times R^{2p} \times R^s$. Then

$$\frac{\mathrm{meas}_{n+m+s+2\nu}(\mathcal{W} \cap \mathcal{U}_\sigma)}{\mathrm{meas}_{n+m+s+2\nu}(\mathcal{U}_\sigma)} \to 1 \quad \left(\sigma + \frac{\delta^*}{\sigma} \downarrow 0\right). \tag{4.15}$$

Here σ controls the smallness of f, g, $\sigma^{-1}h$ whereas $\sigma^{-1}\delta^*$ controls the smallness of $\tilde{f}$, $\tilde{g}$, $\sigma^{-1}\tilde{h}$.

Remark 2. Sevryuk [323, 324] and Quispel & Sevryuk [287] announced limits similar to (4.15) as $\sigma + \delta^* \downarrow 0$ (in the notations of [287, 323, 324], as $\varepsilon + \delta \downarrow 0$). This was not quite correct. In fact, in the notations of [287, 323, 324], one had to require $\delta + \varepsilon/\delta \downarrow 0$.

Remark 3. The excitation of elliptic normal modes of Floquet Diophantine invariant n-tori in reversible systems for $n \geq 2$ was first conjectured by Sevryuk [319]. The precise theorems were announced by him in [323] for $m \geq n$, $s = 0$. The method of proving the excitation theorems we have used here was first introduced in [326] (for $s = 0$).

4.1.3 The reversible context ($m = s = 0$ and $n \leq 1$)

Now consider the case where $m = s = 0$ but $n \leq 1$. We have an individual vector field X which is reversible with respect to involution G of type (p, p) [if $n = 0$] or $(p+1, p)$ [if $n = 1$] and possesses an equilibrium [if $n = 0$] or periodic trajectory [if $n = 1$] invariant under G. This equilibrium or periodic trajectory generically persists under small G-reversible perturbations of X, so one need not perturb X: any sufficiently small perturbation of X would have respectively an equilibrium or periodic trajectory of the same kind as X has. It suffices to consider X itself only. We will examine the case $n = 0$, the case $n = 1$ can be treated in a similar manner (with some nonessential modifications).

The Normal Form Lemma

Thus, consider an analytic G-reversible vector field on R^{2p}

$$X = [\Omega z + h(z)]\frac{\partial}{\partial z}, \qquad G : z \mapsto Rz \tag{4.16}$$

where $z \in O_{2p}(0)$, R is an involutive $2p \times 2p$ real matrix whose 1- and (-1)-eigenspaces are p-dimensional, $\Omega R + R\Omega = 0$, $h(z) = O_2(z)$, $h(Rz) \equiv -Rh(z)$, and the eigenvalues of matrix Ω are given by (2.50). The main tool to study the excitation of normal modes in this case is the so called *Birkhoff normal forms*.

The classical theory of nonlinear normal forms initiated mainly by Poincaré [274], Dulac [108], and Birkhoff (see [41]), and then developed further by many others, see an extended bibliography in Bruno [70, 75], deals with vector fields in a neighborhood of an equilibrium or periodic trajectory. This theory describes the simplest ("normal") forms to which one can reduce, by an appropriate coordinate change, the Taylor (or Taylor–Fourier, in the case of a periodic trajectory) expansion of a given vector field around an equilibrium (periodic trajectory). The expansion can be normalized up to order $K \in N$ provided that the eigenvalues of the linearization (respectively the Floquet multipliers, see § 1.1.1) of the field satisfy certain nonresonance conditions dependent on K. If the vector fields in question are compatible with some structure $\mathfrak{S}$ on the phase space then the normalizing coordinate change can often be chosen to respect $\mathfrak{S}$ in a certain sense (see, e.g., recent papers [68, 118, 191]). In the dissipative context, one usually speaks of Poincaré–Dulac normal forms, whereas in the Hamiltonian and reversible realms, one prefers to speak of Birkhoff normal forms. In fact, the normal form theory can be developed around a Floquet n-torus with parallel dynamics of arbitrary dimension $n \in \mathbf{Z}_+$, but, as a rule, normal forms for $n \geq 2$ are more complicated and applied much more rarely than those for $n \leq 1$. It sometimes suffices to consider Taylor (or Taylor–Fourier) expansions intermediate between the initial one and the simplest one, in which case one speaks of *partial* Birkhoff normal forms.

Of innumerable works expounding the normal form theory around an equilibrium or periodic trajectory, we here list the books, surveys, and important papers [1, 10, 12, 13, 19, 27, 38, 39, 41, 70–73, 75, 98, 131, 165, 166, 247, 248, 314, 315, 334, 335, 341]. Normal forms around Floquet invariant n-tori with parallel dynamics for $n \geq 2$ have been developed in [26, 39, 42, 47, 48, 64, 72, 97, 309] for dissipative systems (see also Sections 4.3 and 5.2 below), in [72, 73, 75] for the Hamiltonian context, and in [323] for the reversible context.

In fact, normal forms can be considered around an *arbitrary* compact smooth invariant manifold [67].

Return to vector field (4.16). By the standard normal form technique, one can easily prove the following statement.

Lemma 4.2 [316, 318, 325] *Let all the eigenvalues (2.50) of matrix Ω be simple. Suppose that for some $K \in \mathbf{N}$, there hold the inequalities*

$$\begin{array}{llll}
\langle \hat\varepsilon, k\rangle \neq 0 & \text{for} \quad 1 \leq |k| \leq 2K+2, & & \\
\langle \hat\varepsilon, k\rangle \neq \varepsilon_j & \text{for} \quad |k| \leq 4K+1, & \nu+1 \leq j \leq N_2, & \\
\langle \hat\varepsilon, k\rangle \neq \varepsilon_{j_1} + \varepsilon_{j_2} & \text{for} \quad |k| \leq 2K, & \nu+1 \leq j_1 \leq j_2 \leq N_2, & \\
\langle \hat\varepsilon, k\rangle \neq \varepsilon_{j_1} - \varepsilon_{j_2} & \text{for} \quad |k| \leq 2K, & \nu+1 \leq j_1 < j_2 \leq N_2, & \\
\langle \hat\varepsilon, k\rangle \neq \beta_{j_1} + \beta_{j_2} & \text{for} \quad |k| \leq 2K, & 1 \leq j_1 \leq j_2 \leq N_3, & \alpha_{j_1} = \alpha_{j_2}, \\
\langle \hat\varepsilon, k\rangle \neq \beta_{j_1} - \beta_{j_2} & \text{for} \quad |k| \leq 2K, & 1 \leq j_1 < j_2 \leq N_3, & \alpha_{j_1} = \alpha_{j_2},
\end{array}$$

where $\hat\varepsilon = (\varepsilon_1, \dots, \varepsilon_\nu)$ and $k \in \mathbf{Z}^\nu$. Then in some coordinate system (u, w) in $\mathbf{R}^{2p}$ ($u \in \mathbf{R}^{2\nu}$, $w \in \mathbf{R}^{2(p-\nu)}$) centered at 0, the differential equations determined by X take the form

$$\begin{aligned}
\dot u_{2j-1} &= -u_{2j}A_j(v) + U_{2j-1}(u, w) \\
\dot u_{2j} &= u_{2j-1}A_j(v) + U_{2j}(u, w) \\
\dot w &= B(v)w + W(u, w)
\end{aligned} \tag{4.17}$$

($1 \leq j \leq \nu$), while involution G takes the form

$$(u_1, u_2, u_3, u_4, \dots, u_{2\nu-1}, u_{2\nu}, w) \mapsto (u_1, -u_2, u_3, -u_4, \dots, u_{2\nu-1}, -u_{2\nu}, \widehat{R}w) \tag{4.18}$$

where $v = (v_1, \dots, v_\nu)$, $v_j = u_{2j-1}^2 + u_{2j}^2$. Here A_j are polynomials in $v_1, \dots, v_\nu$ of degrees $\leq K$ with constant terms ε_j ($1 \leq j \leq \nu$) whereas B is a polynomial function $\mathbf{R}^\nu \to gl(2p-2\nu, \mathbf{R})$ of degree $\leq K$, the remainders U_j and W possess the properties

$$\begin{aligned}
U_j(\sigma u, \sigma^{2K+1}w) &= O(\sigma^{2K+2}) \qquad (\sigma \downarrow 0), \\
W(\sigma u, \sigma^{2K+1}w) &= O(\sigma^{4K+2}) \qquad (\sigma \downarrow 0)
\end{aligned} \tag{4.19}$$

$(1 \le j \le 2\nu)$, and $\widehat{R}$ is an involutive $(2p-2\nu)\times(2p-2\nu)$ matrix whose 1- and (-1)-eigenspaces are $(p-\nu)$-dimensional. Moreover, the coordinate change $z \mapsto (u,w)$ is polynomial.

System (4.17) exemplifies a partial Birkhoff normal form: the normalization with respect to u is much "deeper" than that with respect to w.

The local KAM theorem

Let the hypotheses of Lemma 4.2 be satisfied for $K = 2$ and

$$\begin{aligned} A_j(v) &= \varepsilon_j + \sum_{l=1}^{\nu} L_{jl}v_l + O_2(v) \quad (1 \le j \le \nu), \\ B(v) &= \widehat{\Omega} + F(v) + O_2(v) \end{aligned} \tag{4.20}$$

where $F : \mathbb{R}^\nu \to gl(2p-2\nu, \mathbb{R})$ is a linear function. The eigenvalues of matrix $\widehat{\Omega}$ are

$$\pm\,\delta_1,\dots,\pm\delta_{N_1}, \quad \pm i\varepsilon_{\nu+1},\dots,\pm i\varepsilon_{N_2}, \quad \pm\alpha_1 \pm i\beta_1,\dots,\pm\alpha_{N_3} \pm i\beta_{N_3}. \tag{4.21}$$

Let matrix $M \in GL(2p-2\nu, \mathbb{C})$ diagonalize $\widehat{\Omega}$, i.e., $M\widehat{\Omega}M^{-1} = diag(\lambda_1,\dots,\lambda_{2p-2\nu})$ where entries λ_κ $(1 \le \kappa \le 2p-2\nu)$ are numbers (4.21).

Now suppose that the matrix $L = \{L_{jl}\}_{j,l=1}^{\nu}$ in (4.20) is non-singular: $\det L \ne 0$, and consider linear function

$$\mathbb{R}^\nu \to gl(2p-2\nu, \mathbb{C}), \quad v \mapsto MF(L^{-1}v)M^{-1} = \{[MF(L^{-1}v)M^{-1}]_{\kappa\tau}\}_{\kappa,\tau=1}^{2p-2\nu}.$$

Introduce the notation

$$C_\kappa = \max_{\|v\|=1} \left|\mathrm{Im}[MF(L^{-1}v)M^{-1}]_{\kappa\kappa}\right| \ge 0 \tag{4.22}$$

$(1 \le \kappa \le 2p-2\nu)$ and suppose that the inequalities

$$\langle\hat{\varepsilon}, k\rangle \ne \mathrm{Im}\,\lambda_\kappa \quad \text{for} \quad 1 \le \kappa \le 2p-2\nu, \quad 0 < \|k\| \le C_\kappa, \tag{4.23}$$

$$\langle\hat{\varepsilon}, k\rangle \ne \mathrm{Im}(\lambda_\kappa - \lambda_\tau) \quad \text{for} \quad 1 \le \kappa < \tau \le 2p-2\nu, \; 0 < \|k\| \le C_\kappa + C_\tau \tag{4.24}$$

hold ($k \in \mathbb{Z}^\nu$).

Theorem 4.3 [316, 318, 325] *Let the G-reversible vector field X determining the system of differential equations (4.17) satisfy the conditions above. Then in any neighborhood $\mathcal{O}_{2p}(0)$ of the origin in $\mathbb{R}^{2p}$, the field X possesses a Whitney-smooth ν-parameter family of Floquet Diophantine invariant ν-tori, the frequency vectors of these tori being close to $\hat{\varepsilon} = (\varepsilon_1,\dots,\varepsilon_\nu)$. If $\mathcal{U}_\sigma$ is the (real) σ-neighborhood of the origin on the (2ν)-dimensional plane $\{w = 0\}$ and $\mathcal{W}$ is the union of the images of the tori in question under the projection $\mathbb{R}^{2p} \to \mathbb{R}^{2\nu}$, $(u,w) \mapsto u$, then*

$$\frac{\mathrm{meas}_{2\nu}(\mathcal{W} \cap \mathcal{U}_\sigma)}{\mathrm{meas}_{2\nu}(\mathcal{U}_\sigma)} \to 1 \quad (\sigma \downarrow 0).$$

Remark. The existence of invariant tori near equilibria of reversible vector fields is well known. Various statements similar to Theorem 4.3 can be traced back up to the very discovery of the reversible KAM theory [40].

The small twist KAM theorem

To prove Theorem 4.3, we need the so called *small twist* analogue of Theorem 2.20. In fact, all the KAM theorems presented in Sections 2.3–2.5 (in any context) have their small twist counterparts adapted expressly for studying vector fields near "critical elements" [1] such as equilibria or periodic trajectories (see, e.g., [62, 137, 162, 248, 278, 279, 315, 318, 325]). We will present, in a rather informal manner, the small twist version of the "miniparameter" theorem in the reversible context 1. The small twist analogue of the "main" theorem in the dissipative context is discussed, in an even much more informal fashion, in Section 5.2.

Instead of family (2.60) of G-reversible vector fields, where involution G is given by (2.61), one should consider the family

$$X_\sigma = \{X_\sigma^\mu\}_\mu = \omega(\sigma, y, \mu)\frac{\partial}{\partial x} + \Omega(\sigma, y, \mu)z\frac{\partial}{\partial z} \tag{4.25}$$

where σ is a small parameter and the functions ω and Ω are analytic in σ:

$$\omega(\sigma, y, \mu) = \omega_{(0)} + \sum_{\iota=1}^{\infty}\sigma^\iota\omega_{(\iota)}(y, \mu), \qquad \Omega(\sigma, y, \mu) = \Omega_{(0)} + \sum_{\iota=1}^{\infty}\sigma^\iota\Omega_{(\iota)}(y, \mu)$$

[$\omega_{(0)} \in \mathrm{R}^n$ and $\Omega_{(0)} \in sl(2p, \mathrm{R})$ being constants and $\Omega_{(0)}R + R\Omega_{(0)} = 0$]. All the eigenvalues of the infinitesimally R-reversible matrix $\Omega_{(0)}$ are assumed to be simple, in which case the eigenvalues of matrix $\Omega(\sigma, y, \mu)$ are analytic in σ. In particular, one can expand the positive imaginary parts of the eigenvalues of the matrix $\Omega(\sigma, y, \mu)$ in powers of σ:

$$\omega^N(\sigma, y, \mu) = \omega^N_{(0)} + \sum_{\iota=1}^{\infty}\sigma^\iota\omega^N_{(\iota)}(y, \mu)$$

[$\omega^N_{(0)} \in \mathrm{R}^r$ being a constant]. Note that family (4.25) is integrable.

Now introduce the notations

$$\widehat{\rho}^Q(y, \mu) := \min_{\|e\|=1}\max_{j=1}^{Q}\max_{\|u\|=1}\left|\sum_{|q|=j}\langle D^q\omega_{(1)}(y, \mu), e\rangle\, u^q\right| \tag{4.26}$$

($q \in \mathrm{Z}_+^{m+s}$, $e \in \mathrm{R}^n$, $u \in \mathrm{R}^{m+s}$) where $Q \in \mathrm{N}$ [cf. (2.103)],

$$\widehat{\Xi}^Q_\ell(y, \mu) := \max_{j=1}^{Q}\max_{\|u\|=1}\left|\sum_{|q|=j}\left\langle D^q\omega^N_{(1)}(y, \mu), \ell\right\rangle u^q\right| \tag{4.27}$$

($q \in \mathrm{Z}_+^{m+s}$, $u \in \mathrm{R}^{m+s}$) where $Q \in \mathrm{N}$ and $\ell \in \mathrm{Z}^r$ [cf. (2.104)]. Note that the value $j = 0$ is excluded here. Let the mapping $(y, \mu) \mapsto \left(\omega_{(1)}, \omega^N_{(1)}\right) = (\omega_{(1)}, \varepsilon_{(1)}, \beta_{(1)})$ possess the following properties:

a) there exists $Q \in \mathrm{N}$ such that for any $(y, \mu) \in \Gamma$ the collection of $[(m + s + Q)!/(m + s)!Q!] - 1$ vectors

$$D^q\omega_{(1)}(y, \mu) \in \mathrm{R}^n, \quad q \in \mathrm{Z}_+^{m+s},\ 1 \le |q| \le Q$$

[cf. (2.111)] span R^n, or, equivalently, the image of the map $\omega_{(1)} : \Gamma \to \mathrm{R}^n$ does not lie in any *affine* hyperplane (not necessarily passing through the origin); this condition ensures that $\widehat{\rho}^Q(y, \mu) > 0$ for any $(y, \mu) \in \Gamma$,

b) for each $\ell \in Z^r$, $1 \leq |\ell| \leq 2$, and

$$k \in Z^n, \quad 0 < \|k\| \leq \max_{(y,\mu)\in\Gamma} \frac{\widehat{\Xi}^Q_\ell(y,\mu)}{\widehat{\rho}^Q(y,\mu)},$$

the inequality $\langle \omega_{(0)}, k\rangle \neq \left\langle \omega^N_{(0)}, \ell\right\rangle$ holds.

Theorem 4.4 *Let X_σ be an analytic family of analytic G-reversible vector fields (4.25) on $T^n \times R^{m+2p}$ satisfying the conditions above. Then for any $\gamma^* > 0$ there exist $\sigma^* > 0$ and $\delta^* > 0$ (independent of σ) such that the following holds. Consider a small G-reversible perturbation of family (4.25)*

$$\begin{aligned} \widetilde{X}_\sigma = \{\widetilde{X}^\mu_\sigma\}_\mu &= \left[\omega(\sigma,y,\mu) + \widetilde{f}(\sigma,x,y,z,\mu)\right]\frac{\partial}{\partial x} + \widetilde{g}(\sigma,x,y,z,\mu)\frac{\partial}{\partial y} + \\ & \left[\Omega(\sigma,y,\mu)z + \widetilde{h}(\sigma,x,y,z,\mu)\right]\frac{\partial}{\partial z}. \end{aligned} \tag{4.28}$$

Suppose that $|\widetilde{f}(\sigma,\cdot)|$, $|\widetilde{g}(\sigma,\cdot)|$, and $|\widetilde{h}(\sigma,\cdot)|$ are less than $\sigma^2\delta^*$ in a fixed (independent of γ^* and σ) complex neighborhood of $T^n \times \{0\} \times \Gamma \subset T^n \times R^{2p} \times R^{m+s}$. Then family (4.28) of vector fields possesses, for $0 < \sigma < \sigma^*$, a Whitney-smooth $(m+s)$-parameter family of Floquet Diophantine invariant n-tori close to the unperturbed tori $\{y = \text{const}, z = 0\}$. The Lebesgue measure of the image of the union of these tori (treated as tori in the space $T^n \times Y \times R^{2p} \times P$) under the projection $(x,y,z,\mu) \mapsto (x,y,\mu)$ is larger than $(1-\gamma^*)(2\pi)^n \,\mathrm{meas}_{m+s}\Gamma$.

Otherwise speaking, if the internal and normal frequencies of the unperturbed tori have the form const + $O(\sigma)$, then one should insert the factor σ^2 in the estimate of a perturbation to make the relative measure of the union of the perturbed tori independent of σ.

The proof of this theorem is based on the small twist analogues of Lemma 2.13 and Theorem 2.7. The small twist version of Theorem 2.7 can be proven parallel to Theorem 2.7 itself, one should just insert carefully σ (in appropriate powers) in all the expressions and estimates.

The proof of the local result

Proof of Theorem 4.3. This theorem can be reduced to Theorem 4.4 in the same manner as Theorem 4.1 reduces to Theorem 2.20.

Fix an arbitrary open domain $Y \subset R^\nu$ such that $[L^{-1}\eta]_j > 0$ for all $1 \leq j \leq \nu$, $\eta \in \overline{Y}$ ($\overline{Y}$ denoting the closure of Y and $[L^{-1}\eta]_j$ designating the jth entry of vector $L^{-1}\eta$). Introduce new ("cylindrical") coordinates (χ, η, ζ) in R^{2p} via the formulas

$$\begin{aligned} u_{2j-1} &= \sqrt{\sigma[L^{-1}\eta]_j}\cos\chi_j, \\ u_{2j} &= \sqrt{\sigma[L^{-1}\eta]_j}\sin\chi_j \end{aligned}$$

$(1 \leq j \leq \nu)$, or

$$u = \sqrt{\sigma}\theta(\chi,\eta) \qquad \text{(for short)},$$

and

$$w = \sigma^{(2K+1)/2}\zeta,$$

where $\chi \in \mathrm{T}^{\nu}$, $\eta \in Y$, $\zeta \in \mathrm{R}^{2(p-\nu)}$, σ is a small positive parameter, and $K = 2$. One observes that $v = \sigma L^{-1}\eta$. In coordinates (χ, η, ζ), the system (4.17) takes the form

$$\begin{aligned} \dot{\chi}_j &= A_j(\sigma L^{-1}\eta) + \frac{1}{\sqrt{\sigma[L^{-1}\eta]_j}}(U_{2j}\cos\chi_j - U_{2j-1}\sin\chi_j) \\ \dot{\eta} &= \frac{2LV}{\sqrt{\sigma}} \\ \dot{\zeta} &= B(\sigma L^{-1}\eta)\zeta + \frac{W}{\sigma^{(2K+1)/2}} \end{aligned} \tag{4.29}$$

where

$$V_j = \sqrt{[L^{-1}\eta]_j}\,(U_{2j-1}\cos\chi_j + U_{2j}\sin\chi_j)$$

$(1 \le j \le \nu)$, while involution G given by (4.18) takes the form

$$(\chi, \eta, \zeta) \mapsto (-\chi, \eta, \widehat{R}\zeta). \tag{4.30}$$

Here the arguments of functions U and W are $(\sqrt{\sigma}\theta(\chi,\eta), \sigma^{(2K+1)/2}\zeta)$.

Now note that

$$\begin{aligned} &\tfrac{1}{\sqrt{\sigma}}U\left(\sqrt{\sigma}\theta(\chi,\eta), \sigma^{(2K+1)/2}\zeta\right) = O(\sigma^{(2K+1)/2}) = \sigma^K o(1), \\ &\tfrac{1}{\sigma^{(2K+1)/2}}W\left(\sqrt{\sigma}\theta(\chi,\eta), \sigma^{(2K+1)/2}\zeta\right) = O(\sigma^{(2K+1)/2}) = \sigma^K o(1) \end{aligned}$$

according to (4.19), and

$$\begin{aligned} A(\sigma L^{-1}\eta) &= \hat{\varepsilon} + \sigma\eta + O(\sigma^2), \\ B(\sigma L^{-1}\eta) &= \widehat{\Omega} + \sigma F(L^{-1}\eta) + O(\sigma^2) \end{aligned}$$

according to (4.20). We claim that system (4.29) satisfies all the conditions of Theorem 4.4 for $s = 0$,

the rôle of n, m, p, r being played respectively by ν, ν, $p-\nu$, $r-\nu$,
the rôle of x, y, z being played respectively by χ, η, ζ,
the rôle of $\omega(\sigma, y)$ being played by $A(\sigma L^{-1}\eta)$,
the rôle of $\omega_{(0)}$ being played by $\hat{\varepsilon}$,
the rôle of $\omega_{(1)}(y)$ being played by the *identity* function η,
the rôle of $\Omega(\sigma, y)$ being played by $B(\sigma L^{-1}\eta)$,
the rôle of $\Omega_{(0)}$ being played by $\widehat{\Omega}$,
the rôle of $\Omega_{(1)}(y)$ being played by the *linear* function $F(L^{-1}\eta)$,
the rôle of the perturbations $\tilde{f}$, $\tilde{g}$, $\tilde{h}$ being played respectively by the functions

$$\frac{1}{\sqrt{\sigma[L^{-1}\eta]_j}}(U_{2j}\cos\chi_j - U_{2j-1}\sin\chi_j) \quad (1 \le j \le \nu), \qquad \frac{2LV}{\sqrt{\sigma}}, \qquad \frac{W}{\sigma^{(2K+1)/2}}.$$

To prove this claim, it suffices to verify the corresponding nondegeneracy and nonresonance conditions. This step is easy and straightforward but requires some care and exactness, so we consider it expedient to present here the reasoning in detail. For arbitrary $d \in \mathbb{N}$, denote by $e^j \in \mathbb{Z}^d$ the jth basis vector of $\mathbb{R}^d$ ($1 \le j \le d$): its jth entry is 1 while all the remaining $d-1$ entries vanish. Since $D^{e^j}\eta = e^j$ ($e^j \in \mathbb{Z}^\nu$, $1 \le j \le \nu$), the nondegeneracy condition a) of Theorem 4.4 is met for $Q = 1$. Now calculate the quantities $\widehat{\rho}^1(\eta)$ and $\widehat{\Xi}^1_\ell(\eta)$ given by the expressions (4.26)–(4.27). One has

$$\widehat{\rho}^1(\eta) = \min_{\|e\|=1} \max_{\|u\|=1} \left| \sum_{j=1}^{\nu} \langle e^j, e\rangle u_j \right| = \min_{\|e\|=1} \max_{\|u\|=1} |\langle e, u\rangle| = 1$$

($e^j \in \mathbb{Z}^\nu$, $e \in \mathbb{R}^\nu$, $u \in \mathbb{R}^\nu$),

$$\widehat{\Xi}^1_\ell(\eta) = \max_{\|u\|=1} \left| \sum_{j=1}^{\nu} \left\langle \frac{\partial \omega^N_{(1)}(\eta)}{\partial \eta_j}, \ell \right\rangle u_j \right| = \left\| \nabla \left\langle \omega^N_{(1)}(\eta), \ell \right\rangle \right\|$$

($u \in \mathbb{R}^\nu$), ∇ being the gradient. Recall that $\ell \in \mathbb{Z}^{r-\nu}$, $1 \le |\ell| \le 2$.

On the other hand, $\omega^N_{(0)} + \sigma\omega^N_{(1)}(\eta)$ coincides, up to the terms of the order of σ^2, with the vector constituted by the positive imaginary parts of the eigenvalues of matrix $\widehat{\Omega} + \sigma F(L^{-1}\eta)$. The eigenvalues of this matrix are equal to those of matrix

$$M\left(\widehat{\Omega} + \sigma F(L^{-1}\eta)\right)M^{-1} = diag(\lambda_1, \ldots, \lambda_{2p-2\nu}) + \sigma M F(L^{-1}\eta) M^{-1} \tag{4.31}$$

where $M \in GL(2p-2\nu, \mathbb{C})$ diagonalizes $\widehat{\Omega}$. In turn, the eigenvalues of matrix (4.31) are equal, up to the order of σ^2, to $\lambda_\kappa + \sigma[MF(L^{-1}\eta)M^{-1}]_{\kappa\kappa}$ where $[MF(L^{-1}\eta)M^{-1}]_{\kappa\kappa}$ denotes the κth diagonal element ($1 \le \kappa \le 2p-2\nu$) of matrix $MF(L^{-1}\eta)M^{-1}$ (recall that the numbers $\lambda_1, \ldots, \lambda_{2p-2\nu}$ are pairwise distinct). The vector-valued function $\omega^N_{(1)}(\eta)$ is therefore linear, and if $\left[\omega^N_{(0)}\right]_a$ ($1 \le a \le r-\nu$) is defined as $\operatorname{Im}\lambda_\kappa$ for some κ, then

$$\left[\omega^N_{(1)}(\eta)\right]_a = \operatorname{Im}[MF(L^{-1}\eta)M^{-1}]_{\kappa\kappa}.$$

In particular,

$$\left\|\nabla\left[\omega^N_{(1)}(\eta)\right]_a\right\| = \|\nabla \operatorname{Im}[MF(L^{-1}\eta)M^{-1}]_{\kappa\kappa}\| = \max_{\|\eta\|=1} |\operatorname{Im}[MF(L^{-1}\eta)M^{-1}]_{\kappa\kappa}| = C_\kappa$$

[see (4.22)]. Since for any κ there exists τ such that $\lambda_\tau = -\lambda_\kappa$ ($1 \le \kappa \le 2p-2\nu$ and $1 \le \tau \le 2p-2\nu$), $\left[\omega^N_{(0)}\right]_a$ can also be defined as $-\operatorname{Im}\lambda_\tau$. In this case,

$$\left[\omega^N_{(1)}(\eta)\right]_a = -\operatorname{Im}[MF(L^{-1}\eta)M^{-1}]_{\tau\tau},$$

and $\left\|\nabla\left[\omega^N_{(1)}(\eta)\right]_a\right\| = C_\tau$.

Now consider an arbitrary vector $\ell \in \mathbb{Z}^{r-\nu}$, $1 \le |\ell| \le 2$. Let first $|\ell| = 1$ and $\ell = \delta e^a$ ($\delta = \pm 1$, $1 \le a \le r-\nu$). Let $\left[\omega^N_{(0)}\right]_a$ be defined as $\delta \operatorname{Im}\lambda_\kappa$ for some κ ($1 \le \kappa \le 2p-2\nu$). Then

$$\widehat{\Xi}^1_\ell(\eta) = \left\|\nabla\left[\omega^N_{(1)}(\eta)\right]_a\right\| = C_\kappa$$

and the nonresonance condition b) corresponding to ℓ consists in that

$$\langle \hat{\varepsilon}, k\rangle \ne \left\langle \omega^N_{(0)}, \ell \right\rangle = \operatorname{Im}\lambda_\kappa \quad \text{for} \quad k \in \mathbb{Z}^\nu,\ 0 < \|k\| \le C_\kappa.$$

This is ensured by (4.23). Now let $|\ell| = 2$ and $\ell = \delta_1 e^a + \delta_2 e^b$ ($\delta_1 = \pm 1$, $\delta_2 = \pm 1$, $1 \le a \le r - \nu$, $1 \le b \le r - \nu$, and $a \ne b$ for $\delta_1 = -\delta_2$). Let $\left[\omega^N_{(0)}\right]_a$ be defined as $\delta_1 \operatorname{Im} \lambda_\kappa$ and $\left[\omega^N_{(0)}\right]_b$ as $-\delta_2 \operatorname{Im} \lambda_\tau$ for some κ and τ ($1 \le \kappa \le 2p - 2\nu$ and $1 \le \tau \le 2p - 2\nu$). It is of importance to notice that $\kappa \ne \tau$. Indeed, if $a \ne b$ then $\kappa \ne \tau$ due to obvious reasons. On the other hand, if $a = b$ then $\delta_1 = \delta_2$ and $\operatorname{Im} \lambda_\kappa = -\operatorname{Im} \lambda_\tau$, whence $\kappa \ne \tau$. Thus, $\kappa \ne \tau$ for all the admissible values of a, b, δ_1, and δ_2. One has

$$\widehat{\Xi}^1_\ell(\eta) = \left\| \nabla \left(\delta_1 \left[\omega^N_{(1)}(\eta)\right]_a + \delta_2 \left[\omega^N_{(1)}(\eta)\right]_b \right) \right\| \le C_\kappa + C_\tau$$

and the nonresonance condition b) corresponding to ℓ is implied by

$$\langle \hat{\varepsilon}, k \rangle \ne \left\langle \omega^N_{(0)}, \ell \right\rangle = \operatorname{Im}(\lambda_\kappa - \lambda_\tau) \quad \text{for} \quad k \in \mathbf{Z}^\nu, \ 0 < \|k\| \le C_\kappa + C_\tau.$$

This is ensured by (4.24) [it is clear that if the inequality in (4.24) holds for any $\kappa < \tau$ and $0 < \|k\| \le C_\kappa + C_\tau$ then it also holds for any $\kappa > \tau$ and $0 < \|k\| \le C_\kappa + C_\tau$]. Thus, system (4.29) satisfies all the nondegeneracy and nonresonance conditions of Theorem 4.4 which completes the proof of Theorem 4.3. □

Remark 1. Although nonresonance conditions (4.23)–(4.24) seem to be somewhat unusual, they are essential. Appropriate counterexamples are given in Sevryuk [316, 318] (see also Section 2.6).

Remark 2. Nonresonance conditions (4.23)–(4.24) can be weakened to

$$\begin{aligned} &i\langle \hat{\varepsilon}, k \rangle \ne \lambda_\kappa && \text{for} \quad 1 \le \kappa \le 2p - 2\nu, && 0 < \|k\| \le C_\kappa, \\ &i\langle \hat{\varepsilon}, k \rangle \ne \lambda_\kappa - \lambda_\tau && \text{for} \quad 1 \le \kappa < \tau \le 2p - 2\nu, && 0 < \|k\| \le C_\kappa + C_\tau \end{aligned}$$

($k \in \mathbf{Z}^\nu$), see [316, 318, 325].

One may ask why we succeeded in dispensing with the small twist troubles and the (partial) Birkhoff normal forms in the proof of Theorem 4.1 which pertains to the case $m + s \ge 1$. The point here is as follows. To obtain Floquet Diophantine invariant tori in a dynamical system, one needs *at least a one-dimensional* parameter labeling the tori (cf. § 1.4.1). The suitable nondegeneracy conditions are to be formulated in terms of this parameter. When $m+s \ge 1$, the parameter in question can be chosen to be $(y, \mu) \in \mathbf{R}^{m+s}$. On the other hand, if $m = s = 0$, we need another parameter which was chosen to be $\upsilon \in \mathbf{R}^\nu$ [see (4.17)]. It is this parameter that is "served" by the partial Birkhoff normal form and the small twist scheme. Of course, an analogous parameter is present in the case $m+s \ge 1$ as well, and the family of invariant $(n + \nu)$-tori in Theorem 4.1 is $(m + s + \nu)$-parameter rather than $(m+s)$-parameter, but the nondegeneracy condition can be expressed in terms of the $(m + s)$-dimensional parameter (y, μ) alone. Therefore, in Theorem 4.1, we were able to manage without normalizing the nonlinear terms and applying the small twist technique.

Remark. However, the *first* proof of the excitation of elliptic normal modes for invariant n-tori ($n \ge 2$) in reversible systems did use the partial Birkhoff normal forms and small twist theorems [323].

The Lyapunov–Devaney theorem

The case $\nu = 1$ in Theorem 4.3 is special and does not require the KAM methods. This case is described by the so called Lyapunov–Devaney theorem.

Theorem 4.5 [22, 38, 105, 136, 261, 315, 317, 348, 349] *Let the matrix Ω in (4.16) possess a pair of simple purely imaginary eigenvalues $\pm i\varepsilon$ ($\varepsilon > 0$) and have no eigenvalues of the form $il\varepsilon$, $l \in \mathrm{Z}_+ \setminus \{1\}$. Then through the origin in R^{2p}, there passes an analytic two-dimensional surface S tangent at 0 to the two-dimensional Ω-invariant plane in R^{2p} corresponding to the eigenvalues $\pm i\varepsilon$. The surface $S \setminus \{0\}$ is analytically foliated into closed trajectories of the field X (4.16) which are invariant under the reversing involution G. The period of these trajectories tends to $2\pi/\varepsilon$ as the initial conditions tend to 0.*

Note that the hypotheses of Theorem 4.5 include no nondegeneracy conditions but infinitely many nonresonance conditions. All these conditions are indeed necessary if we wish the surface S to be analytic at 0. The resonances $1 : l$ (i.e., the presence of eigenvalues $\pm i\varepsilon$, $\pm il\varepsilon$) for $l \geq 3$ do not prevent closed trajectories with periods near $2\pi/\varepsilon$ from existing in an arbitrary neighborhood of 0. However, these trajectories constitute a two-dimensional surface which is generically either of class C^{l-3} at 0 (and not of class C^{l-2}) or even not of class C^1 at 0, see Sevryuk [315, 317].

One may also wonder whether elliptic normal modes can excite in the case where $m = s = 0$ and $n \geq 2$. In fact, this is plausible but has not been examined yet in the literature. Note that for $m = s = 0$ and $n \geq 2$, parallel dynamics on the unperturbed n-torus itself does not survive small perturbations (see § 1.4.1). It seems possible, however, that this unperturbed n-torus is surrounded by ("dressed" with) Floquet Diophantine invariant tori of larger dimensions (provided that $N_2 > 0$), and the latter tori persist under small perturbations (see also a discussion in [326]).

The main difficulty here is that Theorem 4.4 is not applicable to the case where $n \geq 1$. Trying to proceed as in the proof of Theorem 4.3, one again obtains a *linear* function $\omega_{(1)} : \mathrm{R}^\nu \to \mathrm{R}^{\nu+n}$ (see the normal form in [323]) which for $n > 0$, of course, cannot satisfy the nondegeneracy condition a) of Theorem 4.4. For $n = 1$ one may invoke some improvements of Theorem 4.4 or the Poincaré section reduction and the analogue of Theorem 4.3 for reversible diffeomorphisms (see § 5.1.7). But for $n \geq 2$, this difficulty seems to be rather serious.

4.1.4 The Hamiltonian context ($n + s \geq 1$)

Now proceed to the excitation of normal modes in Hamiltonian systems. Theorem 4.3, its analogue for periodic trajectories, and Theorem 4.5 can be carried over to the Hamiltonian case without essential modifications, and the Hamiltonian counterparts of these theorems are well known, see, e.g., [36, 39, 137, 249, 250, 279] (we have listed only some works dealing with invariant tori of all the dimensions $2, \ldots, n$ in Hamiltonian systems with n degrees of freedom). The Hamiltonian analogue of Theorem 4.5 is called the Lyapunov center theorem [1, 38, 136, 174, 175, 211, 220, 247, 335, 348, 349]. The situation with Theorem 4.1 is somewhat different.

The point here is that our method of proving Theorem 4.1 cannot be applied straightforwardly to the Hamiltonian case. Indeed, consider again "blow-up" coordinate change

(4.12). It is in fact a "part" of the coordinate change

$$\begin{aligned} x &= x, & & \\ y &= y, & & \\ z_{2j-1} &= \sigma\eta_j \cos\chi_j, & & 1 \le j \le \nu, \\ z_{2j} &= \sigma\eta_j \sin\chi_j, & & 1 \le j \le \nu, \\ z_{2\nu+j} &= \sigma\zeta_j, & & 1 \le j \le 2(p-\nu) \end{aligned} \tag{4.32}$$

(here we assume, in accordance with the Hamiltonian framework, that $x \in \mathrm{T}^n$ and $y \in \mathrm{R}^n$). This change casts the reversing involution (2.61) into involution (4.14) of the same form. On the other hand, the small parameter σ enters the change of variables z only and does not affect variables (x, y). Consequently, neither transformation (4.32) itself nor any similar transformation can put the symplectic structure

$$\sum_{i=1}^{n} dy_i \wedge dx_i + \sum_{j=1}^{\nu} dz_{2j-1} \wedge dz_{2j} + \sum_{j=1}^{p-\nu} dz_{2\nu+j} \wedge dz_{p+\nu+j} \tag{4.33}$$

into, say, a structure proportional to

$$\sum_{i=1}^{n} dy_i \wedge dx_i + \sum_{j=1}^{\nu} d\eta_j \wedge d\chi_j + \sum_{j=1}^{p-\nu} d\zeta_j \wedge d\zeta_{j+p-\nu} \tag{4.34}$$

[cf. (1.13)]. In the variables $(x, y, \chi, \eta, \zeta)$ given by (4.32), the symplectic structure (4.33) is not "uniform", namely,

$$\sum_{i=1}^{n} dy_i \wedge dx_i + \sigma^2 \left(\sum_{j=1}^{\nu} \eta_j d\eta_j \wedge d\chi_j + \sum_{j=1}^{p-\nu} d\zeta_j \wedge d\zeta_{j+p-\nu} \right),$$

which prevents one from applying Theorem 2.19 (although "non-uniform" structures of this kind are sometimes used in the theory of Hamiltonian systems [141]). To obtain a symplectic structure proportional to (4.34), one should "stretch" not only z, but y as well. For $s = 0$, this would destroy our technique completely, because if $s = 0$ then y is the only parameter we use that labels the tori.

One can express the same in other words. The internal frequencies of the unperturbed tori $\{y = const, \eta = const, \zeta = 0\}$ corresponding to system (4.13) are

$$\omega_1(y, \mu), \ldots, \omega_n(y, \mu), \varepsilon_1(y, \mu), \ldots, \varepsilon_\nu(y, \mu).$$

In the reversible realm, these frequencies may be arbitrary. But in the Hamiltonian realm, they are the partial derivatives of a certain function of y, η, μ and should obey the identities of the form $\partial^2/\partial a \partial b = \partial^2/\partial b \partial a$, e.g.,

$$\partial\varepsilon_j/\partial y_i = \partial\omega_i/\partial\eta_j \equiv 0, \quad 1 \le i \le n, \ 1 \le j \le \nu.$$

Of course, these identities are generically not satisfied.

However, our method can be modified to handle the Hamiltonian case as well. Consider an analytic s-parameter family $X = \{X^\mu\}_\mu$ of analytic Hamiltonian vector fields (2.49) on

$T^n \times R^{n+2p}$ ($n + s \geq 1$), the symplectic structure being given by (1.13) and the Hamilton function being equal to (2.48). Assume all the eigenvalues (2.50) of matrix $\Omega(y, \mu)$ to be simple for all y and μ. Introduce the vector-valued functions $\widehat{\omega} = \widehat{\omega}(y, \mu)$ and $\widehat{\omega}^N = \widehat{\omega}^N(y, \mu)$ ranging in $R^{n+\nu}$ and $R^{r-\nu}$, respectively, by formulas (4.3) [recall again that $r = N_2 + N_3$]. Define the quantities $\widehat{\rho}^Q(y, \mu)$ and $\widehat{\Xi}_\ell^Q(y, \mu)$ by (4.4) and (4.5), respectively (with $u \in R^{n+s}$ instead of $u \in R^{m+s}$ and $q \in Z_+^{n+s}$ instead of $q \in Z_+^{m+s}$).

Let the mapping $(y, \mu) \mapsto (\omega, \varepsilon, \beta)$ possess the following properties:

a) there exists $Q \in N$ such that for any y and μ the collection of $(n+s+Q)!/(n+s)!Q!$ vectors

$$D^q \widehat{\omega}(y, \mu) \in R^{n+\nu}, \quad q \in Z_+^{n+s}, \ 0 \leq |q| \leq Q \tag{4.35}$$

[cf. (2.105) and (4.6)] span $R^{n+\nu}$, or, equivalently, the image of the map $\widehat{\omega}$ does not lie in any linear hyperplane passing through the origin (see Lemma 2.14); this condition ensures that $\widehat{\rho}^Q(y, \mu) > 0$ for any y and μ,

b) for each y, μ, ℓ [$\ell \in Z^{r-\nu}$, $1 \leq |\ell| \leq 2$], and k [$k \in Z^{n+\nu}$, $0 < \|k\| \leq \widehat{\Xi}_\ell^Q(y, \mu)/\widehat{\rho}^Q(y, \mu)$], inequality (4.7) holds.

Theorem 4.6 [330] *Let X be an analytic family of analytic Hamiltonian vector fields (2.49) on $T^n \times R^{n+2p}$ satisfying the conditions above. Then for any $\sigma > 0$ there exists a neighborhood $\mathcal{X}$ of the Hamilton function H^μ (2.48) in the space of all analytic and analytically μ-dependent Hamiltonians (2.55) such that for any $\widetilde{H}^\mu \in \mathcal{X}$, the family $\widetilde{X}$ of vector fields determined by $\widetilde{H}^\mu$ possesses a Whitney-smooth $(n + s + \nu)$-parameter family of Floquet Diophantine isotropic invariant $(n + \nu)$-tori lying in the σ-neighborhood of the surface $\{z = 0\}$, the frequency vector of an $(n + \nu)$-torus that is invariant under the field $\widetilde{X}^{\mu^0}$ and lies in the close vicinity of the n-torus $\{y = y^0, z = 0\}$ being close to $\widehat{\omega}(y^0, \mu^0)$.*

Remark. This family of invariant $(n + \nu)$-tori is entirely within the framework of Theorem 2.28 and can be regarded as another illustration to that theorem. The Floquet matrices of the $(n+\nu)$-tori in question have $N_2 - \nu$ pairs of purely imaginary eigenvalues.

Proof of Theorem 4.6. This theorem can be reduced to Theorem 2.19 in the same manner as Theorem 4.1 reduces to Theorem 2.20, but one should invoke the so called *localization* trick heavily exploited (in a somewhat different set-up) in [60, 62, 162]. Consider a perturbed Hamilton function $\widetilde{H}^\mu$ (2.55). Introduce a new parameter $a \in Y \subset R^n$ and the change of variables $y = \hat{y} + a$ ($\hat{y}$ being the "local" variable). We obtain an $(n+s)$-parameter family of Hamiltonian vector fields governed by the Hamilton function $\widetilde{H}^\mu(x, \hat{y} + a, z)$. Of course, this family is just the original perturbed s-parameter family $\widetilde{X}$ "multiplied" by R^n.

There exists a linear coordinate transformation $z = T(a, \mu) z^{\text{new}}$ in R^{2p} which depends analytically on a and μ and reduces the function $\frac{1}{2}\langle z, K(a, \mu) z\rangle$ and the symplectic structure $\sum_{j=1}^p dz_j \wedge dz_{j+p}$ to the normal forms

$$\begin{aligned} &\tfrac{1}{2}\langle T(a, \mu) z^{\text{new}}, K(a, \mu) T(a, \mu) z^{\text{new}}\rangle = \\ &\tfrac{1}{2}\sum_{j=1}^{\nu} \varepsilon_j^\star(a, \mu) \left[\left(z_{2j-1}^{\text{new}}\right)^2 + \left(z_{2j}^{\text{new}}\right)^2 \right] + \tfrac{1}{2}\langle \hat{z}^{\text{new}}, \widehat{K}(a, \mu) \hat{z}^{\text{new}}\rangle \end{aligned}$$

and

$$\sum_{j=1}^{\nu} dz_{2j-1}^{\text{new}} \wedge dz_{2j}^{\text{new}} + \sum_{j=1}^{p-\nu} dz_{2\nu+j}^{\text{new}} \wedge dz_{p+\nu+j}^{\text{new}},$$

respectively (see [120, 161, 178]). Here

$$\varepsilon_j^*(a,\mu) = \pm\varepsilon_j(a,\mu) \quad \text{for} \quad 1 \le j \le \nu,$$

$\hat{z}^{\rm new} = (z_{2\nu+1}^{\rm new}, \ldots, z_{2p}^{\rm new}) \in \mathrm{R}^{2(p-\nu)}$, and the $(2p-2\nu)\times(2p-2\nu)$ matrix $\widehat{K}(a,\mu)$ is symmetric for all values of a and μ. Define the $(2p-2\nu)\times(2p-2\nu)$ matrix $\widehat{J}$ by the same formula (1.14) as the $2p \times 2p$ matrix J, with $I \in SL(p-\nu, \mathrm{R})$ instead of $I \in SL(p, \mathrm{R})$, and define Hamiltonian matrix $\widehat{\Omega}(a,\mu) = \widehat{J}\widehat{K}(a,\mu)$.

Now introduce new coordinates $(x, \xi, \chi, \eta, \zeta)$ in $\mathrm{T}^n \times \mathrm{R}^{n+2p}$ around the n-torus $\{\hat{y} = 0, z = 0\}$ via the formulas

$$\begin{aligned}
x &= x, \\
\hat{y} &= \sigma^2\xi, \\
z_{2j-1}^{\rm new} &= \sigma\sqrt{2\eta_j}\cos\chi_j, && 1 \le j \le \nu, \\
z_{2j}^{\rm new} &= \sigma\sqrt{2\eta_j}\sin\chi_j, && 1 \le j \le \nu, \\
z_{2\nu+j}^{\rm new} &= \sigma\zeta_j, && 1 \le j \le 2(p-\nu),
\end{aligned}$$

or

$$x = x, \quad \hat{y} = \sigma^2\xi, \quad z^{\rm new} = \sigma\theta(\chi, \eta, \zeta) \qquad \text{(for short)},$$

where $\xi \in \mathrm{R}^n$, $\chi \in \mathrm{T}^\nu$, η_j ranges in some fixed open interval $0 < \eta_j^{\min} < \eta_j < \eta_j^{\max}$ $(1 \le j \le \nu)$, $\zeta \in \mathrm{R}^{2(p-\nu)}$, and σ is a small positive parameter [cf. (4.32)]. In these coordinates, the symplectic structure (1.13) takes the form

$$\sigma^2\left(\sum_{i=1}^n d\xi_i \wedge dx_i + \sum_{j=1}^\nu d\eta_j \wedge d\chi_j + \sum_{j=1}^{p-\nu} d\zeta_j \wedge d\zeta_{j+p-\nu}\right) \tag{4.36}$$

[cf. (4.34)], while the new Hamilton function is

$$\begin{aligned}
&\widetilde{H}^\mu(x, a+\sigma^2\xi, \sigma T(a,\mu)\theta(\chi,\eta,\zeta)) = \\
&F(a+\sigma^2\xi,\mu) + \sigma^2\sum_{j=1}^\nu \varepsilon_j^*(a,\mu)\eta_j + \frac{\sigma^2}{2}\langle\zeta, \widehat{K}(a,\mu)\zeta\rangle + \\
&\quad \frac{\sigma^2}{2}\langle T(a,\mu)\theta(\chi,\eta,\zeta),\ [K(a+\sigma^2\xi,\mu) - K(a,\mu)]\,T(a,\mu)\theta(\chi,\eta,\zeta)\rangle + \\
&\quad \Delta\,(x, a+\sigma^2\xi, \sigma T(a,\mu)\theta(\chi,\eta,\zeta),\mu) + \widetilde{\Delta}\,(x, a+\sigma^2\xi, \sigma T(a,\mu)\theta(\chi,\eta,\zeta),\mu) = \\
&F(a,\mu) + \sigma^2\left[\sum_{i=1}^n \omega_i(a,\mu)\xi_i + \sum_{j=1}^\nu \varepsilon_j^*(a,\mu)\eta_j + \tfrac{1}{2}\langle\zeta, \widehat{K}(a,\mu)\zeta\rangle + \right. \\
&\quad \left. \sigma\Delta^*(x,\xi,\chi,\eta,\zeta,a,\mu,\sigma)\right] + \widetilde{\Delta}\,(x, a+\sigma^2\xi, \sigma T(a,\mu)\theta(\chi,\eta,\zeta),\mu),
\end{aligned} \tag{4.37}$$

where remainder Δ^* is analytic in all its arguments for $-\sigma_0 < \sigma < \sigma_0$ (σ_0 being sufficiently small), because $\Delta(x,y,z,\mu) = O_3(z)$ [however, we deal here with the positive values of σ only]. One can drop the irrelevant constant term $F(a,\mu)$ in (4.37) and divide both the new Hamiltonian and the new symplectic structure (4.36) by σ^2. Consequently, if the perturbation $\sigma^{-2}\widetilde{\Delta}$ is small then symplectic structure (4.36) and Hamilton function (4.37) satisfy all the conditions of Theorem 2.19,

the rôle of n, p, r, s being played by $n+\nu, p-\nu, r-\nu, n+s$, respectively,
the rôle of x being played by (x, χ),
the rôle of y being played by (ξ, η),
the rôle of z being played by ζ,
the rôle of μ being played by (a, μ),
the rôle of $\omega(y, \mu)$ being played by

$$\widehat{\omega}^*(a, \mu) = (\omega_1, \ldots, \omega_n, \varepsilon_1^*, \ldots, \varepsilon_\nu^*)|_{(a,\mu)}, \tag{4.38}$$

the rôle of $K(y, \mu)$ and $\Omega(y, \mu)$ being played by $\widehat{K}(a, \mu)$ and $\widehat{\Omega}(a, \mu)$, respectively,
the rôle of $\omega^N(y, \mu)$ being played by

$$\widehat{\omega}^N(a, \mu) = (\varepsilon_{\nu+1}, \ldots, \varepsilon_{N_2}, \beta_1, \ldots, \beta_{N_3})|_{(a,\mu)}, \tag{4.39}$$

the analogue of the function F being linear in (ξ, η),
the analogue of the function Δ vanishing identically,
the rôle of the perturbation $\widetilde{\Delta}$ being played by the function $\sigma\Delta^* + \sigma^{-2}\widetilde{\Delta}$.
Indeed, the value of $\widehat{\rho}^Q$ does not change if one replaces $\widehat{\omega}$ [defined by (4.3)] with $\widehat{\omega}^*$ [defined by (4.38)] in the right-hand side of (4.4) [with $u \in \mathbf{R}^{n+s}$ and $q \in \mathbf{Z}_+^{n+s}$ instead of respectively $u \in \mathbf{R}^{m+s}$ and $q \in \mathbf{Z}_+^{m+s}$].

Now Theorem 2.19 provides us with a Whitney-smooth $(2n+s+\nu)$-parameter family of Floquet Diophantine isotropic invariant $(n+\nu)$-tori for the Hamilton function $\widetilde{H}^\mu(x, \hat{y}+a, z)$. On the other hand, for any fixed value of parameter a, the latter function is just the original perturbed Hamilton function $\widetilde{H}^\mu(x, y, z)$ [given by (2.55)] after the translation $y = \hat{y} + a$. So, for $\widetilde{H}^\mu(x, y, z)$, we obtain a Whitney-smooth $(n+s+\nu)$-parameter family of invariant $(n+\nu)$-tori. This completes the proof. We emphasize that the unperturbed internal $\widehat{\omega}^*$ (4.38) and normal $\widehat{\omega}^N$ (4.39) frequency vectors for Hamiltonian (4.37) do not depend on the action variables (ξ, η) [the so called proper degeneracy, cf. the remark at the end of § 2.3.4]. □

Remark 1. Denote by $\mathcal{U}_\sigma$ the (real) σ-neighborhood of the $(2n+s)$-dimensional set $\mathbf{T}^n \times Y \times \{z=0\} \times P$ on the $(2n+2\nu+s)$-dimensional surface $\{z_{2\nu+1}^{\text{new}} = \cdots = z_{2p}^{\text{new}} = 0\} \subset \mathbf{T}^n \times Y \times \mathbf{R}^{2p} \times P$. Denote also by $\mathcal{W}$ the union of the images of the $(n+\nu)$-tori in question (considered as tori in the space $\mathbf{T}^n \times Y \times \mathbf{R}^{2p} \times P$) under the projection

$$\mathbf{T}^n \times \mathbf{R}^{n+2p+s} \to \mathbf{T}^n \times \mathbf{R}^{n+2\nu+s}, \quad (x, y, z^{\text{new}}, \mu) \mapsto (x, y, z_1^{\text{new}}, \ldots, z_{2\nu}^{\text{new}}, \mu).$$

Let $|\widetilde{\Delta}|$ be less than δ^* in a fixed complex neighborhood of $\mathbf{T}^n \times Y \times \{0\} \times P \subset \mathbf{T}^n \times \mathbf{R}^n \times \mathbf{R}^{2p} \times \mathbf{R}^s$. Then

$$\frac{\operatorname{meas}_{2n+s+2\nu}(\mathcal{W} \cap \mathcal{U}_\sigma)}{\operatorname{meas}_{2n+s+2\nu}(\mathcal{U}_\sigma)} \to 1 \quad \left(\sigma + \frac{\delta^*}{\sigma^2} \downarrow 0\right) \tag{4.40}$$

[cf. (4.15)]. Here σ controls the smallness of $\sigma\Delta^*$ whereas $\sigma^{-2}\delta^*$ controls the smallness of $\sigma^{-2}\widetilde{\Delta}$.

Remark 2. Diophantine isotropic invariant $(n+p)$-tori around invariant n-tori in Hamiltonian systems with $n+p$ degrees of freedom were first studied by Arnol'd [5] (see also [20]). Arnol'd [5] considered the properly degenerate framework as well as the case where the proper degeneracy and the limit one combine.

Remark 3. Bruno [72, Part II] examined in detail the following problem. Consider an analytic s-parameter family $X = \{X^\mu\}_\mu$ of analytic Hamiltonian vector fields and suppose that the field X^{μ_0} possesses a Floquet Diophantine isotropic invariant analytic n-torus T_0 whose Floquet matrix has $N_\star \geq 1$ pairs of purely imaginary eigenvalues. Assume that $s \geq N_\star - 1$. Bruno proved by means of his general theory (see Section 6.3) that under suitable nondegeneracy and nonresonance conditions, for each integer ν in the range $0 \leq \nu \leq N_\star$, family X of vector fields admits $(s - N_\star + \nu + 1)$-parameter analytic families $\mathfrak{T}$ of Diophantine isotropic invariant analytic $(n+\nu)$-tori (not necessarily Floquet), families $\mathfrak{T}$ adjoining the torus T_0. This result agrees well with Theorem 3.7.

The actual motion near T_0 is however much more complicated, because most of the $(n+\nu)$-tori in question are in fact Floquet and belong to Whitney-smooth $(n+s+\nu)$-parameter families of Floquet Diophantine isotropic invariant analytic $(n+\nu)$-tori. For $s \geq 1$, this statement follows from Theorem 4.6, it also gives rise to no doubt for $s = N_\star - 1 = 0$.

Remark 4. While discussing Bruno's result in [287], Quispel & Sevryuk asserted that all the $(n+\nu)$-tori in the $(s - N_\star + \nu + 1)$-parameter families $\mathfrak{T}$ are Floquet. This was not quite correct. Invariant tori constructed within the framework of the Bruno continuation theory are not necessarily Floquet. The statement in [287] was true only when all the eigenvalues of the Floquet matrix of the "central" torus T_0 are purely imaginary or real. Suppose, on the other hand, that the Floquet matrix of the torus T_0 has $N_\star$ pairs of purely imaginary eigenvalues and $N_{\star\star}$ quadruplets $\pm\alpha\pm i\beta$ ($\alpha > 0$, $\beta > 0$). Then, according to Theorem 2.28, each Floquet $(n+\nu)$-torus belongs generically to an $(s - N_\star - N_{\star\star} + \nu + 1)$-parameter analytic subfamily of Floquet $(n+\nu)$-tori (provided that s is sufficiently large).

4.1.5 The measure estimates

To conclude this section, mention the following interesting estimates of the Lebesgue measure of the union of invariant tori near elliptic equilibria of Hamiltonian vector fields. Let $0 \in \mathbb{R}^{2n}$ be an equilibrium point of an analytic Hamiltonian vector field X. Suppose that all the eigenvalues of the linearization of X at 0 are purely imaginary and equal to $\pm\varepsilon_1, \ldots, \pm\varepsilon_n$ where $\varepsilon_1 > 0, \ldots, \varepsilon_n > 0$. Then, under certain nonresonance and nondegeneracy conditions, the field X possesses a Whitney-smooth n-parameter family of Floquet Diophantine Lagrangian invariant n-tori in any neighborhood of 0, the frequency vectors of these tori being close to $\varepsilon = (\varepsilon_1, \cdots, \varepsilon_n)$ [a particular case of the Hamiltonian analogue of Theorem 4.3]. Denote by m_σ the relative measure of the complement of the union of the tori in question in the σ-neighborhood $\mathcal{U}_\sigma$ of 0:

$$m_\sigma = \frac{\mathrm{meas}_{2n}(\mathcal{U}_\sigma \setminus \mathcal{W})}{\mathrm{meas}_{2n}(\mathcal{U}_\sigma)}$$

($\mathcal{W}$ being the union of the tori). Then $m_\sigma = o(1)$ as $\sigma \downarrow 0$.

It turns out that if the vector ε is nonresonant up to order $K \geq 4$ (i.e., $\langle \varepsilon, k \rangle \neq 0$ for $k \in \mathbb{Z}^n$, $1 \leq |k| \leq K$), then $m_\sigma = O(\sigma^{(K-3)/2})$ [278, Sect. 5]. On the other hand, if the vector ε is Diophantine:

$$|\langle \varepsilon, k \rangle| \geq \gamma |k|^{-\tau} \quad \forall k \in \mathbb{Z}^n \setminus \{0\},$$

then m_σ is exponentially small with σ:

$$m_\sigma = O\left(\exp\left\{-\left[\frac{c\gamma}{\sigma}\right]^{1/(\tau+1)}\right\}\right)$$

for some constant $c > 0$ [102, 104] (see also [131]). For the latter estimate, the analyticity of X is very essential. Generalizations to the case of resonant vectors ε are presented in [102] as well.

Similar estimates undoubtedly hold near elliptic equilibria of reversible vector fields [227], but this seems to have not been verified yet in the literature.

4.2 Resonance zones

In this section, we consider the motion in resonance zones between Whitney-smooth families of Floquet Diophantine invariant tori. We will confine ourselves with the Hamiltonian isotropic $(n, 0, 0)$ context (for $n \geq 2$), because the results pertaining to other contexts are rather slender. On the other hand, very ample information has been obtained by now on the motion in resonance zones in Hamiltonian systems. We will dwell on some selected topics only.

Consider a Hamiltonian vector field X^ϵ which depends on a small parameter $\epsilon \geq 0$ and is associated to the Hamilton function

$$H(x, y, \epsilon) = H^0(y) + \epsilon H^1(x, y, \epsilon) \tag{4.41}$$

analytic in x, y, and ϵ. Here $x \in \mathrm{T}^n$, $y \in Y \subset \mathrm{R}^n$ (Y being an open bounded domain in R^n), H^0 is defined in a complex neighborhood of the closure $\overline{Y}$ of Y, and $|H^1(x, y, \epsilon)| < C$ in an ϵ-independent complex neighborhood of $\mathrm{T}^n \times \overline{Y}$ with ϵ-independent constant $C > 0$. The symplectic structure is supposed to be $\sum_{i=1}^n dy_i \wedge dx_i$. The vector field X^0 is governed by Hamiltonian $H^0(y)$ and has the form $X^0 = \omega(y)\partial/\partial x$ where $\omega(y) = \partial H^0(y)/\partial y$. Throughout this section, we will assume that $\omega(y) \neq 0$ in $\overline{Y}$. Each n-torus $\{y = const\}$ is invariant under the flow of X^0 and carries parallel dynamics with frequency vector $\omega(y)$. The coordinates (y, x) are the action-angle variables for integrable Hamiltonian field X^0.

According to Theorem 2.19 and Proposition 2.22, the Hamiltonian system X^ϵ possesses many Diophantine invariant n-tori close to the unperturbed tori $\{y = const\}$ for any function H^1 and all sufficiently small ϵ if and only if the unperturbed Hamiltonian $H^0(y)$ is *nondegenerate in the sense of Rüssmann* ([302–304], see also Herman [153] as well as Sevryuk [327, 329]): the image of the frequency map $\omega : Y \to \mathrm{R}^n$ does not lie in any linear hyperplane passing through the origin. This condition is very weak: the image of the frequency map for an integrable Hamilton function $H^0(y)$ nondegenerate in the sense of Rüssmann can be a variety in R^n of any positive dimension. It can even be a curve.

Example 4.7 ([74, 327, 329], note that paper [74] contains certain inaccuracies corrected in [329]) Let m be an integer in the range $1 \leq m \leq n$. Denote by $u = u(y_1, \ldots, y_m)$ the solution of the equation

$$\textstyle\sum_{j=1}^m j u^{j-1} y_j = u$$

which is defined and analytic in $y_1, \ldots, y_m$ near the point $y_1 = \cdots = y_m = 0$ and vanishes at that point. The local existence and uniqueness of function u are ensured by the Implicit

Function Theorem. If $m = 1$ then $u = y_1$, and if $m = 2$ then $u = y_1(1 - 2y_2)^{-1}$. It is easy to verify that

$$\frac{\partial u}{\partial y_j} = \frac{ju^{j-1}}{1 - \sum_{\iota=2}^{m} \iota(\iota - 1)u^{\iota-2}y_\iota} = ju^{j-1}\frac{\partial u}{\partial y_1} = \frac{\partial(u^j)}{\partial y_1}, \quad 1 \le j \le m \tag{4.42}$$

and that $u = 0$ for $y_1 = 0$. Now consider the Hamiltonian

$$H^0(y) = \int_0^{y_1} u(s, y_2, \ldots, y_m)\, ds + \tfrac{1}{2}\sum_{j=m+1}^{n} y_j^2. \tag{4.43}$$

It follows from (4.42) that the frequencies corresponding to this Hamiltonian are

$$\omega_j(y) = \frac{\partial H^0(y)}{\partial y_j} = \left[\begin{array}{ll} (u(y_1, \ldots, y_m))^j & \text{for } 1 \le j \le m \\ y_j & \text{for } m+1 \le j \le n. \end{array}\right.$$

Thus, Hamiltonian (4.43) is nondegenerate in the sense of Rüssmann and the image of its frequency map is of dimension $n - m + 1$.

Remark 1. Interesting topological conditions (in terms of the so called topological Conley index) for the existence of invariant n-tori in nearly integrable Hamiltonian systems have been found by Plotnikov [272]. To be precise, Plotnikov constructs a one-to-one correspondence between invariant tori (with a fixed Diophantine frequency vector) of the flow governed by Hamiltonian (4.41) for $\epsilon \ne 0$ sufficiently small and critical points of a certain analytic function defined in Y (in our notations). His results imply, in particular, the existence of invariant tori under rather weak nondegeneracy conditions on $H^0(y)$ (close to the Rüssmann nondegeneracy condition).

Remark 2. The existence of invariant tori in a system with Hamiltonian (4.41) for $\epsilon \ne 0$ small under the conditions very close to the Rüssmann nondegeneracy condition was also proven by Cheng & Sun [89] and by Xu, You & Qiu [360].

However, assume $H^0(y)$ to satisfy the much stronger nondegeneracy condition: the Hessian matrix $\partial^2 H^0(y)/\partial y^2 = \partial\omega(y)/\partial y$ is non-singular in $\overline{Y}$, i.e., the frequency map is submersive in $\overline{Y}$ (the so called *nondegeneracy in the sense of Kolmogorov* [4, 179]). In this case, Theorem 2.6 provides us with the following description of the fate of the unperturbed tori under small perturbations of the integrable Hamiltonian: tori $\{y = y^*\}$ with Diophantine frequency vectors

$$|\langle\omega(y^*), k\rangle| \ge \gamma|k|^{-\tau} \quad \forall k \in \mathbb{Z}^n \setminus \{0\} \tag{4.44}$$

survive a perturbation and just undergo a slight deformation [for a fixed $\tau > n - 1$, a torus satisfying (4.44) withstands a perturbation of $H^0(y)$ of the order of γ^2]. The deformation (i.e., the difference between the unperturbed torus and the perturbed one with the same frequencies) is of the order of $\epsilon/\gamma \lesssim \sqrt{\epsilon}$ (if $\gamma \gtrsim \sqrt{\epsilon}$) [257, 278]. All the perturbed invariant n-tori (the so called KAM tori) constitute an n-parameter Whitney-smooth family $\mathfrak{T}_0 = \mathfrak{T}_0(\epsilon)$. The union of these tori is sometimes called the Kolmogorov set.

On the other hand, resonant tori $\{y = y^*\}$ disintegrate under arbitrarily small generic perturbations and give rise to *resonance zones* (or resonance layers) which size ("width") is of the order of $\sqrt{\epsilon}$ [257, 278].[3] These zones contain stochastic trajectories as well as quasi-periodic trajectories filling up a complicated infinite hierarchy of subsidiary ("secondary", "tertiary", "quaternary", ...) Whitney-smooth families of Floquet Diophantine invariant tori of various dimensions (from 1 to n). Each family of tori is described by Theorem 2.28. One of the main "sources" of the invariant tori inside resonance zones is as follows.

A resonant unperturbed n-torus $T_\star = \{y = y^\star\}$ is foliated by quasi-periodic invariant subtori of smaller dimension (say, ν, where $1 \le \nu \le n-1$). It turns out that if these subtori are not only quasi-periodic but even Diophantine then the torus $T_\star$ does not break up completely under a perturbation but, as a rule, gives birth to a finite collection of Floquet Diophantine invariant ν-tori.

Let us proceed to exact formulations. Consider an unperturbed invariant n-torus $T_\star = \{y = y^\star\}$. Let its frequencies $\omega_1(y^\star), \dots, \omega_n(y^\star)$ be rationally dependent:

$$G = \{k \in \mathbb{Z}^n \ : \ \langle \omega(y^\star), k\rangle = 0\} \approx \mathbb{Z}^{n-\nu} \tag{4.45}$$

($1 \le \nu \le n-1$). It is obvious that if $sk \in G$, where $s \in \mathbb{Z} \setminus \{0\}$ and $k \in \mathbb{Z}^n$, then $k \in G$. Consequently, the quotient group $\mathbb{Z}^n/G \approx \mathbb{Z}^\nu$ is torsion-free, and there therefore exist vectors $\varsigma_1, \dots, \varsigma_n \in \mathbb{Z}^n$ such that the $n \times n$ matrix K with columns $\varsigma_1, \dots, \varsigma_n$ is unimodular (i.e., $\det K = 1$) and vectors $\varsigma_{\nu+1}, \dots, \varsigma_n$ generate G [45]. One has $\langle \omega(y^\star), \varsigma_j\rangle = 0$ for $\nu+1 \le j \le n$. Denote $\langle \omega(y^\star), \varsigma_j\rangle$ by ϖ_j for $1 \le j \le \nu$. The numbers $\varpi_1, \dots, \varpi_\nu$ are rationally independent.

In the sequel, we will sometimes write $K = K(T_\star)$ and $\varpi_j = \varpi_j(T_\star)$, $1 \le j \le \nu$. Emphasize however that neither matrix $K \in SL(n, \mathbb{Z})$ nor vector $\varpi \in \mathbb{R}^\nu$ are defined uniquely. E.g., vector ϖ is determined up to changes of the form $\varpi \mapsto A\varpi$ with $A \in GL(\nu, \mathbb{Z})$.

Now consider the well defined canonical coordinate change $x = (K^t)^{-1}x_{\text{new}}$, $y = Ky_{\text{new}}$, where $x_{\text{new}} \in \mathbb{T}^n$, $y_{\text{new}} \in Y_{\text{new}} = K^{-1}Y \subset \mathbb{R}^n$, and K^t denotes the matrix transposed to K (here x, x_{new}, y, and y_{new} are treated as column-vectors). The unperturbed Hamilton function in the new coordinates takes the form $H^0_{\text{new}}(y_{\text{new}}) = H^0(Ky_{\text{new}})$ and the frequency map, the form $\omega_{\text{new}}(y_{\text{new}}) = \omega(Ky_{\text{new}})K$ (here ω and ω_{new} are treated as row-vectors). The resonant n-torus $T_\star$ in question gains the form $\{y_{\text{new}} = y^\star_{\text{new}} = K^{-1}y^\star\}$, its frequency vector being $\omega_{\text{new}}(y^\star_{\text{new}}) = \omega(y^\star)K = (\varpi_1, \dots, \varpi_\nu, \underbrace{0, \dots, 0}_{n-\nu})$. This torus is therefore foliated into quasi-periodic invariant ν-subtori $\{(x_{\text{new}})_{\nu+1} = \mathit{const}, \dots, (x_{\text{new}})_n = \mathit{const}, y_{\text{new}} = y^\star_{\text{new}}\}$ with frequency vector ϖ.

Hereafter we will use the notation $\hat{u} = (u_{\nu+1}, \dots, u_n) \in \mathbb{R}^{n-\nu}$ (or $\mathbb{T}^{n-\nu}$) for any vector $u \in \mathbb{R}^n$ (or $u \in \mathbb{T}^n$). Our analysis above has shown that when studying an individual resonant unperturbed n-torus $\{y = y^\star\}$, one may assume without loss of generality that $\widehat{\omega}(y^\star) = 0$.

4.2.1 Poincaré trajectories

First consider the break-up of "maximally" resonant unperturbed tori ($\nu = 1$). In this case, $\hat{u} = (u_2, \dots, u_n) \in \mathbb{R}^{n-1}$ (or $\mathbb{T}^{n-1}$) for any vector $u \in \mathbb{R}^n$ (or $u \in \mathbb{T}^n$).

---

[3]If $H^0(y)$ is just nondegenerate in the sense of Rüssmann, then the resonance zone width is of the order $\le \epsilon^{1/2Q}$ where $Q \in \mathbb{N}$ is the integer that enters the nondegeneracy condition a) of Theorem 2.19 [327].

Theorem 4.8 (Poincaré [275], see also Kozlov [181–183] for a modern presentation) *Let for some $y^* \in Y$*

$$\omega(y^*) = (\varpi, \underbrace{0, \ldots, 0}_{n-1})$$

where $\varpi > 0$,

$$\det \frac{\partial^2 H^0(y^*)}{\partial y^2} = \det \frac{\partial \omega(y^*)}{\partial y} \neq 0,$$

and $\hat{x}^* = (x_2^*, \ldots, x_n^*)$ is a nondegenerate critical point of the function

$$h(\hat{x}) = h(x_2, \ldots, x_n) = \frac{1}{2\pi} \int_0^{2\pi} H^1(x_1, \hat{x}, y^*, 0)\, dx_1 \in \mathrm{R}$$

(this means that $\partial h(\hat{x}^*)/\partial \hat{x} = 0$ and $\det \partial^2 h(\hat{x}^*)/\partial \hat{x}^2 \neq 0$). Then for all sufficiently small $\epsilon \geq 0$, Hamiltonian vector field X^ϵ with Hamilton function (4.41) possesses a closed trajectory ξ_ϵ of period $2\pi/\varpi$ (called a Poincaré trajectory in the sequel) near the circle $\{\hat{x} = \hat{x}^*, y = y^*\}$. This trajectory depends analytically on ϵ, and ξ_0 coincides with that circle.

Proof. One can assume without loss of generality that $y^* = 0$ and $\hat{x}^* = 0$. Denote by g_ϵ^t the phase flow of field X^ϵ. We will look for points of the form $(0, \hat{x}^a, y^a)$ such that $g_\epsilon^{2\pi/\varpi}(0, \hat{x}^a, y^a) = (2\pi, \hat{x}^a, y^a)$. If $g_\epsilon^t(0, \hat{x}^a, y^a) = (x(t), y(t))$ then

$$\begin{aligned} x_1(t) &= \omega_1(y^a)t + O(\epsilon) \\ \hat{x}(t) &= \hat{x}^a + \hat{\omega}(y^a)t + O(\epsilon) \\ y(t) &= y^a - \epsilon \int_0^t \frac{\partial H^1(\omega_1(y^a)s, \hat{x}^a + \hat{\omega}(y^a)s, y^a, 0)}{\partial x}\, ds + O(\epsilon^2). \end{aligned}$$

Define the function

$$\mathcal{F}(\hat{x}^a, y^a, \epsilon) = \left(\frac{\varpi}{2\pi}\left[x_1\left(\frac{2\pi}{\varpi}\right) - 2\pi\right], \frac{\varpi}{2\pi}\left[\hat{x}\left(\frac{2\pi}{\varpi}\right) - \hat{x}^a\right], \frac{\varpi}{2\pi\epsilon}\left[\hat{y}^a - \hat{y}\left(\frac{2\pi}{\varpi}\right)\right]\right) \in \mathrm{R}^{2n-1}$$

[note that $y_1(t)$ does not enter $\mathcal{F}$]. Then

$$\begin{aligned} \mathcal{F}(\hat{x}^a, y^a, \epsilon) &= \Bigg(\omega_1(y^a) - \varpi + O(\epsilon), \hat{\omega}(y^a) + O(\epsilon), \\ &\quad \frac{\varpi}{2\pi} \int_0^{2\pi/\varpi} \frac{\partial H^1(\omega_1(y^a)s, \hat{x}^a + \hat{\omega}(y^a)s, y^a, 0)}{\partial \hat{x}}\, ds + O(\epsilon) \Bigg). \end{aligned}$$

One has $\mathcal{F}(0,0,0) = 0$ because $\omega_1(0) = \varpi$, $\hat{\omega}(0) = 0$, and $\partial h(0)/\partial \hat{x} = 0$. On the other hand, the Jacobi matrix of the function $\mathcal{F}(\hat{x}^a, y^a, 0)$ at the point $(\hat{x}^a = 0, y^a = 0)$ is equal to

$$\left.\frac{\partial \mathcal{F}(\hat{x}^a, y^a, 0)}{\partial(\hat{x}^a, y^a)}\right|_{\hat{x}^a=0, y^a=0} = \begin{pmatrix} 0 & \partial\omega(0)/\partial y \\ \partial^2 h(0)/\partial \hat{x}^2 & M \end{pmatrix} \tag{4.46}$$

where M is a certain $(n-1) \times n$ matrix. Since $\det \partial\omega(0)/\partial y \neq 0$ and $\det \partial^2 h(0)/\partial \hat{x}^2 \neq 0$, matrix (4.46) is non-singular. By the Implicit Function Theorem, the equation $\mathcal{F}(\hat{x}^a, y^a, \epsilon) = 0$ with respect to $\hat{x}^a$, y^a has a unique solution $\hat{x}^a = \widehat{\mathrm{X}}(\epsilon)$, $y^a = \mathrm{Y}(\epsilon)$ depending analytically on ϵ and such that $\widehat{\mathrm{X}}(0) = 0$, $\mathrm{Y}(0) = 0$. For $\hat{x}^a = \widehat{\mathrm{X}}(\epsilon)$, $y^a = \mathrm{Y}(\epsilon)$, one has $x_1(2\pi/\varpi) = 2\pi$, $\hat{x}(2\pi/\varpi) = \hat{x}^a$, $y_1(2\pi/\varpi) = z(\epsilon)$, and $\hat{y}(2\pi/\varpi) = \hat{y}^a$, function $z(\epsilon)$ depending analytically on ϵ and $z(0) = 0$. The energy conservation law gives

$$H(0, \widehat{\mathrm{X}}(\epsilon), \mathrm{Y}(\epsilon), \epsilon) = H(0, \widehat{\mathrm{X}}(\epsilon), z(\epsilon), \widehat{\mathrm{Y}}(\epsilon), \epsilon). \tag{4.47}$$

Since $\partial H(0,0,0)/\partial y_1 = \partial H^0(0)/\partial y_1 = \omega_1(0) = \varpi \neq 0$, equality (4.47) implies $z(\epsilon) \equiv \mathrm{Y}_1(\epsilon)$, i.e., $y(2\pi/\varpi) = y^a$.

Thus, $g_\epsilon^{2\pi/\varpi}(0, \widehat{\mathrm{X}}(\epsilon), \mathrm{Y}(\epsilon)) = (2\pi, \widehat{\mathrm{X}}(\epsilon), \mathrm{Y}(\epsilon))$. The phase trajectory passing through the point $(0, \widehat{\mathrm{X}}(\epsilon), \mathrm{Y}(\epsilon))$ is the desired orbit ξ_ϵ. □

Remark 1. The function $h(\hat{x})$ is defined on T^{n-1}. Therefore, it generically possesses at least 2^{n-1} nondegenerate critical points [234], and one obtains at least 2^{n-1} Poincaré trajectories $\xi_\epsilon^{(1)}, \xi_\epsilon^{(2)}, \ldots$.

Remark 2. The Poincaré trajectories $\xi_\epsilon^{(1)}(T_\star), \xi_\epsilon^{(2)}(T_\star), \ldots$ corresponding to a "maximally" resonant unperturbed n-torus $T_\star$ exist for $0 < \epsilon < \epsilon_0(T_\star)$. It turns out that, as a rule, $\epsilon_0(T_\star) \to 0$ as $\varpi(T_\star) \downarrow 0$ where $2\pi/\varpi(T_\star)$ is the period of the trajectories in question [351]. Note that for any $\delta > 0$, on each energy level hypersurface $\{H^0 = const\}$, there generically exist only finite number $N(\delta)$ of tori $T_\star$ for which $\varpi(T_\star) > \delta$, and $N(\delta) \to +\infty$ as $\delta \downarrow 0$. The value of $\varpi(T_\star)$ is in fact controlled by the entries of the matrix $K(T_\star)$. Namely, for each $\delta > 0$, there exists a $\Delta > 0$ such that $\varpi(T_\star) > \delta$ whenever the absolute values of all the entries of $K(T_\star)$ are less than Δ [to be more precise, whenever $K(T_\star)$ can be chosen in such a way that the absolute values of all its entries are less than Δ].

On the other hand, if the unperturbed Hamilton function $H^0(y)$ is quasi-convex (see Definition 4.10 below) then for all sufficiently small ϵ ($0 < \epsilon < \epsilon_0$), there are at least n closed trajectories of period $2\pi/\varpi(T_\star)$ near each "maximally" resonant unperturbed torus $T_\star$ [33], ϵ_0 being independent of $T_\star$ (see also [204]). These trajectories are located at a distance $\le O(\epsilon^{1/3})$ from $T_\star$. Moreover, at least one of the trajectories is located at a distance $\le O(\epsilon^{1/2})$ from $T_\star$.

Remark 3. The "maximally" resonant unperturbed n-tori are organized into analytic one-parameter families (the parameter being the period of the corresponding closed trajectories or the value of H^0). Consequently, the Poincaré trajectories of a perturbed vector field X^ϵ are also organized, for any given $\epsilon > 0$, into analytic one-parameter families (the parameter being the period or the value of H). To these families, one can apply the theorem on the excitation of elliptic normal modes. This gives "secondary" Whitney-smooth families of invariant tori of dimensions $2, \ldots, n$ in the resonance zones of X^ϵ. The motions carried by those tori are sometimes called phase oscillations, see Arnol'd [13, Appendix 7].

Remark 4. The analogues of Poincaré trajectories in the reversible $(n, m, 0, 0)$ context 1 for $m \ge n$ have been constructed by Sevryuk [315].

4.2.2 Treshchëv tori

Now consider the case $\nu > 1$. As one should expect, the arithmetical properties of the frequencies $\varpi_1, \ldots, \varpi_\nu$ are of importance for $\nu > 1$.

Theorem 4.9 (Treshchëv [342, 343]) *Let for some $y^* \in Y$*

$$\omega(y^*) = (\varpi_1, \ldots, \varpi_\nu, \underbrace{0, \ldots, 0}_{n-\nu})$$

$(2 \le \nu \le n-1)$ and the frequency vector $\varpi = (\varpi_1, \ldots, \varpi_\nu) \in \mathrm{R}^\nu$ is Diophantine, the $n \times n$ Hessian matrix

$$W = \frac{\partial^2 H^0(y^*)}{\partial y^2} = \frac{\partial \omega(y^*)}{\partial y}$$

is non-singular, and $\hat{x}^* = (x^*_{\nu+1}, \ldots, x^*_n)$ is a nondegenerate critical point of the function

$$h(\hat{x}) = h(x_{\nu+1}, \ldots, x_n) = \frac{1}{(2\pi)^\nu} \int_0^{2\pi} \cdots \int_0^{2\pi} H^1(x_1, \ldots, x_\nu, \hat{x}, y^*, 0)\, dx_1 \cdots dx_\nu \in \mathrm{R}.$$

The latter condition means that $\partial h(\hat{x}^*)/\partial \hat{x} = 0$ and the $(n-\nu) \times (n-\nu)$ Hessian matrix

$$V = \frac{\partial^2 h(\hat{x}^*)}{\partial \hat{x}^2}$$

is non-singular. Finally, denote by $W_{n-\nu}$ the lower right principal $(n-\nu) \times (n-\nu)$ minor of matrix W and assume that none of the eigenvalues of the $(n-\nu) \times (n-\nu)$ matrix $VW_{n-\nu}$ is a non-negative real number. Then for all sufficiently small $\epsilon \ge 0$, Hamiltonian vector field X^ϵ with Hamilton function (4.41) possesses a Floquet Diophantine isotropic invariant analytic ν-torus T_ϵ with frequency vector ϖ (called a Treshchëv torus in the sequel) near the ν-torus $\{\hat{x} = \hat{x}^*, y = y^*\}$. The torus T_ϵ depends analytically on ϵ for $\epsilon > 0$ and smoothly (C^∞) on $\sqrt{\epsilon}$ for $\epsilon \ge 0$, and T_0 coincides with the torus $\{\hat{x} = \hat{x}^*, y = y^*\}$. The torus T_ϵ is hyperbolic for $\epsilon > 0$, i.e., its Floquet matrix has eigenvalue 0 of multiplicity ν and $2(n-\nu)$ eigenvalues $\pm\lambda_1(\epsilon), \ldots, \pm\lambda_{n-\nu}(\epsilon)$ none of which is purely imaginary.

This theorem was proven by Treshchëv [342]. The case $\nu = n-1$ was recently examined in detail by Cheng [87] and Rudnev & Wiggins [295]. Theorem 4.9 is also discussed in [183].

Remark 1. The condition $\det V \ne 0$ is in fact superfluous because it follows from the condition $\det(VW_{n-\nu}) \ne 0$.

Remark 2. Since the frequencies $\varpi_1, \ldots, \varpi_\nu$ are rationally independent, function $h(\hat{x})$ can be defined as

$$h(\hat{x}) = \lim_{\vartheta \to +\infty} \frac{1}{\vartheta} \int_0^\vartheta H^1(x_1 + \varpi_1 t, \ldots, x_\nu + \varpi_\nu t, \hat{x}, y^*, 0)\, dt$$

(the limit in the right-hand side does not depend on $x_1, \ldots, x_\nu$), see [12, 13].

Remark 3. The function $h(\hat{x})$ is defined on $\mathrm{T}^{n-\nu}$. Therefore, it generically possesses at least $2^{n-\nu}$ nondegenerate critical points [234]. However, some of them may violate the hyperbolicity condition (the spectrum of the corresponding matrix $VW_{n-\nu}$ may contain positive real numbers). Conjecturally, such points give rise[4] to non-hyperbolic invariant ν-tori of X^ϵ for small $\epsilon > 0$, but this very plausible statement has not been proven yet.

[4] in a much weaker sense, however, than it is described in Theorem 4.9 pertaining to the hyperbolic case. One must not expect that, e.g., non-hyperbolic invariant ν-tori depend continuously on ϵ.

Remark 4. For any fixed $\epsilon > 0$, the Treshchëv ν-tori are organized into Whitney-smooth ν-parameter families (for a proof in the case $\nu = n - 1$, see [295]). Each such family is constituted by Treshchëv ν-tori born at resonant unperturbed n-tori $T_\star = \{y = y^\star\}$, $y^\star \in \Xi \subset \overline{Y}$, subject to the following conditions:

a) for any $y^\star \in \Xi$, the torus $T_\star$ is foliated by Diophantine invariant subtori of dimension ν with frequency vector $\varpi = \varpi(T_\star) \in \mathbf{R}^\nu$:

$$|\langle \varpi(T_\star), k\rangle| \geq \gamma(T_\star)|k|^{-\tau} \quad \forall k \in \mathbf{Z}^\nu \setminus \{0\},$$

$\tau > \nu - 1$ being a fixed constant;

b) there exists a constant $\gamma_0 > 0$ such that $\gamma(T_\star) \geq \gamma_0$ for any $y^\star \in \Xi$;

c) there exists a constant $\Delta > 0$ such that for any $y^\star \in \Xi$, the matrix $K(T_\star)$ can be chosen in such a way that the absolute values of all its entries are no greater than Δ.

One easily sees that Ξ is a closed subset in $\overline{Y}$ and $meas_\nu\, \Xi > 0$ for any prescribed sufficiently small γ_0 and $1/\Delta$.

Such ν-parameter families of Treshchëv ν-tori can be regarded as new illustrations to Theorem 2.28.

Disintegration of resonant unperturbed tori is another example of situations where small twist KAM theorems are to be invoked. The conjectural general small twist theorem deals with families of vector fields of the form

$$\begin{gathered}\left[\sum_{\iota=0}^{M}\sigma^\iota\omega_{(\iota)}(y,\mu)+\sigma^{l_1}f(\sigma,x,y,z,\mu)+\sigma^{L_1}\tilde{f}(\sigma,x,y,z,\mu)\right]\frac{\partial}{\partial x}+\\ \left[\sigma^{l_2}g(\sigma,x,y,z,\mu)+\sigma^{L_2}\tilde{g}(\sigma,x,y,z,\mu)\right]\frac{\partial}{\partial y}+\\ \left[\sum_{\iota=0}^{N}\sigma^\iota\Omega_{(\iota)}(y,\mu)z+\sigma^{l_3}h(\sigma,x,y,z,\mu)+\sigma^{L_3}\tilde{h}(\sigma,x,y,z,\mu)\right]\frac{\partial}{\partial z}\end{gathered} \tag{4.48}$$

where $x \in \mathbf{T}^n$, $y \in Y \subset \mathbf{R}^m$, $z \in \mathcal{O}_{2p}(0)$, $\mu \in P \subset \mathbf{R}^s$ is an external parameter, $\sigma \geq 0$ is a small parameter, $f = O(z)$, $g = O(z)$, $h = O_2(z)$, and $\tilde{f}$, $\tilde{g}$, $\tilde{h}$ are small perturbations (the fixed numbers $l_1 < L_1$, $l_2 < L_2$, $l_3 < L_3$ are not necessarily integers). These vector fields are also assumed to be compatible with some structure $\mathfrak{S}$ on the phase space depending on the context at hand, while the functions $\omega_{(\iota)}$, $0 \leq \iota \leq M$, and $\Omega_{(\iota)}$, $0 \leq \iota \leq N$, are let satisfy appropriate nondegeneracy and nonresonance conditions. The simplest particular case was considered in § 4.1.3 [cf. (4.25)] where, roughly speaking, $\omega_{(0)}$ and $\Omega_{(0)}$ are constants, while the nondegeneracy condition is imposed on $\omega_{(1)}(y,\mu)$. This case is enough for studying invariant tori near equilibria. The classical theory of properly degenerate Hamiltonians [5, 20] (cf. the remark at the end of § 2.3.4) deals with the case where $s = p = 0$, the function $\omega_{(0)}$ does not depend on some y_j, and $\omega_{(0)}(y) + \sigma\omega_{(1)}(y)$ is nondegenerate. On the other hand, the excitation of normal modes in the reversible $(n, 0, p, 0)$ context 1 with $n \geq 2$ requires small twist theorems where the nondegeneracy condition is imposed on at least $\omega_{(1)}$ and $\omega_{(2)}$ which seems to have not been explored yet in the literature. The main difficulty in the proof of Theorem 4.9 is that the latter is reduced to examining a vector field of the form (4.48) [for $s = 0$] where $\Omega_{(0)} \equiv 0$ [342]. A general theory of invariant n-tori for vector fields (4.48) has not been developed yet.

4.2.3 Stability of the action variables

Now discuss briefly stochastic trajectories in the resonance zones. Consider an arbitrary solution $x(t)$, $y(t)$ of Hamiltonian system X^ϵ with Hamilton function (4.41). Over a time interval of the order of 1, functions $x(t)$ and $y(t)$ are respectively ϵ-close to $x(0)+\omega(y(0))t$ and $y(0)$. The question of fundamental importance is how the variation of the action variables $|y(t)-y(0)|$ behaves over *long* time intervals.

Generally speaking, the variation $|y(t)-y(0)|$ can become of the order of 1 over a time interval of the order of ϵ^{-1}, even if the unperturbed Hamiltonian $H^0(y)$ is nondegenerate in the sense of Kolmogorov. For instance, cf. Nekhoroshev [254], consider the system X^ϵ with Hamiltonian

$$H(x,y,\epsilon) = \tfrac{1}{2}(y_1^2 - y_2^2) + \epsilon \sin(x_1 - x_2), \tag{4.49}$$

here $n=2$ and $Y = \{y \in \mathbb{R}^2 \;:\; 0 < r_0 < \|y\| < R_0\}$. One has $\omega(y) = (y_1, -y_2)$ and $\det \partial\omega(y)/\partial y \equiv -1$, so that $H^0(y) = \frac{1}{2}(y_1^2 - y_2^2)$ in this example is nondegenerate in the sense of Kolmogorov. System X^ϵ admits the special solutions

$$x_1(t) = x_2(t) = -\tfrac{1}{2}(\epsilon t^2 + 2ct), \quad y_1(t) = -(\epsilon t + c), \quad y_2(t) = \epsilon t + c \tag{4.50}$$

(c being a constant) for which $|y(t)-y(0)| = 2\epsilon|t|$ for all t. Nevertheless, the evolution of the action variables in *generic* nearly integrable Hamiltonian systems turns out to proceed much more slowly, some features of the motion in the dimensions $n=2$ and $n \geq 3$ being drastically different.

Besides the Rüssmann and Kolmogorov nondegeneracy conditions, one exploits the so called *isoenergetic nondegeneracy* in the theory of nearly integrable Hamiltonian vector fields. The unperturbed Hamiltonian $H^0(y)$ is said to be isoenergetically nondegenerate if the $(n+1)\times(n+1)$ symmetric matrix

$$\mathcal{Q}(y) := \begin{pmatrix} \partial^2 H^0(y)/\partial y^2 & \partial H^0(y)/\partial y \\ \partial H^0(y)/\partial y & 0 \end{pmatrix} = \begin{pmatrix} \partial\omega(y)/\partial y & \omega(y) \\ \omega(y) & 0 \end{pmatrix}$$

is non-singular in $\overline{Y}$. The condition $\det \mathcal{Q}(y) \neq 0$ means that for each energy level hypersurface $\{H^0 = h_* = const\}$, the map

$$\{H^0 = h_*\} \to \mathbb{RP}^{n-1}, \qquad y \mapsto \Pi\omega(y)$$

is submersive on $\{H^0 = h_*\}$. This condition is much stronger than the Rüssmann nondegeneracy but independent of the Kolmogorov nondegeneracy. One can verify the independence of the conditions $\det \partial\omega(y)/\partial y \neq 0$ and $\det \mathcal{Q}(y) \neq 0$ very easily by looking at the following examples:

$$F_1(y) = \tfrac{1}{2}\textstyle\sum_{j=1}^{n-1} y_j^2 + y_n \quad \text{and} \quad F_2(y) = \sum_{j=1}^n a_j \log y_j$$

where $a_1, \ldots, a_n$ are constants and $\sum_{j=1}^n a_j = 0$. For $H^0(y) = F_1(y)$, one has $\det \partial\omega(y)/\partial y \equiv 0$ and $\det \mathcal{Q}(y) \equiv -1$. For $H^0(y) = F_2(y)$, one has

$$\det \frac{\partial\omega(y)}{\partial y} = (-1)^n \prod_{j=1}^n \frac{a_j}{y_j^2} \quad \text{and} \quad \det \mathcal{Q}(y) = \left(\det \frac{\partial\omega(y)}{\partial y}\right) \sum_{j=1}^n a_j \equiv 0.$$

For $n=2$ these examples are given in [103, 335].

The determinant det $Q(y)$ was introduced by Arnol'd [4, 5] and is sometimes called the *Arnol'd determinant* [204].

If $H^0(y)$ is isoenergetically nondegenerate then for any function $H^1(x, y, \epsilon)$, the relative $(2n-1)$-dimensional measure of the union of the perturbed invariant n-tori (KAM tori) on *each* energy level hypersurface $\{H = const\}$ tends to 1 as $\epsilon \downarrow 0$ [5, 13, 59, 102–104, 162].[5] On the other hand, for $n = 2$ two-dimensional invariant tori divide three-dimensional energy hypersurfaces. Consequently, if $n = 2$ and $H^0(y)$ is isoenergetically nondegenerate then trajectories in the resonance zones turn out to be *locked* between the KAM tori on each energy level hypersurface (provided that ϵ is sufficiently small). This implies that the variations $|y(t) - y(0)|$ remain small (of the order of $\sqrt{\epsilon}$) over infinite time intervals for all the initial conditions $x(0)$, $y(0)$.

The isoenergetic nondegeneracy condition is very essential here, as the example of Hamiltonian (4.49) shows. For this Hamiltonian, one has det $Q(y) = y_1^2 - y_2^2 = 2H^0(y)$, so that the isoenergetic nondegeneracy fails at the straight lines $y_1 = \pm y_2$ ("superconductivity channels").

On the other hand, for $n \geq 3$ the n-dimensional KAM tori no longer divide the $(2n-1)$-dimensional energy level hypersurfaces. Hence, the resonance zones on each hypersurface form a connected web and stochastic trajectories may wander along this web in an arbitrary manner. The variations $|y(t) - y(0)|$ can therefore become large for sufficiently large t, however small $\epsilon > 0$ is. This phenomenon was discovered by Arnol'd [7] and has been called the *Arnol'd diffusion*. It does take place in *generic* nearly integrable Hamiltonian systems with $n \geq 3$ degrees of freedom, see Chierchia & Gallavotti [90, 94].[6]

Nevertheless, the diffusion rate turns out to be *exponentially small* with ϵ for generic *analytic* Hamiltonian systems (the Nekhoroshev theorem) [251, 254, 255]. To be more precise, if the unperturbed Hamilton function $H^0(y)$ meets certain conditions known as *steepness*, then there exist constants $a > 0$, $b > 0$, $R_* > 0$, $K_* > 0$, and $\epsilon_* > 0$ such that the solution $x(t)$, $y(t)$ of system X^ϵ governed by Hamiltonian (4.41) satisfies the inequality

$$|y(t) - y(0)| \leq R_* \epsilon^b \quad \text{for} \quad |t| \leq \exp(K_* \epsilon^{-a}) \tag{4.51}$$

for any initial conditions $x(0) \in \mathrm{T}^n$ and $y(0) \in Y$ provided that $0 < \epsilon \leq \epsilon_*$ [29, 31, 103, 132, 134, 204–206, 208–210, 251, 254, 255, 283]. This phenomenon is sometimes referred to as *effective stability* of the action variables (cf. [103, 131, 227]).

The steepness conditions are very weak: the nonsteep functions $H^0(y)$ constitute a set of infinite codimension in an appropriate functional space [253, 254]. The numbers a and b in (4.51) are called the *stability exponents*. They depend on n and the so called *steepness indices* of the function $H^0(y)$ and tend to zero as $n \to +\infty$.

For the definition and discussion of the concept of steepness, the reader is referred to [163, 251, 253–255]. Here we mention only that among all steep functions $H^0(y)$, the "steepest" ones are so called convex and quasi-convex functions [254].

[5] In fact, the latter property is implied by the following condition: the image under $\omega : Y \to \mathbb{R}^n$ of each connected component of each energy level hypersurface does not lie in any linear hyperplane in $\mathbb{R}^n$ passing through the origin.

[6] However, cf. a discussion on "a priori stability" and "a priori instability" in [207].

Definition 4.10 *An unperturbed Hamilton function $H^0(y)$ is said to be convex if there exists a constant $c > 0$ such that*

$$\left\langle \eta, \frac{\partial \omega(y)}{\partial y}\eta \right\rangle \geq c\|\eta\|^2 \tag{4.52}$$

for all $y \in \overline{Y}$ and $\eta \in \mathbb{R}^n$. A function $H^0(y)$ is said to be quasi-convex if there exists a constant $c > 0$ such that inequality (4.52) holds whenever $y \in \overline{Y}$, $\eta \in \mathbb{R}^n$, and $\langle \omega(y), \eta \rangle = 0$.

One easily verifies that quasi-convexity of an integrable Hamilton function means convexity of the corresponding energy level hypersurfaces.

It is clear that convexity implies the Kolmogorov nondegeneracy. There exist quasi-convex Hamiltonians which are degenerate in the sense of Kolmogorov, e.g.,

$$H^0(y) = \tfrac{1}{2}\textstyle\sum_{j=1}^{n-1} y_j^2 + y_n$$

in $Y = \{y \in \mathbb{R}^n \ : \ \|y\| < R_0\}$, but one can easily verify that quasi-convexity (and a *fortiori* convexity) implies the isoenergetic nondegeneracy [204]. For quasi-convex unperturbed Hamiltonians, the Nekhoroshev estimate (4.51) turns out to hold for $a = b = 1/(2n)$, see Lochak, Neĭshtadt, Pöschel, Niederman, Delshams & Gutiérrez [103, 205, 208–210, 283]. Moreover, one can take

$$a = \frac{\mu}{2n}, \qquad b = \frac{1-\mu}{2} + \frac{\mu}{2n}$$

for any $0 < \mu \leq 1$ [205, 283]. These values of the stability exponents seem to be optimal.

On the other hand, near resonant unperturbed tori the estimates of the stability exponents can be improved considerably (the so called *stabilization via resonance* [204, 205, 207], cf. also [208–210]). We will not discuss this effect here.

Although nonsteep functions are "infinitely degenerate", steepness is independent of the Kolmogorov nondegeneracy: the function

$$H^0(y) = \tfrac{1}{2}\left(\textstyle\sum_{j=1}^{n-1} y_j^2 - y_n^2\right) \tag{4.53}$$

is nondegenerate in the sense of Kolmogorov [$\det \partial\omega(y)/\partial y \equiv -1$] but not steep in the domain $Y = \{y \in \mathbb{R}^n : \ 0 < r_0 < \|y\| < R_0\}$ [cf. (4.49)]. Indeed, the system X^ϵ with Hamiltonian $H(x, y, \epsilon) = H^0(y) + \epsilon \sin(x_{n-1} - x_n)$ admits special solutions with $x_n(t) = x_{n-1}(t) = -\frac{1}{2}(\epsilon t^2 + 2ct)$, $y_n(t) = -y_{n-1}(t) = \epsilon t + c$ (c being a constant) for which $|y(t) - y(0)| = 2\epsilon|t|$ [cf. (4.50)]. In the case of two degrees of freedom ($n = 2$), the isoenergetic nondegeneracy implies quasi-convexity and a *fortiori* steepness. For $n \geq 3$, steepness is independent of the isoenergetic nondegeneracy as well. For instance, function (4.53) is isoenergetically nondegenerate in the domain

$$Y = \{y \in \mathbb{R}^n \ : \ r_0^2 + 1 < y_1^2 + \cdots + y_{n-2}^2 < R_0^2, \ y_{n-1}^2 + y_n^2 < r_0^2\}$$

[$\det Q(y) = 2H^0(y) \geq 1$ in $\overline{Y}$] but not steep in this domain.

For a detailed discussion of the Arnol'd diffusion, the reader is referred to, e.g., [17, 94–96, 159, 201, 202, 204, 207, 212, 216, 254, 268, 269, 357]. The so called *whiskers*, or separatrix manifolds, of the hyperbolic invariant tori of dimensions less than n play the central rôle in the diffusion mechanisms.

Remark 1. Exponential estimates on the diffusion rate can simetimes be obtained also for highly degenerate (even linear) unperturbed Hamilton functions $H^0(y)$ provided that a perturbation "removes" the degeneracy [206]. To be more precise, the perturbed Hamilton function (4.41) in [206] is assumed to have the form $H(x, y, \epsilon) = H^0(y) + \epsilon H^1_{(1)}(y) + \epsilon^2 H^1_{(2)}(x, y)$ with $H^0(y)$ linear and $H^1_{(1)}(y)$ convex.

Remark 2. A Hamiltonian system is said to be *quasi-ergodic* if on each energy level hypersurface, there is an everywhere dense trajectory. The considerations above show that for two-degree-of-freedom Hamiltonian vector fields, it is a typical property to be not quasi-ergodic (and, moreover, everywhere dense trajectories exist on *none* of the energy hypersurfaces). The same conclusion turns out to hold for Hamiltonian systems with $n \geq 3$ degrees of freedom as well [361]. To see this, one should consult the Hamiltonian coisotropic $(N, p, 0)$ context with $N = 2n - 2$ and $p = (N - 2)/2 = n - 2$ (see §§ 1.3.3 and 1.4.2). In this context, one has invariant $(2n - 2)$-tori which divide $(2n - 1)$-dimensional energy level hypersurfaces (cf. Theorem 1.5). Under appropriate conditions, these tori constitute a set of large $(2n - 1)$-dimensional measure on each energy hypersurface, and this property occurs typically [361].

On the other hand, the symplectic structure in the Hamiltonian coisotropic context is always nonexact (see § 5.1.3 below). Non-quasi-ergodicity of a typical Hamiltonian system with $n \geq 3$ degrees of freedom and an exact symplectic structure is still an open question (we are grateful to J.N. Mather for a discussion on this problem).

Remark 3. The Nekhoroshev theory admits the local counterpart. For instance, the rate of moving away from an elliptic equilibrium of a Hamiltonian vector field is exponentially small provided that the collection of the corresponding eigenfrequencies is Diophantine [131] (cf. also § 4.1.5 above). In the latter paper, a generalization to the case of a resonant collection of eigenfrequencies is presented as well.

4.2.4 Stickiness of the invariant tori

Another aspect of long time stability in nearly integrable analytic Hamiltonian systems is the so called *stickiness* of the KAM tori. It turns out that under the conditions of the Kolmogorov nondegeneracy and quasi-convexity, all the trajectories starting at a distance of the order of $\varrho < \varrho^*$ from a Diophantine perturbed n-torus remain close to this torus for an exceedingly long time of the order of

$$\exp\left\{\exp\left[\left(\frac{\varrho^*}{\varrho}\right)^{1/(\tau+1)}\right]\right\}$$

(provided that ϵ is sufficiently small). Here $\varrho^* > 0$ is a certain ϵ-independent constant and τ is the exponent entering the Diophantine inequalities (4.44). One says that the rate of moving away from a KAM torus is *superexponentially small.* The reader is referred to Morbidelli & Giorgilli [239] for the precise formulations and proof, see [241] as well.

It is interesting to note that in the first papers devoted to the stickiness of the KAM tori, an exponential estimate only was obtained, see [237, 270].

It also turns out that the relative measure of the union of invariant n-tori (close to the unperturbed tori $\{y = const\}$) in a ϱ-neighborhood of any given KAM torus differs from 1 by an exponentially small quantity of the order of $\exp(-\varrho^{-1})$ [239, 241]. Thus, for any sufficiently small fixed $\epsilon > 0$, each KAM torus T (belonging to the Whitney-smooth n-parameter family $\mathfrak{T}_0$ of Diophantine n-tori) is a "density point" of a certain "secondary" Whitney-smooth n-parameter family $\mathfrak{T}_1(T)$ of Diophantine n-tori, cf. Section 5.2. This phenomenon has not been well understood yet but can be rephrased as follows.

Consider an unperturbed n-torus $\{y = y^\star\}$ whose frequency vector $\omega(y^\star)$ satisfies the Diophantine condition (4.44) with fixed $\tau > n - 1$ and some $\gamma = \gamma(y^\star) > 0$. Fix also the perturbation size $\epsilon > 0$ (to be sufficiently small). If $\gamma(y^\star) \sim \sqrt{\epsilon}$ then the torus $\{y = y^\star\}$ will survive the perturbation. The question now is what will happen to this torus if $\gamma(y^\star)$ is very small. The answer seems to be as follows: such torus will generally break up, but it will survive provided that $\omega(y^\star)$ is sufficiently close to a vector ω' satisfying the inequalities

$$|\langle \omega', k\rangle| \geq \gamma' |k|^{-\tau} \qquad \forall k \in \mathbb{Z}^n \setminus \{0\}$$

with $\gamma' \sim \sqrt{\epsilon}$. In other words, each Diophantine unperturbed n-torus not only survives itself, but also "helps" all the nearby tori to survive.

It would be interesting to find out what happens to unperturbed tori $\{y = \mathit{const}\}$ whose frequency vectors ω meet more general Diophantine condition (1.42) with various $\gamma > 0$.

Every torus $\widetilde{T}$ in a secondary family $\mathfrak{T}_1(T)$ seems to be a density point of a "tertiary" Whitney-smooth n-parameter family $\mathfrak{T}_2(\widetilde{T})$ of Diophantine n-tori, and so on. This hierarchy is discussed by Morbidelli & Giorgilli [134, 238, 240, 241].

Analogous exponential estimates of the relative measure of stochastic regions near *lower-dimensional* invariant tori in Hamiltonian systems were obtained recently by Jorba & Villanueva [172].

Remark 1. For nonanalytic (but sufficiently smooth) Hamiltonian systems, all the exponential [say, $\exp(\epsilon^{-a})$] and superexponential [say, $\exp(\exp(\varrho^{-a}))$] stability estimates are replaced with estimates of the form ϵ^{-N} or ϱ^{-N}. For C^∞ systems, N can be taken arbitrarily large, whereas for systems of class C^r, one has $N = N(r)$. Analyticity is known to be very essential for the existence of exponentially small effects.

Remark 2. One also considers the concept of an almost (or nearly) invariant torus [103, 237], i.e., a manifold in the phase space which is diffeomorphic to $\mathbb{T}^n$ and in fact not invariant, but possesses the stickiness property: a trajectory starting very close to this manifold remains close for a very long time.

Remark 3. Besides a hierarchy of Whitney-smooth families of invariant tori, resonance zones are known to contain the so called Aubry–Mather sets, or *cantori*, i.e., invariant sets diffeomorphic to Cantor subsets of a torus. Cantori are the analogues of usual Diophantine invariant tori for non-Diophantine (but quasi-periodic) frequency vectors or for systems far from integrable ones. Both invariant tori and cantori can be described by a unified *variational principle* [266, 267] (see also [20]). We will not discuss here the destruction of invariant tori as the perturbation size increases and the birth of cantori. The reader is referred to Chenciner, MacKay, Meiss, Percival, Mather & Forni [83, 212, 215, 216, 218, 224, 229, 269] for brilliant surveys.

4.2.5 Peculiarities of the reversible context

To conclude this section, discuss whether the results above on exponential or superexponential stability in Hamiltonian systems can be carried over to the reversible $(n, n, 0, 0)$ context 1, namely, vector fields

$$X^\epsilon = [\omega(y) + \epsilon f(x, y, \epsilon)]\frac{\partial}{\partial x} + \epsilon g(x, y, \epsilon)\frac{\partial}{\partial y}$$

reversible with respect to the involution $(x, y) \mapsto (-x, y)$ [i.e., with f even in x and g odd in x]. The answer seems to be negative due to the following reason. One of the crucial facts in the proofs of all the (global) exponential stability theorems is a remarkable *coupling* between resonances and the averaged evolution of the action variables in nearly integrable Hamiltonian systems, cf. [12, 254].

Namely, consider the Hamiltonian vector field X^ϵ associated to the Hamilton function (4.41):

$$X^\epsilon = \left[\omega(y) + \epsilon\frac{\partial H^1(x, y, \epsilon)}{\partial y}\right]\frac{\partial}{\partial x} - \epsilon\frac{\partial H^1(x, y, \epsilon)}{\partial x}\frac{\partial}{\partial y}.$$

Choose a resonant unperturbed torus $T_\star = \{y = y^\star\}$. To this torus, one assigns group $G \approx \mathbb{Z}^{n-\nu}$ of integer vectors $k \in \mathbb{Z}^n$ orthogonal to $\omega(y^\star)$ [see (4.45)] and the averaged motion

$$y(t) = y(0) - \epsilon\Phi(x)t, \qquad \Phi(x) = \lim_{\vartheta\to+\infty}\frac{1}{\vartheta}\int_0^\vartheta \frac{\partial H^1(x + \omega(y^\star)t, y^\star, 0)}{\partial x}\,dt \in \mathbb{R}^n$$

$(1 \le \nu \le n-1)$. Note that the torus $T_\star$ is foliated by invariant quasi-periodic subtori of dimension ν, and $\Phi(x^1) = \Phi(x^2)$ whenever points $(x^1, y^\star)$ and $(x^2, y^\star)$ lie on one and the same subtorus [12, 13].

Lemma 4.11 (Coupling Lemma) *The vector $\Phi(x)$ lies in the $(n-\nu)$-dimensional linear plane in $\mathbb{R}^n$ spanned by $k \in G$.*

Proof. Let

$$H^1(x, y^\star, 0) = \sum_{k\in\mathbb{Z}^n} c_k e^{i\langle k,x\rangle},$$

then

$$\frac{\partial H^1(x, y^\star, 0)}{\partial x} = i\sum_{k\in\mathbb{Z}^n} c_k e^{i\langle k,x\rangle}k. \tag{4.54}$$

Observe that the coefficient of the kth harmonic in (4.54) is proportional to k. The averaging procedure consists in eliminating all the harmonics $e^{i\langle k,x\rangle}$ except those for which $\langle\omega(y^\star), k\rangle = 0$. Thus,

$$\Phi(x) = i\sum_{k\in G} c_k e^{i\langle k,x\rangle}k,$$

which completes the proof. □

Obviously, this lemma has no reversible analogue, because a generic odd vector-valued function $g(x, y^\star, 0)$ is not the gradient of a scalar function (for $n \ge 2$). Consequently, in contrast to the KAM theory, the Nekhoroshev theory will hardly be developed for reversible systems (cf. also a discussion in [227]).

On the other hand, the *local* counterpart of the Nekhoroshev theory (see Remark 3 at the end of § 4.2.3) is of essentially different nature and does admit a reversible analogue [227].

4.3 Quasi-periodic bifurcation theory

Let us return again to Whitney-smooth families of invariant tori in the dissipative context. Consider an analytic s-parameter family $X = \{X^\mu\}_\mu$ of analytic vector fields, $\mu \in P \subset \mathbb{R}^s$ being an external parameter. According to Theorem 2.25 (see Section 2.7), the following situation occurs typically: for μ in a Cantor-like subset P_γ of P of positive measure, X^μ possesses a Floquet Diophantine invariant analytic n-torus V_μ which depends on $\mu \in P_\gamma$ in a Whitney-C^∞ way. For μ in the complement $P \setminus P_\gamma$, vector field X^μ also has an invariant n-torus V_μ but the latter is, generally speaking, of finite smoothness only and the induced dynamics on V_μ is not parallel. For any $\mu \in P$, the torus V_μ is normally hyperbolic [67, 115, 158, 356] which for $\mu \in P_\gamma$ means just that none of the eigenvalues of the Floquet matrix of V_μ is purely imaginary.

The quasi-periodic bifurcation theory describes the phenomena occurring when, roughly speaking, the normal hyperbolicity of V_μ vanishes for some values of parameter μ. Since loss of the normal hyperbolicity is a "degeneracy of codimension one" (see Arnol'd [10, 12, 15]), it does happen generically at some hypersurfaces in the parameter space. Studies of quasi-periodic motions near these hypersurfaces provide us with another class of examples of an interaction between different Whitney-smooth families of invariant tori.

Up to now, the best examined quasi-periodic bifurcation patterns are the so called quasi-periodic saddle-node bifurcation, period-doubling bifurcation, and Hopf bifurcation. All these three types of quasi-periodic bifurcations are of codimension one and do resemble the well known saddle-node, period-doubling, and Hopf (or Neĭmark–Sacker) bifurcations for periodic trajectories (cf. [1, 10, 12, 16, 32, 58, 98, 166, 220, 297]), which correspond respectively to

(i) the coalescence of two trajectories while at the bifurcation instant, one of the Floquet multipliers (see § 1.1.1) of the trajectory is equal to 1,

(ii) the passage of a multiplier of the trajectory through the value of -1,

(iii) the passage of a pair of complex conjugate multipliers through the unit circle.

In fact, these "periodic" bifurcations are just simple particular cases ($n = 1$) of the quasi-periodic bifurcations. In the present section, we consider briefly the quasi-periodic Hopf bifurcation. For a more detailed presentation of this bifurcation pattern, as well as a description of the two other patterns and various applications, we refer the reader to Braaksma, Broer & Huitema [47, 48, 54–56]. For a Hamiltonian counterpart also see Hanßmann [143].

The quasi-periodic Hopf bifurcation concerns the case where the Floquet matrix of a torus has a pair of complex conjugate purely imaginary eigenvalues. This requires the codimension of the tori in the phase space to be no less than two. For simplicity we will confine ourselves with the case of codimension two. We will also exploit the following "complex notation" for vector fields on $\mathbb{R}^2$: if $z \in \mathbb{C}$, then by $w\partial/\partial z$ with $w \in \mathbb{C}$ we will denote the vector field that determines the differential equation $\dot{z} = w$. Making use of such notation, we will *not* assume w to be a *holomorphic* function in z whatsoever. To emphasize this, we will write $w = w(z, \bar{z})$ (keeping in mind that actually w is defined in $\mathbb{R}^2$ rather than in $\mathbb{C}^2$).

First consider a family of vector fields on $T^n \times R^2$

$$Y = \{Y^\mu\}_\mu = [\omega^* + f(x,z,\bar{z},\mu)]\frac{\partial}{\partial x} + [i\omega^{N*}z + h(x,z,\bar{z},\mu)]\frac{\partial}{\partial z}$$

where $x \in T^n$, $z \in \mathcal{O}_C(0) \subset C$, $\mu \in \mathcal{O}_s(0) \subset R^s$, $\omega^* \in R^n$, $\omega^{N*} \in R$ (and $\omega^{N*} > 0$), the function f is real-valued, and $f(x,z,\bar{z},0) = O(z,\bar{z})$, $h(x,z,\bar{z},0) = O_2(z,\bar{z})$. The field Y^0 admits a Floquet invariant n-torus $\{z = 0\}$ with parallel dynamics. The (internal) frequency vector and Floquet matrix of this torus are respectively equal to ω^* and

$$\begin{pmatrix} 0 & -\omega^{N*} \\ \omega^{N*} & 0 \end{pmatrix}.$$

The eigenvalues of this matrix are $\pm i\omega^{N*}$, so that the torus $\{z = 0\}$ is *not* normally hyperbolic, while its normal frequency is ω^{N*}. Our first purpose is to reduce Y^μ around $\{z = 0\} \times \{\mu = 0\}$ to a normal form (cf. § 4.1.3), which turns out to require stronger Diophantine conditions on the internal and normal frequencies (ω^*, ω^{N*}) than the usual ones (1.4), (2.24) we used throughout.

Lemma 4.12 (Normal Form Lemma) [47, 48] (also cf. Lemma 5.9) *Let for some* $L \in N$, $\tau > n - 1$, *and* $\gamma > 0$ *vector* ω^* *and number* ω^{N*} *satisfy the Diophantine condition*

$$|\langle \omega^*, k\rangle + \ell\omega^{N*}| \geq \gamma|k|^{-\tau} \quad \forall k \in Z^n \setminus \{0\} \ \forall \ell \in Z, \ |\ell| \leq L+1. \tag{4.55}$$

Then there exists an analytic parameter preserving normalizing transformation

$$\Phi : T^n \times C \times R^s \to T^n \times C \times R^s, \quad \Phi : (x,z,\mu) \mapsto (x + \varphi(x,z,\bar{z},\mu), z + \psi(x,z,\bar{z},\mu), \mu)$$

(the function φ *being real-valued), which is tangent to the identity transformation [i.e.,* $\varphi = O_2(z,\bar{z}) + O(\mu)$, $\psi = O_2(z,\bar{z}) + O(\mu)$*] and such that the family* $Y_{\rm norm} = \Phi_* Y$ *has the form*

$$\begin{aligned} Y_{\rm norm} = \{Y^\mu_{\rm norm}\}_\mu &= [\omega^* + F(v,\mu) + O_L(z,\bar{z},\mu)]\frac{\partial}{\partial x} \\ &+ [i\omega^{N*}z + H(v,\mu)z + O_{L+1}(z,\bar{z},\mu)]\frac{\partial}{\partial z} \end{aligned}$$

where $v = z\bar{z} = |z|^2$, *the function* F *is real-valued,* $F(0,0) = 0$, *and* $H(0,0) = 0$.

Remark. The transformation Φ can be chosen to be Whitney-smoothly dependent on ω^* and ω^{N*}. Otherwise speaking, if we include ω^* and ω^{N*} as parameters into the system, in the corresponding parameter directions the transformation can be extended as a C^∞-diffeomorphism to a neighborhood of the Cantor set determined by the Diophantine conditions (4.55).

Now consider the truncated normal form

$$X = \{X^\mu\}_\mu = [\omega^* + F(v,\mu)]\frac{\partial}{\partial x} + [i\omega^{N*} + H(v,\mu)]\, z\frac{\partial}{\partial z} \tag{4.56}$$

which has a T^{n+1}-symmetry. This truncation will be treated in the following standard way. We will call family (4.56) with $F(0,0) = 0$, $H(0,0) = 0$ a *local Hopf family* if the following two conditions are met:

1) the mapping

$$\mathbf{R}^s \ni \mu \mapsto (F(0,\mu), \operatorname{Re} H(0,\mu), \operatorname{Im} H(0,\mu)) \in \mathbf{R}^n \times \mathbf{R} \times \mathbf{R}$$

is submersive at $\mu = 0$ (so that $s \geq n+2$);

2) $\partial \operatorname{Re} H(0,0)/\partial v \neq 0$.

If this is the case, by the Inverse Function Theorem we can choose a reparametrization of the μ-space of the form

$$\mu \mapsto (\omega(\mu), \omega^N(\mu), \lambda(\mu), \nu(\mu))$$

where

$$\omega(\mu) = \omega^\star + F(0,\mu) \in \mathbf{R}^n, \quad \omega^N(\mu) = \omega^{N\star} + \operatorname{Im} H(0,\mu) \in \mathbf{R}, \quad \lambda(\mu) = \operatorname{Re} H(0,\mu) \in \mathbf{R}$$

and $\nu(\mu) \in \mathbf{R}^{s-n-2}$. Treating ω, ω^N, λ as independent variables and suppressing ν for simplicity, we arrive at the $(n+2)$-parameter family of vector fields

$$\begin{aligned} X^{\omega,\omega^N,\lambda} = & \left[\omega + v\widehat{F}(v,\omega,\omega^N,\lambda)\right]\frac{\partial}{\partial x} + \\ & \left[i\omega^N + iv\widehat{H}_1(v,\omega,\omega^N,\lambda) + \lambda + v\widehat{H}_2(v,\omega,\omega^N,\lambda)\right] z\frac{\partial}{\partial z}. \end{aligned} \tag{4.57}$$

Here ω, ω^N, and λ range respectively near $\omega^\star$, $\omega^{N\star}$, and 0, while all the three functions $\widehat{F}$, $\widehat{H}_1$, and $\widehat{H}_2$ are real-valued, with $\widehat{H}_2(0,\omega^\star,\omega^{N\star},0) \neq 0$ due to condition 2).

For all the values of ω, ω^N, λ, the manifold $\{z=0\}$ is a Floquet invariant n-torus of field $X^{\omega,\omega^N,\lambda}$ with parallel dynamics. The (internal) frequency vector and Floquet matrix of this torus are respectively equal to ω and

$$\begin{pmatrix} \lambda & -\omega^N \\ \omega^N & \lambda \end{pmatrix}.$$

The eigenvalues of this matrix are $\lambda \pm i\omega^N$, so that the normal frequency of the torus $\{z=0\}$ is ω^N while the normal hyperbolicity breaks at $\lambda = 0$. This torus is attracting for $\lambda < 0$ and repelling for $\lambda > 0$.

Besides, for

$$\lambda \widehat{H}_2(0,\omega^\star,\omega^{N\star},0) < 0$$

(so that, over one of the two open half λ-axes), the vector field $X^{\omega,\omega^N,\lambda}$ (4.57) admits a Floquet invariant analytic $(n+1)$-torus with parallel dynamics. This torus is given by the equation

$$\lambda + v\widehat{H}_2(v,\omega,\omega^N,\lambda) = 0 \iff v = \Upsilon(\omega,\omega^N,\lambda) = -\frac{\lambda}{\widehat{H}_2(0,\omega,\omega^N,0)} + O_2(\lambda). \tag{4.58}$$

The "width" of torus (4.58) in the $(z,\bar{z})$-direction is proportional to $\sqrt{|\lambda|}$, and it collapses to the n-torus $\{z=0\}$ as $\lambda \to 0$. One also says that torus (4.58) *branches off* from $\{z=0\}$ at $\lambda = 0$. The (internal) frequency vector of this torus is equal to

$$\mathbf{R}^{n+1} \ni \left(\omega + \Upsilon\widehat{F}(\Upsilon,\omega,\omega^N,\lambda),\ \omega^N + \Upsilon\widehat{H}_1(\Upsilon,\omega,\omega^N,\lambda)\right) = (\omega,\omega^N) + O(\lambda),$$

while its 1×1 Floquet matrix is equal to

$$\frac{\partial}{\partial v}\left(2v\left[\lambda + v\widehat{H}_2(v,\omega,\omega^N,\lambda)\right]\right)\Big|_{v=\Upsilon(\omega,\omega^N,\lambda)} =$$
$$2\Upsilon\left[\widehat{H}_2(\Upsilon,\omega,\omega^N,\lambda) + \Upsilon\frac{\partial \widehat{H}_2(\Upsilon,\omega,\omega^N,\lambda)}{\partial v}\right] = -2\lambda + O_2(\lambda).$$

To verify this, it suffices to observe that $X^{\omega,\omega^N,\lambda}$ (4.57) implies the equation

$$\dot{v} = 2v\left[\lambda + v\widehat{H}_2(v,\omega,\omega^N,\lambda)\right]$$

for v. Thus, the $(n+1)$-torus (4.58) is repelling over the half axis $\lambda < 0$ (where the n-torus $\{z = 0\}$ is attracting) and attracting over the half axis $\lambda > 0$ (where the n-torus $\{z = 0\}$ is repelling). Recall that torus (4.58) in fact exists only for $\lambda > 0$ or only for $\lambda < 0$, depending on the sign of $\widehat{H}_2(0,\omega^\star,\omega^{N\star},0)$.

The question now is what happens to the n-tori $\{z = 0\}$ and the $(n+1)$-tori (4.58) if one perturbs the integrable family (4.57) to a nearby family $\widetilde{X}^{\omega,\omega^N,\lambda}$. Of course, one can predict beforehand that the perturbed family of vector fields will have an $(n+2)$-parameter Whitney-smooth family of Floquet Diophantine invariant analytic n-tori and that of $(n+1)$-tori, both families being described by Theorem 2.25. The problem is how these two families arrange and how finitely smooth "phase-lock" invariant n- and $(n+1)$-tori (see § 1.2.1) are located between the analytic Diophantine tori. The answer runs as follows.

If $n = 0$, a perturbation in fact affects nothing, and the perturbed family of vector fields is even topologically equivalent to the unperturbed family, the latter thus turns out to be structurally stable. In this case, any family of vector fields close to X (4.57) can be chosen as the "unperturbed" one. This is what one should expect, because nondegenerate equilibria and periodic trajectories persist under small perturbations of the vector field and this persistence requires no Diophantine-like conditions (and no normal hyperbolicity). The bifurcation in question for $n = 0$ is usually called "the limit cycle birth/annihilation bifurcation", the Poincaré–Andronov, or Hopf bifurcation. A detailed description of the bifurcation pattern and references to the basic papers by Andronov, Leontovich-Andronova, and Hopf can be found in Marsden & McCracken [220] and Arnol'd [12, § 33]. A similar bifurcation pattern is known for other contexts also. For instance, a large body of literature is devoted to the Hamiltonian Hopf bifurcation, see, e.g., van der Meer [344, 345] or Golubitsky, Marsden, Stewart & Dellnitz [136] and references therein. The reversible Hopf bifurcation has been described by Arnol'd & Sevryuk [22, 315, 317]. In the case of Hamiltonian Hopf and reversible Hopf bifurcations, two pairs of purely imaginary eigenvalues coalesce at the bifurcation instant.[7] There are also results on the *equivariant* Hamiltonian Hopf [99, 346, 350] and reversible Hopf [350] bifurcations, where a certain group of symmetries is present.

If $n = 1$, the family of closed trajectories $\{z = 0\}$ persists under small perturbations of the vector fields as well. The family of invariant 2-tori branching off from these periodic

[7] More general patterns, where k pairs ($k \geq 2$) of purely imaginary eigenvalues coalesce at the bifurcation instant, are called the Hopf bifurcation at k-fold resonances in Hamiltonian [177] and reversible [176] systems. This is a degeneracy of codimension $k - 1$, and it requires at least $k - 1$ external parameters to be described properly.

trajectories also persists, but those tori, generally speaking, lose their analyticity and parallelity of the dynamics. To be more precise, over a Cantor set in the parameter space, the 2-torus is analytic and Diophantine, whereas over the open complement of that set, the 2-torus is only finitely smooth and the induced dynamics is "phase-lock" (and therefore not parallel). In fact, such "phase-lock" tori contain a finite number of hyperbolic closed trajectories. The smoothness class of the torus and the periods of these trajectories tend generically to infinity as the torus approaches the "central" $(n+2)$-parameter family of closed trajectories and collapses. This bifurcation is called the Neĭmark excitation, the Neĭmark–Sacker bifurcation, or the Hopf bifurcation for cycles. It is more often described in an equivalent way in terms of the bifurcation of a fixed point 0 of a diffeomorphism A as a pair of two complex conjugate eigenvalues of the linearization of A at 0 passes through the unit circle. We again refer the reader to Marsden & McCracken [220] and Arnol'd [12, §§ 34–35] for a detailed presentation and the bibliography. A similar bifurcation pattern is known for other contexts also. For instance, see Lahiri, Bhowal, Roy & Sevryuk [188] for the reversible Neĭmark–Sacker bifurcation.

Now proceed to the case $n \geq 2$. It turns out that the "Diophantine part" of the n-tori $\{z = 0\}$ persists under small perturbations of the initial integrable family X (4.57) of vector fields. To be precise, consider a small compact neighborhood Γ of point (ω^*, ω^{N*}) in $\mathbb{R}^{n+1}$ and a small segment $I \subset \mathbb{R}$ with 0 inside I. We suppose that Γ is small enough not to intersect the hyperplane $\{\omega^N = 0\}$. Let for $\tau > n-1$ fixed and $\gamma > 0$

$$\begin{aligned} \mathbb{R}^{n+1} \supset \Gamma_\gamma := \{ (\omega, \omega^N) \in \Gamma \; : \; & \forall k \in \mathbb{Z}^n \setminus \{0\} \; \forall \ell \in \mathbb{Z}, \, |\ell| \leq 2, \; |\langle \omega, k\rangle + \ell \omega^N| \geq \gamma |k|^{-\tau} \\ & \text{and} \quad \operatorname{dist}((\omega, \omega^N), \partial\Gamma) \geq \gamma \} \end{aligned} \tag{4.59}$$

[cf. § 1.5.2, (2.29) in § 2.3.1, Section 5.2, (6.1) in § 6.1.1, and (6.3) in § 6.1.2], here $\partial\Gamma$ is the boundary of Γ. The set Γ_γ is a Whitney-smooth foliation of closed segments. Also let

$$\mathbb{R}^{n+2} \supset (\Gamma \times I)_\gamma := \{(\omega, \omega^N, \lambda) \in \Gamma \times I \; : \; (\omega, \omega^N) \in \Gamma_\gamma \quad \text{and} \quad \operatorname{dist}(\lambda, \partial I) \geq \gamma\},$$

here ∂I is the two-element set consisting of the ends of I. One observes that $meas_{n+2}(\Gamma \times I)_\gamma = [1 - O(\gamma)]\, meas_{n+2}(\Gamma \times I)$ as $\gamma \downarrow 0$, as before (cf. Chapter 2).

Theorem 4.13 [47, 48] *Let Γ, I, and γ be small enough. Then for each analytic family $\tilde{X}$ of analytic vector fields sufficiently close to family X (4.57) in the real analytic topology, there exists a C^∞ mapping*

$$\Phi : \mathbb{T}^n \times \mathbb{R}^2 \times \Gamma \times I \to \mathbb{T}^n \times \mathbb{R}^2 \times \mathbb{R}^{n+2}$$

with the following properties.

(i) *The restriction of Φ to $\mathbb{T}^n \times \{0\} \times \Gamma \times I$ is a C^∞ diffeomorphism onto its image, C^∞-near-the-identity map. Moreover, Φ preserves the projection to the parameter space:*

$$\begin{aligned} \Phi(x, z, \bar{z}, \omega, \omega^N, \lambda) &= (\chi(x, z, \bar{z}, \omega, \omega^N, \lambda), \zeta(x, z, \bar{z}, \omega, \omega^N, \lambda), \phi(\omega, \omega^N, \lambda)), \\ & \quad \chi \in \mathbb{T}^n, \; \zeta \in \mathbb{C}, \; \phi \in \mathbb{R}^{n+2}, \end{aligned}$$

and is affine in $(\operatorname{Re} z, \operatorname{Im} z) \in \mathbb{R}^2$ *and analytic in* $x \in \mathbb{T}^n$.

(ii) *The further restriction of Φ to $\mathbb{T}^n \times \{0\} \times (\Gamma \times I)_\gamma$ conjugates X to $\tilde{X}$. Thus, for any $(\omega, \omega^N, \lambda) \in (\Gamma \times I)_\gamma$, the manifold*

$$\{\Phi(x, 0, 0, \omega, \omega^N, \lambda) \; : \; x \in \mathbb{T}^n\} \tag{4.60}$$

is a Diophantine invariant analytic n-torus of the field $\widetilde{X}^{\phi(\omega,\omega^N,\lambda)}$ with frequency vector ω.

(iii) *Moreover, Φ preserves the normal linear behavior of family $\widetilde{X}$ on $\mathbb{T}^n\times\{0\}\times(\Gamma\times I)_\gamma$. In particular, for any $(\omega,\omega^N,\lambda)\in(\Gamma\times I)_\gamma$, the torus (4.60) is Floquet and the eigenvalues of its Floquet 2×2 matrix are equal to $\lambda\pm i\omega^N$.*

Thus, the set of invariant n-tori $\{z=0\}$ of the family X whose internal and normal frequencies (ω,ω^N) satisfy the Diophantine conditions pointed out in (4.59) is structurally stable under small perturbations of the family, although the normal hyperbolicity breaks for $\lambda=0$. The allowed size of the perturbation depends on γ.

What is the fate of, say, *resonant* tori $\{z=0\}$ as one perturbs X to $\widetilde{X}$? Omitting some details, the answer is as follows. There is a monotonically increasing C^∞ function $\varrho:\mathbb{R}_+\to\mathbb{R}_+$ with $\varrho(0)=0$ which is infinitely flat at 0 and possesses the following property. For given $\alpha>0$, there exist positive constants $C_1=C_1(\alpha)$ and $C_2=C_2(\alpha)$ such that for each $(\omega,\omega^N,\lambda)\in\mho_\alpha$ with

$$\mho_\alpha := \bigcup_{(\omega^0,\omega^{N0})\in\Gamma_\gamma}\{(\omega,\omega^N,\lambda)\in\Gamma\times I\ :\ C_2^{-1}(\alpha)\,\varrho\left(|\omega-\omega^0|+|\omega^N-\omega^{N0}|\right)<|\lambda| \leq C_1(\alpha)\} \tag{4.61}$$

the vector field $\widetilde{X}^{\phi(\omega,\omega^N,\lambda)}$ has an invariant n-torus of smoothness C^α which is close to the torus $\{z=0\}$. The constants C_1 and C_2 grow as α decreases, so that $\mho_{\alpha_1}\subset\mho_{\alpha_2}$ for $\alpha_1>\alpha_2$.

Thus, each segment $\{\omega=\omega^0,\omega^N=\omega^{N0},\lambda\in I\}$ with $(\omega^0,\omega^{N0})\in\Gamma_\gamma$ in the parameter space is "fattened" by the sand-glass-like "tube"

$$|\lambda|>C_2^{-1}\,\varrho\left(|\omega-\omega^0|+|\omega^N-\omega^{N0}|\right) \tag{4.62}$$

with an infinitely narrow "waist" at $\lambda=0$ [47, 48, 64], see Figure 4.1. Each of the two halves of the "tube" (4.62) [determined by $\lambda<0$ or $\lambda>0$] is called *a disc* [64], it is indeed diffeomorphic to an open $(n+2)$-dimensional ball. Over the ϕ-image of the segment itself, the perturbed family $\widetilde{X}$ possesses a Floquet Diophantine invariant analytic n-torus, and over the ϕ-image of the "tube", family $\widetilde{X}$ has an invariant C^α-smooth n-torus. The induced dynamics on that torus is generally not parallel.

The unperturbed tori $\{z=0\}$ that are too close to the critical hyperplane $\{\lambda=0\}$ (in the product of the phase space and the parameter space) and do not satisfy suitable Diophantine conditions do *not* survive the perturbation *even as invariant manifolds* (at least as soon as we are interested in invariant tori of a predetermined finite smoothness). Indeed, to obtain the subset in the parameter space that corresponds to persistent tori, one should *cut out* the set

$$\Diamond_\alpha := \bigcap_{(\omega^0,\omega^{N0})\in\Gamma_\gamma}\{(\omega,\omega^N,\lambda)\in\Gamma\times I\ :\ |\lambda|\leq C_2^{-1}(\alpha)\,\varrho\left(|\omega-\omega^0|+|\omega^N-\omega^{N0}|\right)\}, \tag{4.63}$$

i.e., the intersection of a "Cantor collection" of "resonance holes"

$$|\lambda|\leq C_2^{-1}\,\varrho\left(|\omega-\omega^0|+|\omega^N-\omega^{N0}|\right) \tag{4.64}$$

(cf. [47, 48, 64]). Similar low dimensional sets in, e.g., Chenciner [81, 82, 84] or Iooss & Los [167] are called "bubbles". Note that both sets $\mho_\alpha$ (4.61) and $\Diamond_\alpha$ (4.63) depend also on γ. As $\gamma\downarrow 0$, the measure $\mathit{meas}_{n+2}\,\Diamond_\alpha$ decreases faster than any power of γ.

The subset of the parameter space over which $\widetilde{X}$ possesses invariant $(n+1)$-tori (with a prescribed smoothness) is also the ϕ-image of the union of "discs". However, in this case, those discs exist on one side of the bifurcation hyperplane $\{\lambda = 0\}$ only and are no longer centered around the rays $\omega = \omega^0$, $\omega^N = \omega^{N0}$.

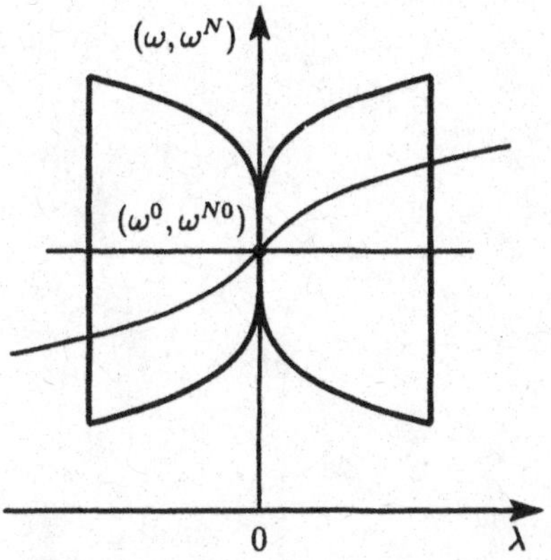

Figure 4.1: A "tube" of "good" parameter values in the quasi-periodic Hopf bifurcation and a Hopf–Landau family (the curve) inside a quasi-periodic Hopf family.

Remark 1. Some generalizations of the theory above, e.g., to the cases where the additional rotational symmetry is present or the Floquet nature of the unperturbed tori is not required, are given in Broer & Takens [64]. In this paper, in particular, the so called skew Hopf bifurcation is studied (concerning the latter, see also Broer [57] and Wagener [352]).

Remark 2. One sometimes also considers a bifurcation of invariant $(n+1)$-tori $\mathcal{V}_\mu$ from invariant n-tori V_μ for $n \geq 2$ without the quasi-periodic dynamics assumption. In such a case, one prefers to speak of the Hopf–Landau bifurcation. This bifurcation has been examined by Chenciner & Iooss [85, 86], Sell [313], and Flockerzi [116], cf. also Bibikov [39]. The external parameter μ in the Hopf–Landau bifurcation is usually considered to be one-dimensional. However, even in this set-up, the n-torus V_μ is assumed to be Diophantine at the critical parameter value $\mu = \mu^\star$, when V_μ loses the normal hyperbolicity. Moreover, although $V_{\mu^\star}$ is not assumed to be Floquet, one can assign a certain kind of normal frequency to $V_{\mu^\star}$, and the whole collection of the internal frequencies of $V_{\mu^\star}$ and this normal frequency should satisfy some Diophantine conditions. In fact, a Hopf–Landau family can be seen as a curve inside a quasi-periodic Hopf family (see Figure 4.1). This curve: 1) at bifurcation, hits a Diophantine non-hyperbolic n-torus and 2) is transversal to the hyperplane $\{\lambda = 0\}$.

Chapter 5

Conclusions

5.1 Analogues and non-analogues for diffeomorphisms

All the theory of quasi-periodic motions in vector fields can be *mutatis mutandis* carried over to dynamical systems with discrete time (diffeomorphisms). Being far from the intention to present here the discrete analogues of all the definitions and theorems, we will confine ourselves with pointing out the main differences from the case of vector fields (differential equations).

5.1.1 Definitions

Let T be an invariant n-torus of a smooth diffeomorphism $A : M \to M$, manifold M being $(n+m)$-dimensional. We say that A on T induces *parallel* (or *conditionally periodic*, or *Kronecker*, or *linear*) motion, evolution, dynamics, or cascade, if there exists a diffeomorphism $T \to \mathrm{T}^n$ transforming the restriction $A|_T$ to a constant shift $x \mapsto x + \omega$ on the standard n-torus T^n. The numbers $\omega_1, \omega_2, \ldots, \omega_n$ are called *(internal) frequencies* of the motion (evolution, dynamics, or cascade) on T, but also of the invariant torus T itself. The frequency vector ω belongs in fact to T^n rather than to R^n.

A parallel motion on an invariant n-torus T with frequency vector ω is called *quasi-periodic* or *nonresonant* if the $n+1$ numbers $\omega_1, \omega_2, \ldots, \omega_n, \pi$ are rationally independent, i.e., if for all $k \in \mathrm{Z}^n \setminus \{0\}$ and $k_0 \in \mathrm{Z}$ one has $\langle \omega, k\rangle + 2\pi k_0 \neq 0$. In this case the torus T itself also is said to be quasi-periodic. Otherwise an invariant torus T with parallel dynamics is called *resonant*.

Quasi-periodic tori are densely filled up by each of the orbits contained therein. However, one should not think, by analogy with the case of vector fields, that resonant tori are foliated by invariant subtori of smaller dimension. In the case of discrete dynamics, resonant tori are foliated by *finite collections* $\widetilde{T}_1, \widetilde{T}_2, \ldots, \widetilde{T}_s$ of subtori of smaller dimension, the cascade $A|_T$ *permuting cyclically* these subtori: $A(\widetilde{T}_i) = \widetilde{T}_{i+1}$ for $1 \le i \le s-1$ and $A(\widetilde{T}_s) = \widetilde{T}_1$. Of course, the value $s = 1$ is possible.

An invariant n-torus with parallel dynamics is said to be *Diophantine* if for some constants $\tau > 0$ and $\gamma > 0$ the corresponding frequency vector ω satisfies the infinite system of inequalities

$$|\langle \omega, k\rangle + 2\pi k_0| \ge \gamma |k|^{-\tau}$$

for all $k \in \mathrm{Z}^n \setminus \{0\}$ and $k_0 \in \mathrm{Z}$, cf. (1.1) [and (1.9) as well]. Diophantine tori are quasi-periodic, but not *vice versa*. For $\tau > n$ the set of all frequency vectors $\omega \in \mathrm{T}^n$ that are

Diophantine in the above sense has positive Lebesgue measure [337]. Note that this set is totally disconnected: the hyperplanes $\langle\omega,k\rangle \equiv 0 \mod 2\pi$, $k \in \mathbb{Z}^n \setminus \{0\}$, densely fill up its complement (cf. § 1.5.2).

An invariant n-torus T with parallel dynamics is called *Floquet* if near T, one can introduce coordinates $(x \in \mathbb{T}^n, y \in \mathbb{R}^m)$ in which the torus T itself gets the equation $\{y = 0\}$ while the mapping A takes the form

$$A : (x, y) \mapsto (x + \omega + O(y), \Omega y + O_2(y))$$

with $\Omega \in GL(m, \mathbb{R})$ independent of $x \in \mathbb{T}^n$, cf. (1.2). If this is the case, matrix Ω is called the *Floquet matrix* of torus T.

The *normal frequencies* of a Floquet invariant torus with parallel dynamics are the *arguments* of the eigenvalues λ of its Floquet matrix with positive imaginary parts. So, if $\lambda = \varrho e^{i\phi}$, $\varrho > 0$, $0 < \phi < \pi$, then the normal frequency associated to λ is ϕ (plus an integral multiple of 2π).

All the normal frequencies $\omega_1^N, \omega_2^N, \ldots, \omega_r^N$ of a given Floquet invariant n-torus with parallel dynamics $(0 \le r \le m)$ constitute the normal frequency vector ω^N which belongs in fact to $\mathbb{T}^r$ rather than to $\mathbb{R}^r$. The "mixed" Diophantine condition on the internal $\omega \in \mathbb{T}^n$ and normal $\omega^N \in \mathbb{T}^r$ frequency vectors has the form

$$|\langle\omega, k\rangle + \langle\omega^N, \ell\rangle + 2\pi k_0| \ge \gamma |k|^{-\tau} \quad \forall k \in \mathbb{Z}^n \setminus \{0\} \;\; \forall k_0 \in \mathbb{Z} \;\; \forall \ell \in \mathbb{Z}^r, \; |\ell| \le 2,$$

cf. (1.4). This condition is not homogeneous, and one cannot consider the ratios $\Pi(\omega, \omega^N)$ of the frequencies instead of frequencies themselves (cf. Section 2.4). Hence, the regularity of Whitney-smooth families of Floquet Diophantine invariant tori for diffeomorphisms is generally less than that for vector fields. Of course, the presence of the term $2\pi k_0$ in all the Diophantine inequalities in the case of discrete dynamics is a consequence of the fact that all the internal and normal frequencies here are determined up to adding an integral multiple of 2π.

The discrete counterparts of purely imaginary eigenvalues of the Floquet matrix are eigenvalues lying on the unit circle.

For vector fields, the dimensions $n = 0$ and $n = 1$ of invariant n-tori are exceptional (no small divisors). For mappings, only the dimension $n = 0$ is exceptional.

5.1.2 Contexts

In the *dissipative context*, one considers just arbitrary diffeomorphisms of a certain manifold (phase space). In § 1.2.1, we presented an example with Poincaré return maps.

In the *volume preserving context*, one should consider only *exact* volume elements: $\sigma = d\tau$, where τ is a certain $(N-1)$-form (N being the dimension of the phase space). The diffeomorphisms A in question are globally volume preserving. To be more precise, the volume preservation condition $A^*\sigma = \sigma$ can be rewritten, in the case of an exact volume element $\sigma = d\tau$, as $d(A^*\tau - \tau) = 0$. The latter equality means that the form $A^*\tau - \tau$ is closed. A volume preserving mapping A is said to be *globally volume preserving* if the form $A^*\tau - \tau$ is not only closed but even exact.

The situation with the discrete counterpart of the Hamiltonian context is similar (note that in the case of diffeomorphisms, one prefers to speak on the *symplectic context*). Namely, in the symplectic *isotropic* context, one should consider only *exact* symplectic structures: $\omega^2 = d\zeta$, where ζ is a certain 1-form. A diffeomorphism A of a symplectic manifold is said to be *symplectic* if $A^*\omega^2 = \omega^2$. In the case of an exact symplectic structure $\omega^2 = d\zeta$, this condition can be rewritten as $d(A^*\zeta - \zeta) = 0$. The latter equality means that the form $A^*\zeta - \zeta$ is closed. A symplectic mapping A is said to be *exact symplectic* if the form $A^*\zeta - \zeta$ is not only closed but even exact [18, 152, 154, 201, 287, 328, 361]. The diffeomorphisms in question in the symplectic isotropic context are exact symplectic.

Recall that a manifold equipped with an exact symplectic structure is also said to be exact symplectic [18, 201].

However, in the symplectic *coisotropic* context, the concept of exact symplecticity turns out to be too restrictive, and one should invoke the weaker concept of r-exact symplecticity ($r \in \mathrm{N}$) introduced by Quispel & Sevryuk [287]. Namely, a symplectic structure ω^2 on a $2N$-dimensional manifold is said to be *r-exact* ($1 \le r \le N$) if the rth exterior power $(\omega^2)^r$ of ω^2 is exact: $(\omega^2)^r = d\zeta$, where ζ is a certain $(2r-1)$-form. It is clear that an r-exact symplectic structure is also r_1-exact for any $r+1 \le r_1 \le N$. If a diffeomorphism A of a $2N$-dimensional manifold equipped with an r-exact symplectic structure ω^2 $[(\omega^2)^r = d\zeta]$ is symplectic, then for any $R \in \mathrm{N}$, $r \le R \le N$, the $(2R-1)$-form $(A^*\zeta - \zeta) \wedge (\omega^2)^{R-r}$ is closed. A symplectic mapping A is said to be *R-exact symplectic* if the latter $(2R-1)$-form is not only closed but even exact. It is clear that an R-exact symplectic mapping is also R_1-exact for any $R+1 \le R_1 \le N$. Exact symplectic structures and diffeomorphisms are just 1-exact symplectic ones in this terminology. The diffeomorphisms in question in the symplectic coisotropic context are r-exact symplectic (see § 5.1.4 below).

A manifold equipped with an r-exact symplectic structure is also said to be r-exact symplectic.

Studies of the occurrence of quasi-periodic motions in the symplectic context were also initiated by Kolmogorov (see [21]).

In the *reversible context*, the diffeomorphisms A in question are reversible with respect to the fixed involution G of the phase space. This means that G conjugates mapping A with its inverse A^{-1}, i.e., $GAG = A^{-1}$. In other words, an invertible mapping A is said to be *reversible* with respect to G if G casts the forward orbit of any point u under A into the backward orbit of the point $G(u)$. As in the case of vector fields, this definition is applicable to any diffeomorphism G (of course, in this general case one should write $GAG^{-1} = A^{-1}$) but only the involutive case $G^2 = id$ is usually considered.

5.1.3 Peculiarities of the symplectic and reversible contexts

Whereas the dissipative and volume preserving contexts for diffeomorphisms are parallel to their analogues for vector fields, the symplectic and reversible contexts for diffeomorphisms differ considerably from their vector field counterparts.

In the case of the symplectic context, one should consider exact symplectic structures and diffeomorphisms in the isotropic subcontext and r-exact symplectic structures and diffeomorphisms in the coisotropic subcontext. The symplectic structure in the Hamiltonian or symplectic *coisotropic* context is necessarily *nonexact*. The reason is that a quasi-periodic invariant torus T of a Hamiltonian flow or symplectic diffeomorphism is automatically *isotropic* (and, in particular, of dimension no greater than the number of degrees of freedom), if the symplectic structure is *exact*.

Indeed, let $\dim T = n$ and in some coordinates $x \in \mathrm{T}^n$ on T, let the induced motion be described by the equation $\dot{x} = \omega$ or the mapping $x \mapsto x + \omega$, respectively, the frequency vector ω being nonresonant. Then the restriction $\omega^2|_T$ of the symplectic structure ω^2 to T has constant (i.e., x-independent) coefficients. Let X_1 and X_2 be arbitrary *constant* vector fields on T. On the one hand, $\omega^2(X_1, X_2) = const$. On the other hand, if the form ω^2 is exact, so is $\omega^2|_T$. The latter form is therefore a sum of terms of the form $[\partial F(x)/\partial x_{j_1}]dx_{j_1} \wedge dx_{j_2}$, and

$$\int_{\mathbb{T}^n} \omega^2(X_1, X_2)dx = 0.$$

Hence, $\omega^2(X_1, X_2) \equiv 0$, i.e., $\omega^2|_T = 0$ and the torus T is isotropic.

The reversible settings for vector fields and diffeomorphisms differ more drastically. Namely, the reversible context for diffeomorphisms is much richer than that for vector fields.

Indeed, a reversible *vector field* is characterized by the type of the reversing involution. On the other hand, any reversible *mapping* has in fact infinitely many reversing involutions (provided that no power of this mapping is the identity transformation) [271, 287, 292]. If a mapping A is reversible with respect to an involution G, then one can easily verify that all the mappings $A^jG = GA^{-j}$ for $j \in \mathbb{Z}$ are also involutions which reverse A. All these involutions enjoy in fact equal rights. The equality $A = (A^{j+1}G)(A^jG)$ shows that the mapping A can be decomposed into the product of two involutions in infinitely many ways.

Thus, at first sight, a G-reversible mapping A is to be characterized by pointing out the types of all the involutions A^jG, and the reversible context for diffeomorphisms is "infinitely many times" as complicated as that for vector fields. Fortunately, the types of only two of these reversing involutions, namely G and AG, are to be taken into account. The reason is that the fixed point submanifolds $\mathrm{Fix}(A^jG)$ of the involutions A^jG can be obtained from those of G and AG by taking their images under powers of A [271, 287, 292]:

$$\mathrm{Fix}(A^{2j}G) = A^j[\mathrm{Fix}\, G], \qquad \mathrm{Fix}(A^{2j+1}G) = A^j[\mathrm{Fix}(AG)].$$

So, if the involution G is of type (q_-, q_+), so is A^jG for any even $j \in \mathbb{Z}$, and if the involution AG is of type (Q_-, Q_+), so is A^jG for any odd $j \in \mathbb{Z}$.

To summarize, G-reversible vector fields X can be classified according to the types of their reversing involutions G, and G-reversible mappings A according to the types of the *two* involutions G and AG. The reversible context for diffeomorphisms turns out to be "twice" as delicate as that for vector fields.

5.1.4 Results

It is not hard to give the discrete time analogues of statements (α)–(η) in § 1.4.1 [note that the counterparts of statements (ζ) and (η) are much more complicated than statements (ζ) and (η) themselves because in the case of reversible diffeomorphisms, one should take into account the types of two involutions]. We will formulate here the analogue of statement (ϵ) [concerning the Hamiltonian coisotropic context] only.

(ϵ') [287] The symplectic coisotropic (n, p, s) context: coisotropic invariant n-tori of s-parameter families of symplectic diffeomorphisms of a $2(n-p)$-dimensional symplectic manifold ($n \geq 3$, $0 < p < n/2$). Let the symplectic structure ω^2 be $(p+1)$-exact. Then, depending on the global properties of ω^2 on the phase space, either 1) a *generic* family of $(p+1)$-exact symplectic diffeomorphisms has no coisotropic invariant n-tori with parallel dynamics, or **2)** a *typical* family of $(p+1)$-exact symplectic diffeomorphisms possesses $(n-2p+s)$-parameter Cantor-like families of Diophantine coisotropic invariant n-tori.

This statement is a conjecture. In fact, the symplectic coisotropic context and the reversible context 2 for mappings have not been examined yet in the literature at all whereas all the remaining five contexts have.

Remark. Quispel & Sevryuk [287] and Sevryuk [328] formulated the above statement (ϵ') for exact symplectic structures and mappings rather than for $(p+1)$-exact ones. This in fact made no sense (see § 5.1.3). The paper [287] contained also the correct version of statement (ϵ').

Below we present the discrete time counterparts of Theorems 2.25–2.29. All the diffeomorphisms and their dependence on external parameters are assumed to be analytic. Let $n \geq 1$.

Theorem 5.1 (dissipative context) *For any $p \geq 0$ and $s \geq 1$, a typical s-parameter family of diffeomorphisms of an $(n+p)$-dimensional manifold possesses a Whitney-smooth s-parameter family of Floquet Diophantine invariant n-tori (at most one torus per each parameter value). If the number of the pairs of complex conjugate eigenvalues of the Floquet matrices of these tori is r and $s \geq n+r$, then the Whitney-smooth s-parameter family of n-tori consists generically of $(s-n-r)$-parameter analytic subfamilies.*

Theorem 5.2 (volume preserving context with $p \geq 2$) *For any $p \geq 2$ and $s \geq 1$, a typical s-parameter family of globally volume preserving diffeomorphisms of an $(n+p)$-dimensional manifold possesses a Whitney-smooth s-parameter family of Floquet Diophantine invariant n-tori (at most one torus per each parameter value). If the number of the pairs of complex conjugate eigenvalues of the Floquet matrices of these tori is r and $s \geq n+r$, then the Whitney-smooth s-parameter family of n-tori consists generically of $(s-n-r)$-parameter analytic subfamilies.*

Theorem 5.3 (volume preserving context with $p = 1$) *For any $s \geq 0$, a typical s-parameter family of globally volume preserving diffeomorphisms of an $(n+1)$-dimensional manifold possesses a Whitney-smooth $(s+1)$-parameter family of Floquet Diophantine invariant n-tori. The Floquet 1×1 matrix of each of these tori is unit. If $s \geq n-1$, then the Whitney-smooth $(s+1)$-parameter family of n-tori consists generically of $(s-n+1)$-parameter analytic subfamilies.*

Theorem 5.4 (symplectic isotropic context) *For any $p \geq 0$ and $s \geq 0$, a typical s-parameter family of exact symplectic diffeomorphisms of a $2(n+p)$-dimensional exact symplectic manifold possesses a Whitney-smooth $(n+s)$-parameter family of Floquet Diophantine isotropic invariant n-tori. The Floquet $(n+2p) \times (n+2p)$ matrix of each of these tori has eigenvalue 1 of multiplicity n, while the remaining $2p$ eigenvalues occur in pairs (λ, λ^{-1}). If the number of distinct values of the arguments ϕ in the range $0 < \phi < \pi$ of the eigenvalues of the Floquet matrices of these tori is r and $s \geq r$, then the Whitney-smooth $(n+s)$-parameter family of n-tori consists generically of $(s-r)$-parameter analytic subfamilies.*

Theorem 5.5 (reversible context 1) *For any $m \geq 0$, $p \geq 0$, $q \geq 0$ and $s \geq \max\{1-m, 0\}$, a typical s-parameter family of diffeomorphisms decomposable into the product of two involutions of types $(n+p, m+p+q)$ and $(n+p+q, m+p)$ [of an $(n+m+2p+q)$-dimensional manifold] possesses a Whitney-smooth $(m+s)$-parameter family of Floquet Diophantine invariant n-tori. These tori are also invariant under both the involutions. The Floquet $(m+2p+q) \times (m+2p+q)$ matrix of each of these tori has eigenvalue 1 of multiplicity m (if $m > 0$), eigenvalue -1 of multiplicity q (if $q > 0$), while the remaining $2p$ eigenvalues occur in pairs (λ, λ^{-1}). If the number of distinct values of the arguments ϕ in the range $0 < \phi < \pi$ of the eigenvalues of the Floquet matrices of these tori is r and $s \geq n-m+r$, then the Whitney-smooth $(m+s)$-parameter family of n-tori consists generically of $(s+m-n-r)$-parameter analytic subfamilies.*

Of course, all the invariant tori in Theorems 5.1–5.5 are analytic.

Remark. Volume preserving context with $p = 1$ for diffeomorphisms was considered by Herman (see [361]) and Xia [358], see also Cheng & Sun [88] and Delshams & de la Llave [101] for the particular case of volume preserving mappings with a three-dimensional phase space (note that [101] analyzed invariant 2-tori in divergence-free *flows* with a three-dimensional phase space as well). Xia also generalized his result to invariant n-tori of diffeomorphisms A of $T^n \times R^m$ $(m \leq n)$ possessing the following *intersection property* (cf. Section 2.2): each n-torus homological to $\{R^m \ni y = const\}$ intersects its A-image [359].

For a complete proof of Theorem 5.4 in the simplest case of symplectic isotropic $(n, 0, 0)$ context, we refer the reader to Arnol'd & Avez [17], see also Lazutkin [201]. *Lower-dimensional* invariant tori in individual exact symplectic diffeomorphisms [the symplectic isotropic $(n, p, 0)$ context with $p \geq 1$] were constructed by Zehnder [363, 364], cf. also Bel'kovich [28].

Lower-dimensional invariant tori in individual reversible diffeomorphisms [the reversible $(n, m, p, q, 0)$ context 1 with $p \geq 1$] have been examined by Sevryuk & Quispel in [318] (for $m = n$ and $q = 0$), [322, 324] (for $m \geq n$ and $q = 0$), [287] (for $m \geq n$ and q arbitrary), and [326] (for arbitrary n, m, and q). Related are also papers [34, 187, 188, 294, 332] devoted to bifurcations of one-parameter families of invariant Diophantine circles in reversible mappings of R^4 (in fact, these papers describe the so called reversible Neĭmark–Sacker bifurcation and its generalizations, see Section 4.3 above). The excitation of elliptic normal modes in reversible diffeomorphisms has been studied in [287, 324, 326]. The authors are unaware of any works considering the excitation of elliptic normal modes in exact symplectic diffeomorphisms.

The Bruno continuation theory (cf. Chapter 3 and Section 6.3) seems to have not been developed yet for diffeomorphisms.

5.1.5 Proofs

The proofs of the "main" theorems in the discrete time case are run entirely parallel to those in the case of vector fields. As to the "relaxed" and "miniparameter" theorems, the case of diffeomorphisms turns out to be slightly different. The point is that the mappings compatible with the structure $\mathfrak{S}$ (an exact volume element, exact symplectic structure, or involution) on the phase space do *not* constitute a linear space, in contrast to vector fields.

This causes some difficulties in proving the "relaxed" and "miniparameter" theorems in the volume preserving, symplectic, and reversible contexts.

Suppose that we have a family $\widetilde{A}^\mu$ of $\mathfrak{G}$-compatible diffeomorphisms

$$\widetilde{A}^\mu : \begin{pmatrix} x \\ y \\ z \end{pmatrix} \mapsto \begin{pmatrix} x + \omega(y,\mu) + f(x,y,z,\mu) + \widetilde{f}(x,y,z,\mu) \\ y + g(x,y,z,\mu) + \widetilde{g}(x,y,z,\mu) \\ \Omega(y,\mu)z + h(x,y,z,\mu) + \widetilde{h}(x,y,z,\mu) \end{pmatrix} \tag{5.1}$$

where $x \in \mathrm{T}^n$, $y \in Y \subset \mathrm{R}^m$, $z \in \mathcal{O}_K(0)$, $f = O(z)$, $g = O(z)$, $h = O_2(z)$, and the perturbations $\widetilde{f}$, $\widetilde{g}$, $\widetilde{h}$ are small (of the order of $\epsilon > 0$). We will say that $\widetilde{A}^\mu$ are diffeomorphisms with linear part $(x+\omega, \Omega z)$. Now one wishes to embed this family into an extended family $\widetilde{A}^{\mu,\nu}_{\text{new}}$ of $\mathfrak{G}$-compatible diffeomorphisms with an additional parameter ν and *prescribed* linear parts

$$(x + \omega^{\text{new}}, \Omega^{\text{new}} z) = (x + \omega^{\text{new}}(y,\mu,\nu), \Omega^{\text{new}}(y,\mu,\nu)z)$$

provided by the unfolding theory of linear operators, cf. [9, 10, 12, 119, 120, 161, 178, 321, 333] (see also [230, 353]). Of course, $\omega^{\text{new}}(y,\mu,0) \equiv \omega(y,\mu)$, $\Omega^{\text{new}}(y,\mu,0) \equiv \Omega(y,\mu)$, and one has to achieve $\widetilde{A}^{\mu,0}_{\text{new}} = \widetilde{A}^\mu$.

We cannot just substitute ω^{new} and Ω^{new} for respectively ω and Ω in (5.1), because in this way we would obtain mappings incompatible with $\mathfrak{G}$ (cf. the proof of Theorem 2.11 in § 2.4.6 and that of Theorem 2.20 in § 2.5.7). The remedy is as follows. Let A^μ_0 be "linear" diffeomorphisms with linear parts $(x+\omega, \Omega z)$:

$$A^\mu_0 : \begin{pmatrix} x \\ y \\ z \end{pmatrix} \mapsto \begin{pmatrix} x + \omega(y,\mu) \\ y \\ \Omega(y,\mu)z \end{pmatrix}$$

and $A^{\mu,\nu}_{0,\text{new}}$ "linear" diffeomorphisms with linear parts $(x + \omega^{\text{new}}, \Omega^{\text{new}} z)$. These diffeomorphisms are compatible with $\mathfrak{G}$ and $A^{\mu,0}_{0,\text{new}} = A^\mu_0$.

In the volume preserving and symplectic contexts, we can now exploit the fact that all the $\mathfrak{G}$-compatible mappings constitute a group, so

$$\widetilde{A}^{\mu,\nu}_{\text{new}} = A^{\mu,\nu}_{0,\text{new}} (A^\mu_0)^{-1} \widetilde{A}^\mu. \tag{5.2}$$

In the reversible context, the involutions $\widetilde{A}^\mu G$ and $A^\mu_0 G$ have the same linear parts $(-x+\omega, \Omega R z)$ where

$$G : \begin{pmatrix} x \\ y \\ z \end{pmatrix} \mapsto \begin{pmatrix} -x \\ y \\ Rz \end{pmatrix}$$

is the reversing involution (R being an involutive $K \times K$ matrix). According to a generalization of the Bochner theorem, the involutions $\widetilde{A}^\mu G$ and $A^\mu_0 G$ are conjugated by some diffeomorphisms $\mathfrak{K}_\mu$ with identity linear parts (x, z):

$$\mathfrak{K}_\mu : \begin{pmatrix} x \\ y \\ z \end{pmatrix} \mapsto \begin{pmatrix} x + O(z) + O(\epsilon) \\ y + O(z) + O(\epsilon) \\ z + O_2(z) + O(\epsilon) \end{pmatrix}$$

(recall that ϵ is the perturbation size), i.e., $\widetilde{A}^\mu G = \mathfrak{K}_\mu^{-1} A^\mu_0 G \mathfrak{K}_\mu$. Consequently, $\widetilde{A}^\mu = \mathfrak{K}_\mu^{-1} A^\mu_0 G \mathfrak{K}_\mu G$, and one can set

$$\widetilde{A}^{\mu,\nu}_{\text{new}} = \mathfrak{K}_\mu^{-1} A^{\mu,\nu}_{0,\text{new}} G \mathfrak{K}_\mu G. \tag{5.3}$$

One straightforwardly verifies that the families $\widetilde{A}^{\mu,\nu}_{\rm new}$ defined by (5.2) or (5.3) possess all the desired properties.

For peculiarities of the proofs of discrete time "converse" KAM theorems (cf. Propositions 2.21–2.24), the reader is referred to Sevryuk [327].

5.1.6 Periodically invariant tori

For diffeomorphisms, it makes sense to consider also "periodically invariant" tori with parallel dynamics, which are invariant under some power of the mapping in question rather than under the mapping itself. For instance, resonant invariant tori of integrable exact symplectic and reversible diffeomorphisms often give rise, under small perturbations, to periodically invariant tori of smaller dimensions. However, the theory of Whitney-smooth families of periodically invariant tori in mappings has not been created yet.

5.1.7 The Poincaré section reduction

The theorems on the existence of invariant tori for vector fields can sometimes be reduced to those for diffeomorphisms by considering Poincaré sections for closed trajectories (cf. § 1.2.1). Such a reduction can be performed in all the contexts.

Let a vector field X on an N-dimensional manifold equipped with a volume element σ be divergence-free. Let ξ be a closed trajectory of X. Consider an arbitrary $(N-1)$-dimensional Poincaré section Σ to ξ and the corresponding Poincaré return map $A : \Sigma \to \Sigma$. Define the $(N-1)$-form $\sigma' = (i_X\sigma)|_\Sigma$ on Σ.

Proposition 5.6 *The form σ' is an exact volume element on Σ, and the diffeomorphism A is globally volume preserving with respect to σ'.*

Now let a vector field X be Hamiltonian with N degrees of freedom and Hamilton function H. Let $\xi \subset \{H = h\}$ be a closed trajectory of X lying on the energy level hypersurface $\{H = h\}$. Consider an arbitrary $(2N-2)$-dimensional "isoenergetic" Poincaré section $\Sigma \subset \{H = h\}$ to ξ and the corresponding "isoenergetic" Poincaré return map $A : \Sigma \to \Sigma$. Define the 2-form $(\omega^2)' = \omega^2|_\Sigma$ on Σ (ω^2 being the symplectic structure on the phase space).

Proposition 5.7 [1, 13, 201] *The form $(\omega^2)'$ is an exact symplectic structure on Σ, and the diffeomorphism A is exact symplectic with respect to $(\omega^2)'$.*

Finally, let a vector field X be reversible with respect to involution G of the phase space. Let ξ be a G-invariant closed trajectory of X. As is well known, ξ contains exactly two fixed points a and b of the reversing involution G. Consider an arbitrary $(N-1)$-dimensional G-invariant Poincaré section Σ to ξ at one of these two points (N being the phase space dimension) and the corresponding Poincaré return map $A : \Sigma \to \Sigma$. Define the involution $G' = G|_\Sigma$ of Σ. If involution G is of type (q_-, q_+) then involution G' is of type $(q_- - 1, q_+)$.

Proposition 5.8 [11, 22, 105, 106, 315] *The diffeomorphism A is reversible with respect to G'.*

In any of these cases, an invariant n-torus of the Poincaré return map A generates an invariant $(n+1)$-torus of the original vector field X.

Sometimes, on the other hand, KAM theorems for diffeomorphisms can be reduced to those for vector fields (see, e.g., Douady [107] or Lazutkin [201]).

In this book, we do not consider *nonautonomous* differential equations *periodic in time*. For such equations, one can also develop the KAM theory (with all the contexts). The latter, however, immediately reduces to the theory of quasi-periodic motions in diffeomorphisms via exploring the corresponding Poincaré map.

One can also obtain KAM theorems for nonautonomous differential equations *quasi-periodic* in time, i.e., those of the form $\dot{u} = F(u, t\omega_1, t\omega_2, \ldots, t\omega_n)$ with $u \in \mathbb{R}^N$ and $F : \mathbb{R}^N \times \mathbb{T}^n \to \mathbb{R}^N$ (cf. a discussion on quasi-periodic forcing in §§ 1.2.3 and 1.5.1). For recent results in the Hamiltonian isotropic context, see [170, 172].

5.2 Density points of quasi-periodicity

Here we proceed to give an application of the dissipative "main" Theorem 2.3 regarding generic integrable families of vector fields with a normally hyperbolic (see [67, 115, 158, 356]) Floquet Diophantine invariant torus. We know already that such a torus is not isolated in the product of the phase space and the parameter space, since it is contained in a "Cantor set" (Whitney-smooth family) of Diophantine invariant tori, see §§ 2.3.1, 2.4.1, and 2.5.2. What will be added in this section is that the corresponding parameter value is a *Lebesgue density point* of the projection of the union of the tori to the parameter space. This result was announced earlier in Broer, Huitema & Takens [62, § 7] (see also [162]).

We confine ourselves with the case where $p = 0$, considering a family of vector fields $X = \{X^\mu\}_\mu$ on $\mathbb{T}^n$ with

$$X^\mu = \sum_{j=1}^{n} [\omega_{0j} + f_j(x, \mu)] \frac{\partial}{\partial x_j} = [\omega_0 + f(x, \mu)] \frac{\partial}{\partial x} \tag{5.4}$$

where $x \in \mathbb{T}^n$, $\mu \in P \subset \mathbb{R}^s$, and $f(x, \mu_0) \equiv 0$ for some $\mu_0 \in P$, so that X^{μ_0} is constant with frequency vector $\omega_0 \in \mathbb{R}^n$. For $s \geq n$ we will show that if ω_0 is Diophantine, then generically μ_0 is a Lebesgue density point of the set of parameter values μ for which X^μ exhibits quasi-periodic dynamics. Without loss of generality, one can put $\mu_0 = 0$.

To this purpose, we first develop an x-independent normal form for X which is formal in the μ-direction, cf. § 4.1.3. The method we will exploit uses induction over the order in μ and closely follows Poincaré [274, 275], also see, e.g., Takens, Broer, Braaksma & Huitema [48, 52, 63, 64, 341]. At each induction step we have to solve a 1-bite small divisor problem as met in Section 1.5. The normalizing transformations $\Phi : \mathbb{T}^n \times P \to \mathbb{T}^n \times P$ will have the form

$$\Phi : (x, \mu) \to (x + \varphi(x, \mu), \mu) \tag{5.5}$$

and will be tangent to the identity transformation to an order in μ suitable for the induction step at hand. We have

Lemma 5.9 (A quasi-periodic normal form, cf. Lemma 4.12) *Let $X = \{X^\mu\}_\mu$ be an analytic family of vector fields (5.4) with $\omega_0 \in \mathbb{R}^n_\gamma$ (see § 1.5.2) and $f(x, 0) \equiv 0$:*

$$|\langle \omega_0, k \rangle| \geq \gamma |k|^{-\tau} \quad \forall k \in \mathbb{Z}^n \setminus \{0\},$$

here $\tau > n-1$ and $\gamma > 0$ are constants, cf. (1.37). Then, given $N \in \mathbb{Z}_+$, there exists an analytic transformation Φ (5.5) such that $\overline{X}^\mu = \Phi_* X^\mu$ has the form

$$\overline{X}^\mu(\xi) = [\omega_0 + \bar{f}(\mu) + O_{N+1}(\mu)] \frac{\partial}{\partial \xi} \tag{5.6}$$

($\xi \in \mathbb{T}^n$) as $\mu \to 0$ uniformly in ξ. Moreover, $\bar{f}(0) = 0$.

Proof. For $N = 0$ there is nothing to prove, so let us assume that the conclusions of the theorem hold for $N-1$. Then $X = \{X^\mu\}_\mu$ has the form

$$X^\mu(x) = \left[\omega_0 + \bar{f}(\mu) + \tilde{f}(x,\mu)\right] \frac{\partial}{\partial x}$$

with $\tilde{f}(x,\mu) = O_N(\mu)$ as $\mu \to 0$ uniformly in x and $\bar{f}(0) = 0$. The desired normalizing transformation Φ now will have the form (5.5) with

$$\varphi(x,\mu) = \mu^N u(x,\mu),$$

where the right-hand side abbreviates the multi-index expression

$$\sum_{|l|=l_1+l_2+\cdots+l_s=N} \mu_1^{l_1}\mu_2^{l_2}\cdots\mu_s^{l_s} u_{l_1,l_2,\ldots,l_s}(x,\mu), \qquad l \in \mathbb{Z}_+^s.$$

Writing $\xi = x + \varphi(x,\mu)$, we obtain

$$\begin{aligned} \dot{\xi} &= \left(id + \frac{\partial\varphi(x,\mu)}{\partial x}\right)\dot{x} \\ &= \left(id + \frac{\partial\varphi(x,\mu)}{\partial x}\right)\left[\omega_0 + \bar{f}(\mu) + \tilde{f}(x,\mu)\right] \\ &= \omega_0 + \bar{f}(\mu) + \tilde{f}(x,\mu) + \frac{\partial\varphi(x,\mu)}{\partial x}\omega_0 + O_{N+1}(\mu) \end{aligned}$$

as $\mu \to 0$ uniformly in x. The conclusion of the theorem now holds for N if

$$\frac{\partial\varphi(x,\mu)}{\partial x}\omega_0 + \tilde{f}(x,\mu) \equiv \mu^N c \pmod{O_{N+1}(\mu)} \quad \text{as } \mu \to 0$$

for a suitable constant vector c. Setting $\tilde{f}(x,\mu) = \mu^N g(x,\mu)$, we arrive at the following equation in u and c:

$$\frac{\partial u(x,\mu)}{\partial x}\omega_0 \equiv -g(x,\mu) + c \pmod{O(\mu)} \quad \text{as } \mu \to 0,$$

to be understood monomial-wise as above. This is another 1-bite small divisor problem, cf. Section 1.5. If

$$g(x,\mu) = \sum_{k\in\mathbb{Z}^n} g_k(\mu) e^{i\langle x,k\rangle}$$

then $c = g_0(0)$ while

$$u(x,\mu) \equiv h(\mu) - \sum_{k\neq 0} \frac{g_k(\mu)}{i\langle\omega_0,k\rangle} e^{i\langle x,k\rangle} \pmod{O(\mu)}$$

for an arbitrary x-independent vector $h = h(\mu)$, cf. (1.35)–(1.36). □

Remark 1. We took $\omega_0 \in R^n_\gamma$ without specifying the constants $\tau > n-1$ and $\gamma > 0$ from the definition of R^n_γ, see § 1.5.2 (note that τ enters the definition of R^n_γ implicitly). In this case there is no harm in taking ω_0 from the union of sets R^n_γ for all possible τ and γ.

Remark 2. One can easily adapt the above proof so that it preserves certain structures, like the symplectic structure or the action of a symmetry group. This approach uses graded Lie algebras, compare Broer, Braaksma & Huitema [48, 52].

We now arrive at the main result of this section. Observe that form (5.6) is nearly integrable, the perturbation being controlled by μ. Taking $N \geq 1$, let us apply the "main" Theorem 2.3 in the dissipative context (also consult Theorem 6.1 in Section 6.1) and see what its consequences are.

The fact that 0 is a Lebesgue density point of the parameter set with quasi-periodic tori can be expressed as follows (cf. [114, 242]). For any $\epsilon_* > 0$, there exists an arbitrarily small neighborhood Σ of 0 in P such that the subset $\Sigma_{\text{q-p}} \subset \Sigma$, defined by

$$\Sigma_{\text{q-p}} = \{\mu \in \Sigma \,:\, T^n \times \{\mu\} \text{ is } X^\mu\text{-quasi-periodic}\},$$

has relative measure at least $1 - \epsilon_*$ in the sense that

$$meas_s\, \Sigma_{\text{q-p}} \geq (1-\epsilon_*)\, meas_s\, \Sigma.$$

Theorem 5.10 (A density point of quasi-periodicity) *Suppose that the Jacobi matrix $\partial \bar{f}(0)/\partial\mu$ is of rank n (so that $s \geq n$). Then the point $\mu = 0$ is a Lebesgue density point of the set*

$$\{\mu \in P \,:\, T^n \times \{\mu\} \text{ is } X^\mu\text{-quasi-periodic}\}.$$

Proof. Consider the integrable truncation

$$Y^\mu(x) = [\omega_0 + \bar{f}(\mu)]\,\frac{\partial}{\partial x}$$

of (5.6). Using the submersivity of $\bar{f}$ at 0, we carry out a local reparametrization $\mu \mapsto (\sigma(\mu), \nu(\mu))$, where $\sigma(\mu) = \omega_0 + \bar{f}(\mu)$, and for simplicity suppress ν.

Now we need the so called *small twist* version of Theorem 2.3, cf. § 4.1.3. To be precise, consider *in abstracto* two vector fields

$$Y^\sigma(x) = \sigma\frac{\partial}{\partial x} \quad \text{and} \quad \tilde{Y}^\sigma(x) = [\sigma + g(x,\sigma)]\frac{\partial}{\partial x}$$

for any small g with σ Diophantine (namely, $\sigma \in R^n_\gamma$):

$$|\langle \sigma, k\rangle| \geq \gamma |k|^{-\tau} \quad \forall k \in Z^n \setminus \{0\},$$

here $\tau > n-1$ and $\gamma > 0$ are constants, cf. (1.37). If we now apply *carefully* the "main" result of the dissipative context, taking into account all the estimates contained in the proof of Theorem 2.3 (or Theorem 6.1), to this situation, we will get the following. Given a suitable neighborhood Γ of ω_0 in R^n, we consider a complex neighborhood

$$O = \text{cl}\,((T^n + \kappa) \times (\Gamma + \rho))$$

of $\mathrm{T}^n \times \Gamma$, determined by $\kappa > 0$ and $0 < \rho \leq 1$, see § 1.5.2 for the notations. Then, if $\gamma \leq \rho$, we find a certain constant $\delta_* > 0$ independent of Γ, ρ, and γ (cf. § 6.1.2) which determines a set of "admissible" real analytic terms g as

$$|g|_O := \max_{(x,\sigma)\in O} |g| < \gamma\delta_*. \tag{5.7}$$

For each "admissible" g, the corresponding perturbation $\widetilde{Y} = \widetilde{Y}^\sigma(x)$ possesses a "Cantor set" (Whitney-smooth family) of quasi-periodic tori whose projection on the parameter space is a C^∞-diffeomorphic image of the set $\Gamma'_\gamma = \Gamma' \cap \mathrm{R}^n_\gamma$ where $\Gamma' = \{\sigma \in \Gamma : \mathrm{dist}(\sigma, \partial\Gamma) \geq \gamma\}$, see [62, 162] ($\partial\Gamma$ is the boundary of Γ). Moreover, this image is a C^∞-small perturbation of Γ'_γ and still contained in Γ. For Γ we may take $\Gamma = \omega_0 + \Delta$, where Δ is some ball centered at $0 \in \mathrm{R}^n$, and one easily shows that

$$meas_n\, \Gamma'_\gamma = [1 - O(\gamma R^{-1})]\, meas_n\, \Gamma \tag{5.8}$$

as $\gamma R^{-1} \downarrow 0$ uniformly in the radius R of Δ, cf. [337]. Moreover, for the perturbed parameter set this estimate also holds, with the same remark on the uniformity.

In the present case, the perturbation term g is of a special kind since $|g(x,\sigma)| = O(|\sigma - \omega_0|^{N+1})$ as $\sigma \to \omega_0$. In order to exploit this, for any small $\epsilon_* > 0$ we let $\Sigma^{\epsilon_*} = \omega_0 + \epsilon_*\Delta$ play the above rôle of Γ and also take $\rho = \epsilon_*$, the ball Δ being fixed. Then for $\kappa > 0$ we get a complex neighborhood

$$O_{\epsilon_*} = \mathrm{cl}\left((\mathrm{T}^n + \kappa) \times (\Sigma^{\epsilon_*} + \epsilon_*)\right).$$

Next we set $\gamma = \epsilon_*^N/\delta_*$ ($\leq \epsilon_*$ for ϵ_* sufficiently small provided that $N \geq 2$), where $\delta_* = \delta_*(\kappa)$ is fixed *as above*! The condition (5.7) on g then becomes

$$|g|_{O_{\epsilon_*}} < \epsilon_*^N$$

which now is satisfied for a suitable κ and all $\epsilon_* > 0$ sufficiently small. The estimate (5.8) and the subsequent remark moreover give

$$meas_n(\Sigma^{\epsilon_*}_{\mathrm{q-p}}) \geq [1 - O(\epsilon_*^{N-1})]\, meas_n(\Sigma^{\epsilon_*}) \tag{5.9}$$

as $\epsilon_* \downarrow 0$. This does imply that $\sigma = \omega_0$ is a density point of the parameter set in R^n with X-quasi-periodic tori, see the beginning of this proof. □

Remark 1. For a similar result see Pöschel [278, Sect. 5, Cor. 3]. There it is shown that in the Hamiltonian setting an elliptic equilibrium generically is a density point of the union of quasi-periodic Lagrangian tori. An even stronger result has been obtained using analyticity by Delshams & Gutiérrez [102, 104], see § 4.1.5.

Remark 2. It is easy to generalize Theorem 5.10 to various contexts, e.g., to the Hamiltonian isotropic $(n, 0, s)$ context (see the previous remark).

Remark 3. One may wonder what happens if we apply the Nekhoroshev–Lochak–Neĭshtadt optimality [209, 210] of N to given ϵ_* (cf. § 4.2.3 above)? Probably estimate (5.9) will become infinitely flat in ϵ_* as $\epsilon_* \downarrow 0$. It may be even more useful to invoke the Morbidelli–Giorgilli optimality [239] (cf. § 4.2.4 above).

5.3 Invariant tori and stability of motion

The existence of Whitney-smooth families of invariant tori in the phase space of a dynamical system often leads to certain stability properties of the motion (not to be confused with the *structural* stability we have mainly dealt with in this book). For instance, the action variables $y(t)$ in a nearly integrable Hamiltonian system with two degrees of freedom remain close to their initial values $y(0)$ for ever provided that the unperturbed Hamilton function $H^0(y)$ is isoenergetically nondegenerate. In the case of $n \geq 3$ degrees of freedom, the evolution of the action variables takes place generically but its rate is exponentially small with the perturbation magnitude provided that $H^0(y)$ is steep (see § 4.2.3). Moreover, the existence of invariant tori often results in the stability of *critical elements* [1] of the system in question, such as equilibria or periodic trajectories. This was discovered at the very beginning of the history of the KAM theory [3–5, 243]. It turns out that

a) a generic elliptic equilibrium 0 of a Hamiltonian vector field with two degrees of freedom is Lyapunov stable (i.e., for any $\sigma > 0$ there exists a $\varrho > 0$ such that an orbit starting at a distance $\leq \varrho$ from 0 remains for ever at a distance $\leq \sigma$ from 0);

b) a generic elliptic periodic trajectory of a Hamiltonian vector field with two degrees of freedom is stable (in the same sense as a generic elliptic equilibrium);

c) an elliptic G-invariant equilibrium of a vector field in $\mathbb{R}^4$ reversible with respect to involution G of type (2, 2) is Lyapunov stable "with probability 1/2" (for a precise meaning of this expression, see below in § 5.3.2).

The key idea in the proofs of all these statements is to apply the elliptic normal modes excitation theorem to an equilibrium or periodic trajectory in question. We will discuss in detail elliptic equilibria of Hamiltonian and reversible vector fields in $\mathbb{R}^4$.

5.3.1 An elliptic equilibrium of a two-degree-of-freedom Hamiltonian vector field

Let $0 \in \mathbb{R}^4$ be an equilibrium point of a Hamiltonian vector field X governed by Hamilton function H. One can assume without loss of generality that $H(0) = 0$. Suppose that the eigenvalues of the linearization of X at 0 are $\pm\varepsilon_1$, $\pm\varepsilon_2$ where $\varepsilon_1 > 0$, $\varepsilon_2 > 0$. If $\langle \varepsilon, k\rangle \neq 0$ for $k \in \mathbb{Z}^2$, $1 \leq |k| \leq 4$ (i.e., $\varepsilon_1 \neq \varepsilon_2$, $\varepsilon_1 \neq 2\varepsilon_2$, $\varepsilon_1 \neq 3\varepsilon_2$, $2\varepsilon_1 \neq \varepsilon_2$, and $3\varepsilon_1 \neq \varepsilon_2$), then one can choose a coordinate system (u_1, v_1, u_2, v_2) in $\mathbb{R}^4$ centered at 0 in which the symplectic structure gets the form $du_1 \wedge dv_1 + du_2 \wedge dv_2$ while the Hamiltonian H takes the fourth-order Birkhoff normal form

$$H(u, v) = \textstyle\sum_{i=1}^2 \alpha_i \eta_i + \sum_{i,j=1}^2 \beta_{ij}\eta_i\eta_j + O_5(u, v), \tag{5.10}$$

where $\eta_i = \frac{1}{2}(u_i^2 + v_i^2)$, $|\alpha_i| = \varepsilon_i$ $(i = 1,\ 2)$, and $\beta_{12} = \beta_{21}$ [41, 70, 71, 75, 248, 335]. If $\alpha_1\alpha_2 > 0$ then the Hamilton function H attains a local extremum at the origin, and the latter is Lyapunov stable due to obvious reasons. The case interesting from the stability viewpoint is $\alpha_1\alpha_2 < 0$. To analyze the stability of the origin for $\alpha_1\alpha_2 < 0$, introduce angular coordinates $(\chi_1, \chi_2) \in \mathbb{T}^2$ via the formulas

$$u_i = \sqrt{2\eta_i}\cos\chi_i, \quad v_i = \sqrt{2\eta_i}\sin\chi_i, \quad i = 1,\ 2, \tag{5.11}$$

then the symplectic structure takes the form $d\eta_1 \wedge d\chi_1 + d\eta_2 \wedge d\chi_2$. Consider the truncated Hamiltonian

$$H^0(\eta) = \sum_{i=1}^2 \alpha_i \eta_i + \sum_{i,j=1}^2 \beta_{ij}\eta_i\eta_j$$

which determines the vector field

$$X^0 = \sum_{i=1}^2 [\alpha_i + 2(\beta_{i1}\eta_1 + \beta_{i2}\eta_2)] \frac{\partial}{\partial \chi_i}.$$

Each set $\{\eta_i = const \geq 0 \ (i = 1,\ 2)\}$ is an invariant torus of the field X^0 with parallel dynamics. For $\eta_1 > 0$, $\eta_2 > 0$ these tori are two-dimensional and their frequency vector is $\omega = \omega(\eta) = (\omega_1, \omega_2)$,

$$\omega_i(\eta) = \alpha_i + 2(\beta_{i1}\eta_1 + \beta_{i2}\eta_2).$$

On the (η_1, η_2)-plane, we therefore obtain two one-parameter families of curves: the energy level conics $\{H^0(\eta) = const\}$ and the straight lines $\{\omega_1(\eta)/\omega_2(\eta) = (\alpha_1/\alpha_2) + const\}$ corresponding to tori with a fixed rotation number. The straight line $\{\omega_1(\eta)/\omega_2(\eta) = \alpha_1/\alpha_2\}$ passes through the point $(0, 0)$.

Remark. The equation

$$\omega_1(\eta)/\omega_2(\eta) = \alpha_1/\alpha_2 \quad \Longleftrightarrow \quad (\beta_{11}\alpha_2 - \beta_{21}\alpha_1)\eta_1 + (\beta_{12}\alpha_2 - \beta_{22}\alpha_1)\eta_2 = 0$$

does define a straight line if and only if $(\beta_{11}\alpha_2 - \beta_{21}\alpha_1)^2 + (\beta_{12}\alpha_2 - \beta_{22}\alpha_1)^2 > 0$. Here we assume this inequality to be satisfied.

The tangent to the curve $\{H^0(\eta) = 0\}$ at the point $(0, 0)$ has the equation $\alpha_1\eta_1 + \alpha_2\eta_2 = 0$. This tangent coincides therefore with the straight line $\{\omega_1(\eta)/\omega_2(\eta) = \alpha_1/\alpha_2\}$ if and only if $Q \neq 0$ where

$$Q := \alpha_1(\beta_{12}\alpha_2 - \beta_{22}\alpha_1) - \alpha_2(\beta_{11}\alpha_2 - \beta_{21}\alpha_1) = \begin{vmatrix} \beta_{11} & \beta_{12} & \alpha_1 \\ \beta_{21} & \beta_{22} & \alpha_2 \\ \alpha_1 & \alpha_2 & 0 \end{vmatrix}$$

$$= -(\beta_{11}\alpha_2^2 - 2\beta_{12}\alpha_1\alpha_2 + \beta_{22}\alpha_1^2).$$

Note that the condition $(\beta_{11}\alpha_2 - \beta_{21}\alpha_1)^2 + (\beta_{12}\alpha_2 - \beta_{22}\alpha_1)^2 > 0$ mentioned above is implied by the condition $Q \neq 0$. Note also that this condition is equivalent to the following property of normal form (5.10): the sum of fourth-degree terms $\sum_{i,j=1}^2 \beta_{ij}\eta_i\eta_j$ is not divisible by the quadratic part $\sum_{i=1}^2 \alpha_i\eta_i$.

We have proven the following statement.

Lemma 5.11 *If $Q \neq 0$ then in the first quadrant $\{\eta_1 > 0, \eta_2 > 0\}$ of the (η_1, η_2)-plane near the point $(0, 0)$, the straight lines $\{\omega_1(\eta)/\omega_2(\eta) = (\alpha_1/\alpha_2) + const\}$ intersect the curves $\{H^0(\eta) = const\}$ transversally.*

Now return to the "full" Hamiltonian (5.10) governing vector field X. According to the elliptic normal modes excitation theorem (the Hamiltonian analogue of Theorem 4.3), vector field X possesses Diophantine invariant 2-tori close to the tori $\{\eta = const\}$ in any neighborhood of the origin, provided that certain nonresonance and nondegeneracy conditions are met. It turns out that the conditions $\langle \varepsilon, k\rangle \neq 0$ for $k \in \mathbb{Z}^2$, $1 \leq |k| \leq 4$, and $Q \neq 0$ are sufficient. Moreover, these tori are organized into analytic one-parameter families consisting of tori with a fixed rotation number. The condition $Q \neq 0$ guarantees that these families are "transversal" to the energy level hypersurfaces $\{H = const\}$ and that each hypersurface contains a "large" set of tori. Every torus T traps the trajectories on the corresponding hypersurface that start nearer the origin than T, see Figure 5.1[1] (recall that the tori are two-dimensional while the energy level hypersurfaces are three-dimensional). Thus, the origin turns out to be Lyapunov stable.

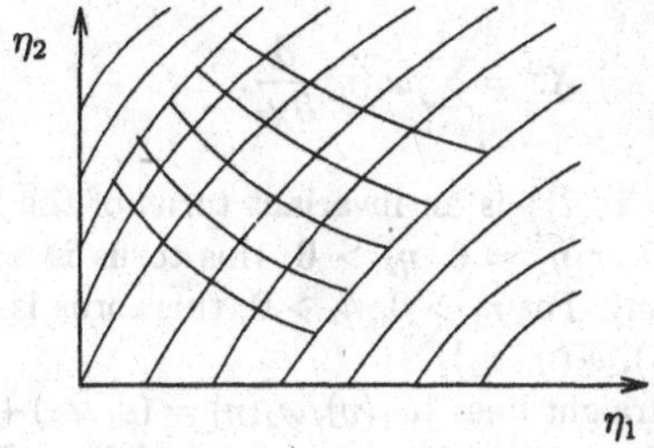

Figure 5.1: Invariant 2-tori near an elliptic equilibrium of a two-degree-of-freedom Hamiltonian system (schematically): the single foliation indicates the hypersurfaces $\{H = const\}$, the intersection of two foliations represents the invariant 2-tori.

In fact, almost no extra work is needed to convert these simple geometric arguments to a rigorous proof of the following theorem.

Theorem 5.12 (Arnol'd [5], see also Moser [247] and Siegel & Moser [335] and a discussion in [70, 164]) *If $\langle \varepsilon, k \rangle \neq 0$ for $k \in \mathbb{Z}^2$, $1 \leq |k| \leq 4$, and $Q \neq 0$ then the origin is a Lyapunov stable equilibrium of the vector field X associated to Hamilton function (5.10).*

Remark. The condition $Q \neq 0$ is the local analogue of the isoenergetic nondegeneracy in two-degree-of-freedom integrable Hamiltonian systems (cf. § 4.2.3).

5.3.2 An elliptic equilibrium of a reversible vector field in $\mathbb{R}^4$

Let now $0 \in \mathbb{R}^4$ be an equilibrium point of a vector field X reversible with respect to involution G of type $(2, 2)$. The origin 0 is assumed to be a fixed point of this involution. Suppose that the eigenvalues of the linearization of X at 0 are $\pm\varepsilon_1$, $\pm\varepsilon_2$ where $\varepsilon_1 > 0$, $\varepsilon_2 > 0$. If $\langle \varepsilon, k \rangle \neq 0$ for $k \in \mathbb{Z}^2$, $1 \leq |k| \leq 4$, then one can choose a coordinate system (u_1, v_1, u_2, v_2) in $\mathbb{R}^4$ centered at 0 in which the reversing involution gets the form $G : (u, v) \mapsto (u, -v)$ while the differential equations determined by the field X take the third-order Birkhoff normal form

$$\dot{u}_i = -v_i \omega_i(\eta) + O_4(u, v), \qquad \dot{v}_i = u_i \omega_i(\eta) + O_4(u, v) \tag{5.12}$$

with

$$\omega_i(\eta) = \varepsilon_i + \gamma_{i1}\eta_1 + \gamma_{i2}\eta_2$$

[1]Cf. Moser [247] and Il'yashenko [164].

($i = 1,\ 2$) and $\eta_i = \frac{1}{2}(u_i^2 + v_i^2)$ [248, 315]. Consider the truncated vector field X^0 which determines the differential equations

$$\dot{u}_i = -v_i\omega_i(\eta), \qquad \dot{v}_i = u_i\omega_i(\eta)$$

($i = 1,\ 2$), this field is still reversible with respect to G. Introduce angular coordinates $(\chi_1, \chi_2) \in \mathrm{T}^2$ via formulas (5.11). In the polar coordinate system (η, χ), involution G gets the form $G : (\eta, \chi) \mapsto (\eta, -\chi)$ while the vector field X^0 takes the form

$$X^0 = \sum_{i=1}^{2} \omega_i(\eta)\frac{\partial}{\partial \chi_i}.$$

Each set $\{\eta_i = const \geq 0\ (i = 1,\ 2)\}$ is an invariant torus of the field X^0 with parallel dynamics. For $\eta_1 > 0,\ \eta_2 = 0$ or $\eta_1 = 0,\ \eta_2 > 0$, this torus is a circle with frequency $\omega_1(\eta_1, 0)$ or $\omega_2(0, \eta_2)$, respectively. For $\eta_1 > 0,\ \eta_2 > 0$, this torus is two-dimensional with frequency vector $\omega = (\omega_1(\eta_1, \eta_2), \omega_2(\eta_1, \eta_2))$.

On the (η_1, η_2)-plane, the straight lines $\{\omega_1(\eta)/\omega_2(\eta) = (\varepsilon_1/\varepsilon_2) + const\}$ correspond to tori with a fixed rotation number. There occur two essentially different generic cases of the arrangement of these straight lines with respect to the axes $\{\eta_2 = 0\}$ and $\{\eta_1 = 0\}$ near the point $(0, 0)$, depending on the sign of

$$\Delta := (\varepsilon_2\gamma_{11} - \varepsilon_1\gamma_{21})(\varepsilon_2\gamma_{12} - \varepsilon_1\gamma_{22}),$$

as is shown in Figure 5.2.

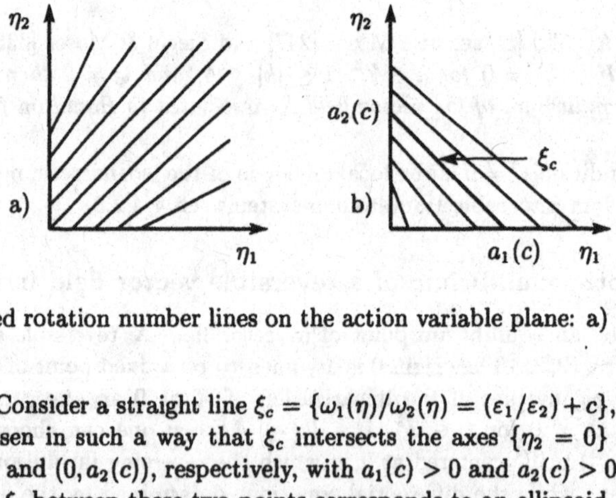

Figure 5.2: Fixed rotation number lines on the action variable plane: a) for $\Delta < 0$ and b) for $\Delta > 0$.

Let $\Delta > 0$. Consider a straight line $\xi_c = \{\omega_1(\eta)/\omega_2(\eta) = (\varepsilon_1/\varepsilon_2) + c\}$, the sign of $c \neq 0$ small being chosen in such a way that ξ_c intersects the axes $\{\eta_2 = 0\}$ and $\{\eta_1 = 0\}$ at points $(a_1(c), 0)$ and $(0, a_2(c))$, respectively, with $a_1(c) > 0$ and $a_2(c) > 0$ (see Figure 5.2). The segment of ξ_c between these two points corresponds to an ellipsoid

$$[\varepsilon_2\gamma_{11} - \varepsilon_1\gamma_{21} + O(c)]\eta_1 + [\varepsilon_2\gamma_{12} - \varepsilon_1\gamma_{22} + O(c)]\eta_2 = b(c)$$

in the phase space R^4. The ellipsoid consists of a continuum of 2-tori and two circles, each torus and circle being invariant under the flow of X^0.

Now return to the "full" vector field X determining equations (5.12). According to the elliptic normal modes excitation theorem (Theorem 4.3), vector field X possesses Diophantine invariant 2-tori close to the tori $\{\eta = const\}$ in any neighborhood of the origin, provided that certain nonresonance and nondegeneracy conditions are met. It turns out that the conditions $\langle \varepsilon, k \rangle \neq 0$ for $k \in \mathrm{Z}^2$, $1 \leq |k| \leq 4$, and $\Delta \neq 0$ are sufficient. Moreover, these tori are organized into analytic one-parameter families consisting of tori with a fixed rotation number. On the other hand, the condition $\varepsilon_1/\varepsilon_2 \notin \mathrm{N}$, $\varepsilon_2/\varepsilon_1 \notin \mathrm{N}$ guarantees that X admits two analytic one-parameter families of invariant circles with periods close to $2\pi/\varepsilon_1$ and $2\pi/\varepsilon_2$ (Theorem 4.5). Clearly, this condition implies also that $\langle \varepsilon, k \rangle \neq 0$ for $k \in \mathrm{Z}^2$, $1 \leq |k| \leq 4$.

It turns out that under the conditions $\varepsilon_1/\varepsilon_2 \notin \mathrm{N}$, $\varepsilon_2/\varepsilon_1 \notin \mathrm{N}$, $\Delta > 0$, in any neighborhood of the origin, vector field X possesses invariant sets consisting of a continuum of Diophantine invariant 2-tori with a fixed rotation number and two periodic trajectories. Each set is diffeomorphic to S^3 and bounds a certain neighborhood of the origin. The latter is therefore Lyapunov stable. We will formulate this statement as a theorem.

Theorem 5.13 (Matveev [225–228]) *If $\varepsilon_1/\varepsilon_2 \notin \mathrm{N}$, $\varepsilon_2/\varepsilon_1 \notin \mathrm{N}$, and $\Delta > 0$ then the origin is a Lyapunov stable equilibrium of the reversible vector field X determining differential equations (5.12).*

The rôle of the condition $\Delta > 0$ (clear in fact from the discussion above) is emphasized by the following remarkable result.

Theorem 5.14 (Matveev [225–227]) *Let $\langle \varepsilon, k \rangle \neq 0$ for $k \in \mathrm{Z}^2$, $1 \leq |k| \leq 4$, and $\Delta < 0$. Then for any $K \in \mathrm{N}$ there exists a G-reversible vector field $\widetilde{X}$ whose K-jet at the origin is arbitrarily close to the K-jet of the field X, the origin being a Lyapunov unstable equilibrium of $\widetilde{X}$.*

Recall that the K-jet means the Taylor expansion up to order K.

Chapter 6

Appendices

6.1 Proof of the dissipative Main Theorem

In this section, we give a complete proof for a special case of the "main" Theorem 2.3 in the dissipative context, namely where there are no normal coordinates ($p = 0$). We briefly discussed a further simplified situation in § 1.2.1 which concerned 2-tori and was based on circle maps. However, our proof is characteristic for all the other contexts mentioned throughout. For a similar proof in the Hamiltonian setting [the Hamiltonian isotropic $(n, 0, 0)$ context], see Pöschel [278].

6.1.1 Formulation of the stability theorem

We will repeat some material of §§ 1.3.2 and 2.3.1 in a slightly another language. Let M be a finite dimensional manifold and P an open domain of a finite dimensional vector space. We consider analytic families $X = \{X^\mu\}_\mu = \{X^\mu(x)\}_\mu$ of analytic vector fields on M parametrized over P, i.e., with $(x, \mu) \in M \times P$. We sometimes regard such a family as a "vertical" vector field on $M \times P$, see §§ 1.2.1 and 1.3.1. Assume a family $X = \{X^\mu\}_\mu$ to possess the following properties:

(i) for all $\mu \in P$, the vector field X^μ has a normally hyperbolic (see [67, 115, 158, 356]) invariant n-torus V_μ depending analytically on μ;

(ii) the dynamics of X^μ on V_μ is parallel for all $\mu \in P$.

If V_μ is Floquet, then the normal hyperbolicity of V_μ means in fact that none of the eigenvalues of the corresponding Floquet matrix is purely imaginary, which is a generic situation. Now the question is what happens to $V = \bigcup_{\mu \in P} V_\mu \times \{\mu\} \subset M \times P$ and its (parallel or even quasi-periodic) dynamics under perturbations to nearby families $\widetilde{X} = \{\widetilde{X}^\mu\}_\mu$. Similarly to examples in § 1.2.1 we here may restrict ourselves to the case where $M = \mathbb{T}^n$. In fact by the center manifold theory, cf. Fenichel [115] or Hirsch, Pugh & Shub [158, Theorem 4.1], V as an invariant manifold always persists under such perturbations (see also [67, 356]), but probably with some loss of differentiability, cf. §§ 1.2.1 and 1.4.1. Consequently, one can always confine oneself with $M = \mathbb{T}^n$ in the finitely differentiable category (moreover, the parallelity of the dynamics may enable one to establish infinite differentiability of the perturbed tori). Nevertheless, we do consider the case $M = \mathbb{T}^n$ and still assume analyticity in the dependence on both x and μ. This simplification is not harmful, since for sufficiently smooth vector fields similar results hold, see [62, 162] (cf. also [151, 243, 277, 278, 306]).

The unperturbed family X of vector fields on T^n has the integrable form

$$X^\mu(x) = \omega(\mu)\frac{\partial}{\partial x}$$

in appropriate coordinates $x = (x_1, x_2, \ldots, x_n)$ modulo 2π. For given $\mu_0 \in P$, we say that X is *nondegenerate* at the torus $\mathrm{T}^n \times \{\mu_0\} \subset \mathrm{T}^n \times P$ if the frequency mapping $\omega : P \to \mathrm{R}^n$ at μ_0 has a surjective derivative (so that $\dim P \geq n$). Now recall from § 1.5.2 and Section 5.2 that for given constants $\tau > n - 1$ (fixed forever) and $\gamma > 0$,

$$\mathrm{R}^n_\gamma := \{\omega \in \mathrm{R}^n \ : \ |\langle \omega, k\rangle| \geq \gamma |k|^{-\tau} \ \ \forall k \in \mathrm{Z}^n \setminus \{0\}\}. \qquad (6.1)$$

If $\Gamma \subset P$ is any subset of P we also write $\Gamma_\gamma = \Gamma \cap \omega^{-1}(\mathrm{R}^n_\gamma)$. If the restriction of the mapping ω to Γ is a submersion then Γ_γ is a Whitney-smooth foliation of smooth manifolds (with boundary) parametrized over a Cantor set. As before (again see § 1.5.2), we shall call such a foliation a "Cantor set". Observe that for $\gamma > 0$ sufficiently small the set $\Gamma_\gamma \subset P$ has positive measure. According to the Inverse Function Theorem, these considerations apply for a sufficiently small neighborhood Γ of μ_0 in P whenever the family X is nondegenerate at the torus $\mathrm{T}^n \times \{\mu_0\}$.

In the perturbation result to follow, the topology from which the neighborhood $\mathcal{X}$ of X will be taken is the *compact-open topology* on the holomorphic extensions of our analytic vector fields.

Theorem 6.1 (Quasi-periodic stability: an example, cf. [62, 162]**)** *For $\mu_0 \in P$, let the analytic integrable family $X = \{X^\mu\}_\mu$ of vector fields on T^n be nondegenerate at the torus $\mathrm{T}^n \times \{\mu_0\}$. Then there exist a neighborhood Γ of μ_0 in P and a neighborhood $\mathcal{X}$ of X in the class of all analytic "vertical" vector fields on $\mathrm{T}^n \times P$ such that for all $\tilde{X} \in \mathcal{X}$, a mapping $\Phi : \mathrm{T}^n \times P \to \mathrm{T}^n \times P$ exists with the following properties:*

(i) *the restriction of Φ to $\mathrm{T}^n \times \Gamma$ is a C^∞ diffeomorphism onto its image, C^∞-near-the-identity map. Moreover, Φ preserves the projection to P and is analytic in $x \in \mathrm{T}^n$;*

(ii) *the further restriction of Φ to $\mathrm{T}^n \times \Gamma_\gamma$ conjugates X to $\tilde{X}$.*

Before we begin the proof, let us see what Theorem 6.1 means. Clearly $\Phi(\mathrm{T}^n \times \Gamma_\gamma)$ is a "Cantor set" consisting of $\tilde{X}$-Diophantine tori. We say that the family X is *quasi-periodically stable*: the restriction $X|_{\mathrm{T}^n \times \Gamma_\gamma}$ is structurally stable (compare the notion of Ω-stability which is the structural stability restricted to the nonwandering set Ω [1]). Moreover, since Φ is at least a C^1-diffeomorphism, the set $\Phi(\mathrm{T}^n \times \Gamma_\gamma)$ again has positive measure. We conclude that in this setting it is a typical or persistent property for families of vector fields on T^n to have such a "Cantor set" of Diophantine tori with positive measure, cf. § 1.4.1. A better formulation, carrying over to the general normally hyperbolic case, projects the union of tori onto the parameter space, there obtaining a set of positive measure.

6.1.2 Preliminary remarks

The vector norms. As in the proof of Theorem 1.7 (see § 1.5.2), we will denote by $|\cdot|$ the maximum (l_∞) norm of independent variables ranging in $\mathbf{R}^n$ and $\mathbf{C}^n$ (so that $|x| = \max_{m=1}^n |x_m|$, $|\omega| = \max_{m=1}^n |\omega_m|$), but *not* in $\mathbf{Z}^n$, where $|\cdot|$ will still denote the l_1-norm: $|k| = \sum_{m=1}^n |k_m|$. On the other hand, for a mapping $f : D \to \mathbf{C}^n$ with $D \subset \mathbf{C}^N$ we will write $|f(\chi)| := \sum_{m=1}^n |f_m(\chi)|$ and

$$\left|\frac{\partial f(\chi)}{\partial \chi}\right| := \max_{\iota=1}^{N} \sum_{m=1}^{n} \left|\frac{\partial f_m(\chi)}{\partial \chi_\iota}\right|.$$

The advantage of this notation system is that if

$$f = \frac{\partial U}{\partial \chi} g$$

with $U : \mathbf{C}^N \to \mathbf{C}^n$ and $g : \mathbf{C}^N \to \mathbf{C}^N$ (i.e., $f = U_* g$ in the case where f and g represent vector fields in local coordinates) then

$$|f| \le \left|\frac{\partial U}{\partial \chi}\right| |g|.$$

Indeed,

$$\begin{aligned} |f| = \sum_{m=1}^{n} |f_m| &= \sum_{m=1}^{n} \left| \sum_{\iota=1}^{N} \frac{\partial U_m}{\partial \chi_\iota} g_\iota \right| \le \sum_{\iota=1}^{N} \left(|g_\iota| \sum_{m=1}^{n} \left| \frac{\partial U_m}{\partial \chi_\iota} \right| \right) \\ &\le \left(\max_{\iota=1}^{N} \sum_{m=1}^{n} \left| \frac{\partial U_m}{\partial \chi_\iota} \right| \right) \sum_{\iota=1}^{N} |g_\iota| = \left| \frac{\partial U}{\partial \chi} \right| |g|. \end{aligned}$$

Moreover, we will write $|f|_D = \sup_{\chi \in D} |f(\chi)|$ whenever this supremum is less than $+\infty$.

Frequencies as parameters. We first reduce our problem to a special case. By the Inverse Function Theorem, near $\mu_0 \in P$ there exists a reparametrization $\mu \mapsto (\omega(\mu), \nu(\mu))$ such that

$$X^{\omega,\nu}(x) = \omega \frac{\partial}{\partial x}.$$

From now on we drop the parameter ν: it turns out that in the proof below it can be easily incorporated again. In this way P is replaced by (an open domain of) $\mathbf{R}^n$.

A real analytic neighborhood. In this reduced setting, let us be more explicit on the form of a real analytic neighborhood $\mathcal{X}$ of X. Let Γ be some compact neighborhood of $\omega_0 = \omega(\mu_0)$ in $\mathbf{R}^n$ and O some compact neighborhood of $\mathbf{T}^n \times \Gamma$ in $(\mathbf{C}/2\pi\mathbf{Z})^n \times \mathbf{C}^n$. Without loss of generality we take O of the form

$$O = \mathrm{cl}\,((\mathbf{T}^n + \kappa) \times (\Gamma + \rho)) \tag{6.2}$$

with constants $\kappa > 0$ and $0 < \rho \le 1$, see § 1.5.2 for the notations. For sufficiently small Γ, κ, and ρ, obviously X has a holomorphic extension to such a set O. Now $\mathcal{X}$ is a compact-open neighborhood of X (intersected with the space of real analytic vector fields we work with) determined by O, γ and a positive constant δ given by the proof below (cf. [65]).

Here $\gamma > 0$ enters the definition (6.1) of R^n_γ and hence that of Γ_γ. To be precise, $\mathcal{X}$ consists of all (families of) vector fields $\tilde{X} = \{\tilde{X}^\omega\}_\omega$ of the form

$$\tilde{X}^\omega(x) = [\omega + f(x,\omega)]\frac{\partial}{\partial x}$$

with real analytic f, defined on O and such that in the supremum norm on O one has

$$|f|_O < \gamma\delta$$

[cf. (5.7)]. If we take $\gamma \leq \rho$ then this constant δ turns out to be independent of Γ, γ, and ρ. Also we announce that now the definition domain of the mapping Φ in the parameter direction is a slight modification of the present set Γ: the rôle of Γ in the formulation of Theorem 6.1 is played by

$$\Gamma' = \{\omega \in \Gamma \; : \; \mathrm{dist}(\omega, \partial\Gamma) \geq \gamma\} \tag{6.3}$$

which, for $\gamma > 0$ sufficiently small, still is a neighborhood as desired ($\partial\Gamma$ is the boundary of Γ).

The idea of the proof. Next let us briefly discuss the idea of the proof. Recall that our aim is to find a map $\Phi : \mathrm{T}^n \times \Gamma'_\gamma \to \mathrm{T}^n \times P$, preserving the projection to P, which conjugates X to $\tilde{X}$, i.e., such that $\Phi_* X = \tilde{X}$. If we write

$$\Phi(\xi,\sigma) = \left(\xi + \tilde{U}(\xi,\sigma), \sigma + \tilde{\Lambda}(\sigma)\right),$$

this conjugacy property translates to

$$\frac{\partial\tilde{U}(\xi,\sigma)}{\partial\xi}\sigma = \tilde{\Lambda}(\sigma) + f\left(\xi + \tilde{U}(\xi,\sigma), \sigma + \tilde{\Lambda}(\sigma)\right) \tag{6.4}$$

[cf. (2.3)]. This is a nonlinear equation in $\tilde{U}$ and $\tilde{\Lambda}$ that should be solved iteratively by a Newtonian procedure. The corresponding linearizations all are 1-bite small divisor problems of the form

$$\frac{\partial\tilde{U}(\xi,\sigma)}{\partial\xi}\sigma = \tilde{\Lambda}(\sigma) + f(\xi,\sigma) \tag{6.5}$$

[cf. (2.4)] as we met in Section 1.5. For a detailed discussion on this Newtonian process we refer to, e.g., Arnol'd [4, 5, 8, 12], Moser [245, 248], Rüssmann [300], or Celletti & Chierchia [78]. An entirely novel technique for proving KAM theorems, based on diagrammatic tree representations of formal Lindstedt series (see [20, 275]) and refined recombination arguments (approximate cancellations of the resonances), was invented recently (for Hamiltonian systems) by Eliasson [110–112], Gallavotti, Gentile & Mastropietro [121–130], and Chierchia & Falcolini [91, 92, 113]. In some of these papers, the new approach is applied just to the so called Thirring model, where a perturbed Hamilton function has the form $H(x,y,\epsilon) = H^0(y) + \epsilon H^1(x)$ [cf. (4.41)], H^0 being sometimes assumed to be a positive definite quadratic form[1] in y and H^1 an even trigonometric polynomial in x.

[1]in which case all the estimates turn out to be uniform in the minimal eigenvalue of that form, and the results are therefore said to be "twistless" [121–123]

6.1.3 Proof

We now start proving the special case mentioned in the previous subsection where $\omega = \mu$. This proof is divided into three parts. First we give its frame presenting the (Newtonian) iteration process which has to provide us with the desired map Φ. Second we give the estimates we need per iteration step and third we are concerned with the convergence of the process.

Frame of the proof

The iteration process. Given a perturbation $\widetilde{X}$ of X, the map Φ solving (6.4) will be obtained as a Whitney-C^∞ limit of a sequence $\{\Phi_j\}_{j\geq 0}$ of real analytic diffeomorphisms. The main ingredient here is the Inverse Approximation Lemma 6.14, compare Zehnder [364] and Pöschel [278], see also § 1.5.2. In order to describe the inductive construction of the Φ_j we first introduce the following notations.

For $j \in \mathbb{Z}_+$, whenever Φ_j is defined, let $x_j = (x_{j1}, x_{j2}, \ldots, x_{jn})$ modulo 2π and $\omega_j = (\omega_{j1}, \omega_{j2}, \ldots, \omega_{jn})$ be the components of the inverse Φ_j^{-1}:

$$(x_j, \omega_j) \xmapsto{\Phi_j} (x, \omega).$$

Abbreviating $\Phi_j^* := (\Phi_j^{-1})_*$ we define $\widetilde{X}_j = \Phi_j^* \widetilde{X}$. Then Φ_j and $\widetilde{X}_j$ take the forms

$$\Phi_j(x_j, \omega_j) = \left(x_j + \widetilde{U}^j(x_j, \omega_j), \omega_j + \widetilde{\Lambda}^j(\omega_j)\right) \tag{6.6}$$

and

$$\widetilde{X}_j^{\omega_j}(x_j) = [\omega_j + f^j(x_j, \omega_j)]\frac{\partial}{\partial x_j}, \tag{6.7}$$

respectively. In view of the Inverse Approximation Lemma 6.14 (see § 6.1.4), we have to take care that both Φ_j and $\widetilde{X}_j$ possess holomorphic extensions to a complex neighborhood D_j of the "Cantor set" $\mathbb{T}^n \times \Gamma'_\gamma$ with $D_0 \subset O$, the neighborhoods D_j shrinking down as $j \to +\infty$ in an appropriate (geometrical) way. As in the case of Theorem 1.7 (see § 1.5.2), application of Lemma 6.14 gives respectively limits Φ_∞ and $\widetilde{X}_\infty$ of class Whitney-C^∞ on the closed set $\mathbb{T}^n \times \Gamma'_\gamma$. The desired mapping Φ defined on $\mathbb{T}^n \times P$ finally will be obtained by the Whitney Extension Theorem 6.15 (again see § 6.1.4) applied to Φ_∞. In the coordinates x_∞, ω_∞ given by Φ_∞^{-1} the vector field $\widetilde{X}_\infty = \Phi_\infty^*(\widetilde{X})$ will have the form

$$\widetilde{X}_\infty^{\omega_\infty}(x_\infty) = \omega_\infty \frac{\partial}{\partial x_\infty}. \tag{6.8}$$

Granted the details, this indeed would prove Theorem 6.1.

Relation between X_j and X_{j+1}. For the moment suppose that we have a sequence $\{\Phi_j\}_{j\geq 0}$ as described above, with $\Phi_0 = id$. In order to see what happens in the iteration step, we write $\Phi_{j+1} = \Phi_j \circ \Psi_j$ for mappings $\Psi_j : D_{j+1} \to D_j$, $j \in \mathbb{Z}_+$:

$$(x_{j+1}, \omega_{j+1}) \xmapsto{\Psi_j} (x_j, \omega_j) \xmapsto{\Phi_j} (x, \omega) \quad \text{and} \quad (x_{j+1}, \omega_{j+1}) \xmapsto{\Phi_{j+1}} (x, \omega).$$

Then for all $j \in \mathbb{Z}_+$

$$\Phi_{j+1} = \Psi_0 \circ \Psi_1 \circ \ldots \circ \Psi_j \quad \text{and} \quad \widetilde{X}_{j+1} = \Psi_j^*(\widetilde{X}_j) \tag{6.9}$$

where $\Psi_j^* := (\Psi_j^{-1})_*$. Since $\Phi_0 = id$ we have $\widetilde{X}_0 = \widetilde{X}$. Using expressions (6.6)–(6.7) we now examine the relationship between $\widetilde{X}_j$ and $\widetilde{X}_{j+1}$.

As long as we are working inside one iteration step, we simplify the notations by the so called '+'-notation system: we write (x,ω) instead of (x_j,ω_j) and (ξ,σ) instead of (x_{j+1},ω_{j+1}); also we replace f^j by f and f^{j+1} by f^+, D_j by D and D_{j+1} by D_+, etc. The map $\Psi : D_+ \to D$, whenever defined, has the form

$$\Psi(\xi,\sigma) = (\xi + U(\xi,\sigma), \sigma + \Lambda(\sigma)), \tag{6.10}$$

compare (6.6). The relation $\widetilde{X}_+ = \Psi^*\widetilde{X}$ (i.e., $\Psi_*\widetilde{X}_+ = \widetilde{X}$) then translates to

$$f^+(\xi,\sigma) + \frac{\partial U(\xi,\sigma)}{\partial \xi}(\sigma + f^+(\xi,\sigma)) = \Lambda(\sigma) + f(\xi + U(\xi,\sigma), \sigma + \Lambda(\sigma)) \tag{6.11}$$

for all $(\xi,\sigma) \in D_+$, compare (2.3) and (6.4).

Determination of the map $\Psi : D_+ \to D$. We now come to the definition of the map $\Psi : D_+ \to D$ (6.10) which has to link Φ to Φ_+. In order to make the Newtonian iteration work, Ψ is determined from the following linearization of (6.11):

$$\frac{\partial U(\xi,\sigma)}{\partial \xi}\sigma = \Lambda(\sigma) + {}_d f(\xi,\sigma) \tag{6.12}$$

[cf. (2.4) and (6.5)], where ${}_d f$, for $d = d_j \in \mathbb{N}$, is the truncation of the Fourier series of f at the order d, to be chosen appropriately later on (compare again the proof of Theorem 1.7 in § 1.5.2). This is the 1-bite small divisor problem referred to above, see (6.5). If

$$f(x,\omega) = \sum_{k\in \mathbb{Z}^n} f_k(\omega) e^{i\langle x,k\rangle},$$

we set $U(\xi,\sigma)$ to be the trigonometric polynomial

$$U(\xi,\sigma) = U_0(\sigma) + \sum_{0<|k|\le d} \frac{f_k(\sigma)}{i\langle \sigma,k\rangle} e^{i\langle \xi,k\rangle}, \tag{6.13}$$

provided the denominators do not vanish, cf. (1.35)–(1.36). Moreover, $U_0(\sigma)$ is arbitrary, while

$$\Lambda(\sigma) = -f_0(\sigma) \quad (= -[f(\cdot,\sigma)]) \tag{6.14}$$

(recall that $[f(\cdot,\sigma)]$ denotes the $\mathbb{T}^n$-average of f). By taking

$$U_0(\sigma) \equiv 0 \tag{6.15}$$

the solution (6.13) becomes unique. As before we need the parameter shift Λ for solvability of equation (6.12), cf. Section 2.1. The solution we have constructed is going to be used in the Inverse Approximation Lemma 6.14. To enable this we at least need U to be a well defined (holomorphic) function on D_+.

Reduction to the case $\gamma = 1$. Before giving the details of the iteration process sketched above, we observe that it is sufficient to prove Theorem 6.1 for the case $\gamma = 1$. To show this just rescale the time t to γt and the parameter ω to $\gamma^{-1}\omega$. Then the parameter domain Γ changes to $\gamma^{-1}\Gamma$ and hence for the sake of uniformity of our results in γ, we now drop the hypothesis that Γ is bounded. The symbols Γ'_γ below are thus to be understood as Γ'_1. Also we may assume that $\rho = 1$.

Estimates for the iteration step

In this part we specify the complex neighborhoods D_j and the orders of truncation d_j as well as estimate $|f^{j+1}|_{D_{j+1}}$ and $|\Psi_{j+1} - id|_{D_{j+1}}$ in terms of $|f^j|_{D_j}$. Also, in view of the Inverse Approximation Lemma 6.14, we estimate the components of $\Phi_{j+1} - \Phi_j$ on D_{j+1}.

Let $\{s_j\}_{j\geq 0}$ be any geometric sequence of positive numbers with ratio less than $\frac{1}{2}$ and next define the geometric sequence $\{r_j\}_{j\geq 0}$ by

$$r_j = \tfrac{1}{2}s_j^{2\tau+2}, \quad j \in \mathbb{Z}_+. \tag{6.16}$$

Then we define domains

$$D_j = (\mathbb{T}^n + \tfrac{1}{2}\kappa + s_j) \times (\Gamma'_\gamma + r_j),$$

see again § 1.5.2 for the notations. In order to ensure that $D_0 \subset O$ we require that

$$0 < s_0 < \min\left\{\tfrac{1}{2}\kappa,\ 2^{1/(2\tau+2)}\right\} \tag{6.17}$$

[see (6.2)], recall that $\rho = 1$. At the end of the proof the sequence $\{s_j\}_{j\geq 0}$ will be fixed such that the iteration process converges. Now in the '+'-notation we further define

$$s_* = \tfrac{1}{2}(s + s_+), \quad r_* = \tfrac{1}{2}(r + r_+)$$

and

$$D_* = (\mathbb{T}^n + \tfrac{1}{2}\kappa + s_*) \times (\Gamma'_\gamma + r_*).$$

Obviously $D_+ \subset D_* \subset D$.

Next recall that the definition of Ψ_j uses an order of truncation d_j, see (6.12)–(6.13). We here specify

$$d_j = \mathrm{Entier}(s_j^{-2}), \quad j \in \mathbb{Z}_+, \tag{6.18}$$

where "Entier" denotes the integral part. Then the denominators in (6.13) are estimated as follows, compare Proposition 1.9 in § 1.5.2.

Proposition 6.2 *For all $\sigma \in \Gamma'_\gamma + r$ and all $k \in \mathbb{Z}^n$ with $0 < |k| \leq d$ one has*

$$|\langle\sigma, k\rangle| \geq \tfrac{1}{2}|k|^{-\tau}.$$

Proof. By definition $d \leq s^{-2} = (2r)^{-1/(\tau+1)}$ and hence, for $0 < |k| \leq d$, we have $|k|^{\tau+1} \leq d^{\tau+1} \leq (2r)^{-1}$ and therefore $2r|k| \leq |k|^{-\tau}$. Now for $\sigma \in \Gamma'_\gamma + r$ there exists $\sigma^* \in \Gamma'_\gamma$ such that $|\sigma - \sigma^*| \leq r$. It then follows for all k with $0 < |k| \leq d$ that

$$|\langle\sigma, k\rangle| \geq |\langle\sigma^*, k\rangle| - |\sigma - \sigma^*|\,|k| \geq |k|^{-\tau} - r|k| \geq \tfrac{1}{2}|k|^{-\tau}.$$

□

Consequently, whenever $f = f^j$ is well defined on $D = D_j$, so is the trigonometric polynomial U, cf. (6.13). Moreover, analyticity of f implies analyticity of U.

Throughout the rest of the proof positive constants will appear that only depend on n, τ, and κ. If we do not need to remember these constants we shall neglect them, and the corresponding inequality sign then will be followed by a dot ($\leq .$). The remaining constants will be denoted by c_0, c_1, c_2, c_3, and c_4.

We now start with a statement on the derivatives of $f = f^j$. Its proof uses the so called Cauchy Estimate, which from now on will be frequently met and therefore spelled out to some detail.

Proposition 6.3 *Let f be holomorphic on D. Then on D_* we have*

$$s\left|\frac{\partial f}{\partial x}\right|,\ r\left|\frac{\partial f}{\partial \omega}\right| \le c_0|f|_D.$$

Proof. By the Cauchy Integral Formula for a point $(x', \omega') \in D_*$ we obtain

$$\frac{\partial f(x',\omega')}{\partial x_\iota} = \frac{1}{2\pi i}\oint \frac{f(x'_1,\ldots,x'_{\iota-1},z,x'_{\iota+1},\ldots,x'_n,\omega')}{(z-x'_\iota)^2}\,dz, \qquad 1 \le \iota \le n,$$

where the integration contour is the circle with radius $s - s_*$ and center x'_ι. Notice that $s - s_* = \frac{1}{2}(s - s_+) > \frac{1}{4}s$ (since $s_+ < \frac{1}{2}s$). Therefore,

$$\left|\frac{\partial f(x',\omega')}{\partial x}\right| \le 4s^{-1}|f|_D.$$

Similarly we get that

$$\left|\frac{\partial f(x',\omega')}{\partial \omega}\right| \le 4r^{-1}|f|_D.$$

□

Next we estimate $\Psi - id$ and some of its derivatives, the mapping Ψ being defined by (6.10) and (6.13)–(6.15). Here the tools are the Paley–Wiener lemma (Lemma 1.8, see § 1.5.2), the Cauchy Estimate and the following lemma.

Lemma 6.4 *Given $\beta > 0$, there exists a constant $c(\beta) > 0$ such that for all $0 < x \le 1$*

$$\sum_{l=1}^{\infty} l^\beta e^{-xl} \le c(\beta)x^{-(\beta+1)}.$$

Proof.

$$\sum_{l=1}^{\infty} l^\beta e^{-xl} = x^{-(\beta+1)}\sum_{l=1}^{\infty} x(xl)^\beta e^{-xl} \le x^{-(\beta+1)}\left(\int_0^\infty t^\beta e^{-t}dt + x(\beta/e)^\beta\right),$$

since $\max_{t\ge 0}(t^\beta e^{-t}) = (\beta/e)^\beta$. □

Proposition 6.5 *Let f be holomorphic on D. Then on D_**

$$\begin{aligned} s^{2\tau+1}|U|,\ s^{2\tau+2}\left|\frac{\partial U}{\partial \xi}\right|,\ rs^{2\tau+1}\left|\frac{\partial U}{\partial \sigma}\right| &\le c_1|f|_D; \\ |\Lambda|,\ r\left|\frac{\partial \Lambda}{\partial \sigma}\right| &\le c_1|f|_D \end{aligned}$$

for some $c_1 \ge c_0$.

Proof. From the Paley–Wiener lemma (Lemma 1.8) it follows for the Fourier coefficients f_k on the domain D

$$|f_k(\sigma)| \le |f|_D e^{-(\frac{1}{2}\kappa+s)|k|}.$$

From this estimate, formulas (6.13), (6.15) and Proposition 6.2 we see that on the domain $(\mathrm{T}^n + \frac{1}{2}\kappa + s_*) \times (\Gamma'_\gamma + r)$

$$|U(\xi,\sigma)| \le \sum_{0<|k|\le d} 2|f|_D|k|^\tau e^{\left(|\operatorname{Im}\xi| - (\frac{1}{2}\kappa+s)\right)|k|} \le 2|f|_D \sum_{k\in\mathbb{Z}^n} |k|^\tau e^{-\frac{1}{4}s|k|},$$

using that $|\operatorname{Im}\xi| \le \frac{1}{2}\kappa + s_*$ and $s - s_* > \frac{1}{4}s$.

In order to estimate the last sum we first realize that

$$\sum_{k\in\mathbb{Z}^n} |k|^\tau e^{-\frac{1}{4}s|k|} \le 2^n \sum_{l=1}^{\infty} l^{\tau+n-1} e^{-\frac{1}{4}sl} < 2^n \sum_{l=1}^{\infty} l^{2\tau} e^{-\frac{1}{4}sl},$$

using that $\tau > n-1$ and exploiting the estimate (1.40), cf. (1.41). Next we apply Lemma 6.4 for $\beta = 2\tau$, which yields

$$\sum_{k\in\mathbb{Z}^n} |k|^\tau e^{-\frac{1}{4}s|k|} \le_\cdot s^{-(2\tau+1)}.$$

Now the estimate on $|U|_{D_*}$ is immediate. Differentiation of (6.13) with respect to ξ similarly gives the inequality on $|\partial U/\partial\xi|$. The inequality on $|\partial U/\partial\sigma|$ can be obtained by the Cauchy Estimate, compare the proof of Proposition 6.3. Similarly, for $\sigma \in \Gamma'_\gamma + r$

$$|\Lambda(\sigma)| = |f_0(\sigma)| \le |f|_D,$$

while with the Cauchy Estimate

$$\left|\frac{\partial\Lambda}{\partial\sigma}\right|_{D_*} \le_\cdot r^{-1}|f|_D.$$

□

Now we estimate $|f^+|_{D_+}$. From the equalities (6.11)–(6.12) it follows that on the domain D_+

$$\left(id + \frac{\partial U}{\partial\xi}\right) f^+ = R_1 + R_2$$

with $R_1(\xi,\sigma) = f(\xi + U(\xi,\sigma), \sigma + \Lambda(\sigma)) - f(\xi,\sigma)$ and $R_2(\xi,\sigma) = f(\xi,\sigma) - {}_d f(\xi,\sigma)$. We then have

Proposition 6.6 *Let f be holomorphic on D and assume that $|f|_D \le \frac{1}{8c_1}s^{2\tau+2}$. Then $\Psi(D_+) \subset D_*$. If, moreover, $ds > 2n$ then*

$$|f^+|_{D_+} \le c_2(I + II),$$

where $I = s^{-(2\tau+2)}|f|_D^2$ and $II = d^n e^{-\frac{1}{4}sd}|f|_D$.

Proof. On behalf of the first conclusion we use Proposition 6.5. Indeed, since $s_* - s_+ = \frac{1}{2}(s - s_+) > \frac{1}{4}s$ and $r_* - r_+ = \frac{1}{2}(r - r_+) > \frac{1}{4}r$ it is sufficient to have that

$$|U|_{D_+} \le \tfrac{1}{4}s \quad \text{and} \quad |\Lambda|_{D_+} \le \tfrac{1}{4}r,$$

which by Proposition 6.5 follows from our assumption:

$$\begin{aligned} |U|_{D_+} &\le c_1 s^{-(2\tau+1)}|f|_D \le \tfrac{1}{8}s < \tfrac{1}{4}s, \\ |\Lambda|_{D_+} &\le c_1|f_D| \le \tfrac{1}{8}s^{2\tau+2} = \tfrac{1}{4}r \end{aligned}$$

[see (6.16)]. A similar argument gives that this assumption implies

$$\left|\frac{\partial U}{\partial\xi}\right|_{D_+} \le c_1 s^{-(2\tau+2)}|f|_D \le \tfrac{1}{8},$$

whence

$$\left|\left(id + \frac{\partial U}{\partial \xi}\right) f^+\right|_{D_+} \geq \left(1 - \left|\frac{\partial U}{\partial \xi}\right|_{D_+}\right) |f^+|_{D_+} \geq \tfrac{7}{8}|f^+|_{D_+}.$$

Next, on behalf of the second conclusion, we estimate each of the terms R_1 and R_2 separately. By the Mean Value Theorem and Propositions 6.3 and 6.5 we have on D_+

$$\begin{aligned} |R_1(\xi,\sigma)| &= |f(\xi + U(\xi,\sigma), \sigma + \Lambda(\sigma)) - f(\xi,\sigma)| \\ &\leq_\cdot \max\left\{\left|\frac{\partial f}{\partial \xi}\right|_{D_*} |U|_{D_+}, \left|\frac{\partial f}{\partial \sigma}\right|_{D_*} |\Lambda|_{D_+}\right\} \\ &\leq_\cdot \max\{s^{-1}|f|_D s^{-(2\tau+1)}|f|_D, r^{-1}|f|_D|f|_D\} \leq_\cdot s^{-(2\tau+2)}|f|_D^2 = I \end{aligned}$$

[see (6.16)], while, again on D_+, by the Paley–Wiener lemma (Lemma 1.8)

$$\begin{aligned} |R_2(\xi,\sigma)| &= |f(\xi,\sigma) - {}_d f(\xi,\sigma)| \\ &\leq \sum_{|k|>d} |f_k(\sigma)||e^{i\langle\xi,k\rangle}| \leq_\cdot |f|_D \sum_{|k|>d} e^{\left(|\operatorname{Im}\xi| - (\frac{1}{2}\kappa + s)\right)|k|} \\ &\leq |f|_D \sum_{|k|>d} e^{-\frac{1}{2}s|k|}, \end{aligned}$$

using that $|\operatorname{Im}\xi| \leq \frac{1}{2}\kappa + s_+$ and $s - s_+ > \frac{1}{2}s$. To estimate the last sum, we proceed as at the end of the proof of Theorem 1.7 in § 1.5.2:

$$\begin{aligned} \sum_{|k|>d} e^{-\frac{1}{2}s|k|} &\leq 2^n \sum_{l=d+1}^{\infty} l^{n-1} e^{-\frac{1}{2}sl} \leq_\cdot \int_d^\infty x^{n-1} e^{-\frac{1}{2}sx} dx \\ &= d^{n-1} e^{-\frac{1}{2}sd} \int_d^\infty \left(\frac{x}{d}\right)^{n-1} e^{-\frac{1}{2}s(x-d)} dx \\ &\leq d^{n-1} e^{-\frac{1}{2}sd} \int_d^\infty \exp\left[\left(\frac{n-1}{d} - \frac{s}{2}\right)(x-d)\right] dx \\ &\leq d^{n-1} e^{-\frac{1}{2}sd} \int_d^\infty e^{-(x-d)/d} dx = d^n e^{-\frac{1}{2}sd}, \end{aligned}$$

using twice that $ds > 2n$. We conclude that

$$|R_2|_{D_+} \leq_\cdot d^n e^{-\frac{1}{2}sd} |f|_D = II.$$

□

We end this part by comparing $\Phi = \Phi_j$ and $\Phi_+ = \Phi_{j+1}$ in a manner suitable for the Inverse Approximation Lemma 6.14. To be more precise, we estimate $|\widetilde{U} - \widetilde{U}^+|$ and $|\widetilde{\Lambda} - \widetilde{\Lambda}^+|$ on the domain D_+ [see (6.6)].

For this purpose, we need an estimate on $\Psi - id$ and its derivatives, which is a direct consequence of Proposition 6.5. Indeed, if we write $\Psi = id + \psi$ we have

Corollary 6.7 *Let f be holomorphic on D. Then on the domain D_+*

$$s^{2\tau+1}|\psi|_{D_+}, \; rs^{2\tau+1}|D\psi|_{D_+} \leq_\cdot |f|_D,$$

$D\psi$ denoting the derivative of ψ.

In the estimates to follow we need the C^1-norm $\|\Phi_j\|_{1,D_j} = \max\{|\Phi_j|_{D_j}, |\mathcal{D}\Phi_j|_{D_j}\}$.

Proposition 6.8 *Under the assumptions of Proposition 6.6*

(i) $\|\Phi_+\|_{1,D_+} \leq_\cdot \|\Phi\|_{1,D} \max\{1 + |\psi|_{D_+}, 1 + |\mathcal{D}\psi|_{D_+}\}$;

(ii) $|\tilde{U} - \tilde{U}^+|_{D_+}, |\tilde{\Lambda} - \tilde{\Lambda}^+|_{D_+} \leq_\cdot \|\Phi\|_{1,D}|\psi|_{D_+}$.

Proof.
(i) $\Phi_+ = \Phi \circ \Psi = (\Phi \circ \Psi - \Phi) + \Phi$, whence

$$|\Phi_+|_{D_+} \leq |\mathcal{D}\Phi|_{D}.|\psi|_{D_+} + |\Phi|_{D_+} \leq \|\Phi\|_{1,D}(1 + |\psi|_{D_+})$$

and

$$\begin{aligned} |D\Phi_+|_{D_+} &\leq |\mathcal{D}\Phi|_{D}.|\mathcal{D}\Psi|_{D_+} + |\mathcal{D}\Phi|_{D_+} \leq |\mathcal{D}\Phi|_{D}.(1 + |\mathcal{D}\psi|_{D_+}) \\ &\leq \|\Phi\|_{1,D}(1 + |\mathcal{D}\psi|_{D_+}), \end{aligned}$$

using the Mean Value Theorem and the Chain Rule.

(ii) This estimate follows from

$$|\Phi_+ - \Phi|_{D_+} = |\Phi \circ \Psi - \Phi|_{D_+} \leq |\mathcal{D}\Phi|_{D}.|\psi|_{D_+} \leq \|\Phi\|_{1,D}|\psi|_{D_+},$$

again using the Mean Value Theorem. □

Convergence

In the following, final Proposition 6.11, by a proper choice of the sequence $\{s_j\}_{j\geq 0}$ we ensure convergence of the iteration process. The "errors" $|f^j|_{D_j}$ as well as the differences $\Phi_j - \Phi_{j+1}$ on the domains D_{j+1} will be reduced superexponentially as $j \to +\infty$, compare the proof of Theorem 1.7 in § 1.5.2.

To start with we further specify the geometric sequence $\{s_j\}_{j\geq 0}$ by

$$s_j = s_0 a^j, \quad j \in \mathbb{Z}_+$$

where, in accordance with (6.17), $0 < a < \frac{1}{2}$ and $0 < s_0 < \min\left\{\frac{1}{2}\kappa, 2^{1/(2\tau+2)}\right\}$. We recall that the geometric sequence $\{r_j\}_{j\geq 0}$ is determined by $r_j = \frac{1}{2}s_j^{2\tau+2}$, $j \in \mathbb{Z}_+$ [see (6.16)]. Also we need an *exponential* sequence $\{\delta_j\}_{j\geq 0}$ of positive numbers:

$$\delta_{j+1} = \delta_j^{1+p} = \delta_0^{(1+p)^{j+1}}, \quad j \in \mathbb{Z}_+$$

with $0 < p < 1$. The δ_j will serve to dominate the "errors" $|f^j|_{D_j}$. As two extra tools we need the following simple technical lemmas.

Lemma 6.9 *Given $\theta > 0$, $A > 0$ and $n > 0$, for all positive $x \geq -\frac{2n}{A}\log\left(\frac{A}{n}\theta^{1/n}\right)$ one has $x^n e^{-Ax} < \theta$.*

Proof. One easily sees that without loss of generality we may assume that $A = n = 1$. Indeed, just scale $\bar{x} = \frac{A}{n}x$, $\bar{\theta} = \frac{A}{n}\theta^{1/n}$ and the general case is reduced to this special one (using that $x > 0$). Thus, it remains to show that $xe^{-x} < \theta$ as soon as $x \geq -2\log\theta$, which is equivalent to $e^{-x/2} \leq \theta$.

Indeed, for all x one has

$$xe^{-x/2} \le \max_{y\in\mathbb{R}}(ye^{-y/2}) = 2/e < 1,$$

whence

$$xe^{-x} = xe^{-x/2}e^{-x/2} < e^{-x/2} \le \theta.$$

□

Lemma 6.10 *Given $p \ge 0$, $\alpha \in \mathbb{R}$ and $\beta \in \mathbb{R}$, for all $j \ge 0$ one has*

$$\frac{\alpha + \beta j}{(1+p)^j} \le |\alpha| + \frac{|\beta|}{\log(1+p)}.$$

Proof. Since $(1+p)^{-j} \le 1$ because of $p \ge 0$ and $j \ge 0$, it suffices to show that $j(1+p)^{-j} \le 1/\log(1+p)$, i.e., $j\log(1+p) \le (1+p)^j$. One has $\max_{y\in\mathbb{R}}(ye^{-y}) = 1/e < 1$, i.e., $y < e^y$ for all y. Taking $y = j\log(1+p)$, we arrive at $j\log(1+p) < (1+p)^j$. □

Proposition 6.11 *Assume that the constants a and p are fixed with*

$$0 < p < 1 \quad \text{and} \quad 0 < a < \tfrac{1}{2}.$$

Then, for $\delta_0 > 0$ sufficiently small, there exist s_0 in the interval

$$0 < s_0 < \min\left\{\tfrac{1}{2}\kappa,\ 2^{1/(2\tau+2)}\right\}$$

and b in the interval

$$0 < b < 1$$

such that the following holds. If $|f^0|_{D_0} < \delta_0$, then for all $j \in \mathbb{Z}_+$

(i) *the assumptions of Proposition 6.6 are satisfied;*

(ii) $|f^j|_{D_j} \le \delta_j$;

(iii) *on the domain D_{j+1}*

$$|\widetilde{U}^j - \widetilde{U}^{j+1}|,\ |\widetilde{\Lambda}^j - \widetilde{\Lambda}^{j+1}| \le c_3\delta_j^b.$$

Proof.

(i) The first assumption of Proposition 6.6 is $|f|_D \le \frac{1}{8c_1}s^{2\tau+2}$. Sufficient for this to be satisfied is that both

$$\delta_0 \le \frac{1}{8c_1}s_0^{2\tau+2} \quad \text{and} \quad \delta_0^p \le a^{2\tau+2}. \tag{6.19}$$

The second assumption of Proposition 6.6 is $ds > 2n$, which will be taken care of later on.

(ii) Regarding the conclusion of Proposition 6.6 it is sufficient that $I,\ II \le \frac{1}{2c_2}\delta_+$. Now, in order to have $I \le \frac{1}{2c_2}\delta_+$, in turn it is sufficient that $\delta^{1-p} \le \frac{1}{2c_2}s^{2\tau+2}$, which is implied by

$$\delta_0^{1-p} \le \frac{1}{2c_2}s_0^{2\tau+2} \quad \text{and} \quad \delta_0^{(1-p)p} \le a^{2\tau+2}. \tag{6.20}$$

Next consider both $II \le \frac{1}{2c_2}\delta_+$ and the condition $ds > 2n$, recalling that $d = \text{Entier}(s^{-2})$ [see (6.18)]. The latter implies that $s^{-1} - 1 < ds \le s^{-1}$. Indeed, if $s > 1$ then $ds = 0 >$

$s^{-1}-1$, whereas if $s \le 1$ then $ds > s^{-1} - s \ge s^{-1} - 1$. So, to meet the condition $ds > 2n$ it is sufficient that for all $j \ge 0$ we have $s_j^{-1} - 1 \ge 2n$, which in turn is implied by

$$0 < a < \tfrac{1}{2} \quad \text{and} \quad s_0 \le \tfrac{1}{2n+1}. \tag{6.21}$$

Moreover it follows from $s^{-1} - 1 < ds \le s^{-1}$ that

$$d^n e^{-sd/2} \le s^{-2n}\sqrt{e}\, e^{-1/(2s)}, \quad \text{whence} \quad II \le s^{-2n}\sqrt{e}\, e^{-1/(2s)}|f|_D.$$

Now apply Lemma 6.9 for $x = s^{-2}$, $A = \frac{1}{2}s$ and $\theta = (2c_2\sqrt{e})^{-1}\delta^p$. This yields that $II \le \frac{1}{2c_2}\delta_+$ is implied by

$$s^{-2} \ge -\frac{4n}{s}\log\left[\frac{s}{2n}\left(\frac{\delta^p}{2c_2\sqrt{e}}\right)^{1/n}\right],$$

which in turn follows from

$$s \le \left\{c_4 + 4n\log\left(\frac{1}{s}\right) + 4p\log\left(\frac{1}{\delta}\right)\right\}^{-1}, \tag{6.22}$$

where we have set

$$c_4 := \max\left\{4n\log\left[2n(2c_2\sqrt{e})^{1/n}\right],\, 0\right\}.$$

Recalling the definitions of sequences $\{s_j\}$ and $\{\delta_j\}$, we can rewrite (6.22) as

$$s_0 a^j \le \left\{c_4 + 4n\log\left(\frac{1}{s_0}\right) + 4nj\log\left(\frac{1}{a}\right) + 4p(1+p)^j\log\left(\frac{1}{\delta_0}\right)\right\}^{-1},$$

or equivalently

$$s_0 a^j \le \frac{1}{(1+p)^j}\left\{\frac{c_4 + 4n\log s_0^{-1} + 4nj\log a^{-1}}{(1+p)^j} + 4p\log\left(\frac{1}{\delta_0}\right)\right\}^{-1}. \tag{6.23}$$

Now apply Lemma 6.10 for $\alpha = c_4 + 4n\log s_0^{-1} > 0$ and $\beta = 4n\log a^{-1} > 0$. This yields that (6.23) is implied by

$$s_0 a^j \le \frac{1}{(1+p)^j}\left\{c_4 + 4n\log\left(\frac{1}{s_0}\right) + \frac{4n\log a^{-1}}{\log(1+p)} + 4p\log\left(\frac{1}{\delta_0}\right)\right\}^{-1},$$

which is in turn ensured by

$$a \le \tfrac{1}{1+p} \tag{6.24}$$

and

$$s_0 \le \left\{c_4 + \frac{4n\log a^{-1}}{\log(1+p)} + 4n\log\left(\frac{1}{s_0}\right) + 4p\log\left(\frac{1}{\delta_0}\right)\right\}^{-1}. \tag{6.25}$$

(iii) Remains the final part of the conclusion. First of all we consider the sequence $\{\|\Phi_j\|_{1,D_j}\}_{j\ge0}$, which we want to bound by a universal constant. Sufficient for this is that

$$\sum_{j=0}^{\infty}\delta_j r_j^{-1} s_j^{-(2\tau+1)} = \sum_{j=0}^{\infty} 2\delta_j s_j^{-(4\tau+3)}$$

[cf. (6.16)] has a universal bound, see Corollary 6.7 and Proposition 6.8. This in turn is implied by

$$\delta_0 s_0^{-(4\tau+3)} < \tfrac{1}{2} \quad \text{and} \quad \delta_0^p a^{-(4\tau+3)} < \tfrac{1}{2}. \tag{6.26}$$

From Corollary 6.7 and Proposition 6.8 we now obtain that $|\widetilde{U}^j - \widetilde{U}^{j+1}|_{D_{j+1}}$ and $|\widetilde{\Lambda}^j - \widetilde{\Lambda}^{j+1}|_{D_{j+1}}$ exhibit superexponential decay as $j \to +\infty$, as soon as $\delta_j s_j^{-(2\tau+1)}$ does. Since $\{\delta_j\}_{j\geq 0}$ is an exponential sequence this imposes no extra condition.

The inequalities

$$\delta_0^{1-b} s_0^{-(2\tau+1)} < 1 \quad \text{and} \quad \delta_0^{p(1-b)} a^{-(2\tau+1)} < 1 \tag{6.27}$$

guarantee that $\delta_j s_j^{-(2\tau+1)} < \delta_j^b$ for all $j \geq 0$ ($b \in]0,1[$ being arbitrary).

The proof is concluded by showing that the conditions (6.17), (6.19)–(6.21) and (6.24)–(6.27) can be simultaneously satisfied. First notice that (6.24) easily follows from the assumptions on a and p. Indeed, since $0 < p < 1$ it follows that $\frac{1}{2} < \frac{1}{1+p}$.

Now fixing a and p as above and choosing $0 < b < 1$ arbitrarily, we are left with conditions on δ_0 and s_0. Observe that all remaining inequalities are of one of the following types:

$$s_0 \leq \vartheta_1, \quad \delta_0 \leq \vartheta_2, \quad \delta_0^{\vartheta_3} \leq \vartheta_4 s_0, \quad s_0 \leq \left\{\vartheta_5 \log\left(\frac{1}{s_0}\right) + \vartheta_6 \log\left(\frac{1}{\delta_0}\right) + \vartheta_7\right\}^{-1},$$

positive constants $\vartheta_1, \vartheta_2, \vartheta_3, \vartheta_4, \vartheta_5, \vartheta_6, \vartheta_7$ possibly depending on a, p, and b. These conditions can all be satisfied by choosing both δ_0 and s_0 small, but in an interdependent way. This follows from the fact that for any $\vartheta > 0$

$$\lim_{\delta_0 \downarrow 0} \delta_0^{\vartheta} \log \delta_0 = 0.$$

□

The proof of Theorem 6.1 finally is completed by taking $\delta = \delta_0$. As said before, we apply the Inverse Approximation Lemma 6.14, see the next subsection, to the mappings Φ_j on the domains D_j, $j \in \mathbb{Z}_+$. First notice that the D_j in the ω_j-direction shrink down to the "Cantor set" Γ'_γ as $j \to +\infty$ in a geometric way. Then conclusion (iii) of Proposition 6.11 gives Φ_∞ as a Whitney-C^∞ mapping on $\mathbb{T}^n \times \Gamma'_\gamma$. Moreover, since the convergence in the x_j-direction is uniform on $\mathbb{T}^n + \frac{1}{2}\kappa$, this limit Φ_∞ is real analytic in x_∞. Also Φ_∞ clearly preserves the projection to the parameter space. Finally consider the limit $\widetilde{X}_\infty = \Phi_\infty^* \widetilde{X}$. From conclusion (ii) of Proposition 6.11 it follows that the "error" has vanished, i.e., $\widetilde{X}_\infty^{\omega_\infty}$ is of the form (6.8), which shows that Φ_∞ does conjugate X to $\widetilde{X}$.

By the Whitney Extension Theorem 6.15, again see the next subsection, we can extend Φ_∞ to a mapping Φ on $\mathbb{T}^n \times \mathbb{R}^n$ such that the conclusions of Theorem 6.1 hold. Moreover, there exists a constant $0 < b_* < 1$, such that for all $j \in \mathbb{N}$ in the C^j-norm on $\mathbb{T}^n \times \Gamma$

$$\|\Phi - id\|_{C^j} = O(|f|_O^{b_*}) \tag{6.28}$$

as $|f|_O \to 0$. In this sense Φ is a C^∞-near-the-identity map (cf. [157]).

6.1.4 Whitney-smoothness: some theory

To make our presentation complete, we here include a brief discussion on the concept of Whitney-differentiability, cf. § 1.5.2. Apart from definitions, a formulation of the Inverse Approximation Lemma is given, which is used to obtain Whitney-differentiable conjugacy results. Here we quote from [160, 278, 338, 355, 364], see also [114, 242].

Let $\beta > 0$ be a fixed order of differentiability. We shall abbreviate

$$\ell = -[1 + \mathrm{Entier}(-\beta)] = \max\{m \in \mathbb{Z} \,:\, m < \beta\} \in \mathbb{Z}_+,$$

in other words, $\ell = \mathrm{Entier}(\beta)$ for $\beta \notin \mathbb{N}$ and $\ell = \beta - 1$ for $\beta \in \mathbb{N}$. One sees $\ell < \beta \leq \ell + 1$.

Definition 6.12 *Let $\Omega \subset \mathbb{R}^n$ be an open domain. The class $C^\beta_{\mathrm{Wh}}(\Omega)$ of Whitney-C^β functions in Ω consists of C^ℓ functions $U : \Omega \to \mathbb{R}$ with uniformly bounded derivatives up to order ℓ, the ℓth derivatives satisfying a Hölder condition with exponent $\beta - \ell$. Otherwise speaking, for any $U \in C^\beta_{\mathrm{Wh}}(\Omega)$ there exists a number $M \geq 0$ such that for all $x \in \Omega$ the inequality*

$$|D^q U(x)| \leq M \tag{6.29}$$

holds for all $q \in \mathbb{Z}^n_+$, $0 \leq |q| \leq \ell$, while for all $x, y \in \Omega$ the inequality

$$|D^q U(x) - D^q U(y)| \leq M|x - y|^{\beta - \ell} \tag{6.30}$$

holds for all $q \in \mathbb{Z}^n_+$, $|q| = \ell$. The norm $\|U\|_\beta$ of functions $U \in C^\beta_{\mathrm{Wh}}(\Omega)$ is the C^β-Hölder norm, defined as the smallest M for which all the inequalities (6.29) and (6.30) are valid.

Remark. For β integer, the class $C^\beta_{\mathrm{Wh}}(\Omega)$ of Whitney-C^β functions defined on Ω does *not* coincide with the class of functions $U : \Omega \to \mathbb{R}$ which are C^β-smooth in the usual sense. First, usual C^β functions are those with continuous derivatives up to order β whereas Whitney-C^β functions (for $\beta \in \mathbb{N}$) are those with *Lipschitz* continuous derivatives up to order $\beta - 1$ (not up to order β). In other words, the $(\beta - 1)^{\mathrm{th}}$ derivatives of usual C^β functions are assumed to have continuous first derivatives whereas the $(\beta - 1)^{\mathrm{th}}$ derivatives of functions $U \in C^\beta_{\mathrm{Wh}}(\Omega)$ are assumed just to satisfy a weaker Lipschitz condition

$$|D^q U(x) - D^q U(y)| \leq \mathrm{const} \cdot |x - y|$$

($q \in \mathbb{Z}^n_+$, $|q| = \beta - 1$) for all $x, y \in \Omega$. Second, the derivatives of usual C^β functions $U : \Omega \to \mathbb{R}$ are not required to be bounded over Ω.

Definition 6.13 *Let $\Omega \subset \mathbb{R}^n$ be a closed set. The elements of the class $C^\beta_{\mathrm{Wh}}(\Omega)$ of Whitney-C^β functions in Ω are collections $U = \{U_q\}_{0 \leq |q| \leq \ell}$ ($q \in \mathbb{Z}^n_+$) of functions defined on Ω and possessing the following property. There exists a number $M \geq 0$ such that for all $x, y \in \Omega$ and all $q \in \mathbb{Z}^n_+$, $0 \leq |q| \leq \ell$, one has*

$$|U_q(x)| \leq M \quad \text{and} \quad |U_q(x) - P_q(x, y)| \leq M|x - y|^{\beta - |q|} \tag{6.31}$$

where

$$P_q(x, y) = \sum_{j=0}^{\ell - |q|} \sum_{|k| = j} \frac{1}{k!} U_{q+k}(y)(x - y)^k, \quad k \in \mathbb{Z}^n_+,$$

is an analogue of the $(\ell - |q|)^{\mathrm{th}}$ Taylor polynomial for U_q [of course, here $k! = k_1! \cdots k_n!$ and $(x - y)^k = (x_1 - y_1)^{k_1} \cdots (x_n - y_n)^{k_n}$]. The norm $\|U\|_\beta$ of collections $U \in C^\beta_{\mathrm{Wh}}(\Omega)$ is defined as the smallest M for which all the inequalities (6.31) hold.

Remark 1. The U_q in Definition 6.13 play the rôle of the partial derivatives of U_0, satisfying the corresponding compatibility relations. Generally speaking, U_q for $|q| \geq 1$ are *not* determined uniquely by U_0.

Remark 2. For $\Omega = \mathbb{R}^n$, these two definitions of $C^\beta_{\mathrm{Wh}}(\mathbb{R}^n)$ coincide and the two corresponding norms $\|\cdot\|_\beta$ are equivalent.

Now let $\Omega \subset \mathbb{R}^n$ be an *either open or closed* subset of $\mathbb{R}^n$ and let $r_j = a\kappa^j$, $j \in \mathbb{Z}_+$, be a fixed geometric sequence with $a = r_0 > 0$ and $0 < \kappa < 1$. Consider the complex domains $W_j = \Omega + r_j$, where, as before,

$$\Omega + r_j := \bigcup_{x \in \Omega} \{z \in \mathbb{C}^n \ : \ |z - x| < r_j\}$$

(cf. § 1.5.2). For $j \in \mathbb{Z}_+$ let also U^j be a real analytic function in domain W_j. The question is when the sequence $\{U^j\}_{j\geq 0}$ has a limit U^∞, defined on Ω, which is of class Whitney-C^β.

Lemma 6.14 (Inverse Approximation Lemma) *Assume that $\beta \notin \mathbb{N}$. Let $\{W_j\}_{j\geq 0}$ and $\{U^j\}_{j\geq 0}$ be as above, such that $U^0 \equiv 0$ while for all $j \geq 1$*

$$|U^j - U^{j-1}|_{W_j} \leq M r_j^\beta$$

for some constant $M \geq 0$. Then there exists a unique function U^∞, defined on Ω, which is of class Whitney-C^β and such that

$$\|U^\infty\|_\beta \leq M c_{\kappa,\beta},$$

the constant $c_{\kappa,\beta} > 0$ depending on n, κ, and β only, whereas

$$\|U^\infty - U^j\|_\alpha \to 0 \quad \text{as} \quad j \to +\infty$$

for all $\alpha < \beta$. Here $\|\cdot\|_\alpha$ and $\|\cdot\|_\beta$ are respectively the C^α- and C^β-norms on Ω.

Remark. This lemma is not valid for β integer. For instance, in the case of open Ω, the $(\beta-1)^{\text{th}}$ derivatives of the limit U^∞ for $\beta \in \mathbb{N}$ are in general no longer Lipschitz continuous, but satisfy a weaker Zygmund condition

$$|D^q U^\infty(x) + D^q U^\infty(y) - 2D^q U^\infty\left(\tfrac{1}{2}(x+y)\right)| \leq const \cdot |x - y|$$

($q \in \mathbb{Z}^n_+$, $|q| = \beta - 1$) for all $x, y \in \Omega$ such that $\frac{1}{2}(x+y) \in \Omega$ [364].

Of course, in the case of closed Ω in Lemma 6.14, $U^\infty - U^j$ should be treated as the collection $\{U^\infty_q - D^q U^j\}_q$.

Whitney-C^β functions defined on closed subsets of $\mathbb{R}^n$ are extendible to Whitney-C^β functions defined on the whole space $\mathbb{R}^n$.

Theorem 6.15 (Whitney Extension Theorem) *For any $\beta > 0$ and closed subset $\Omega \subset \mathbb{R}^n$, there exists a (non-unique) linear extension operator $\mathcal{E} : C^\beta_{\mathrm{Wh}}(\Omega) \to C^\beta_{\mathrm{Wh}}(\mathbb{R}^n)$ such that for each $U = \{U_q\}_q \in C^\beta_{\mathrm{Wh}}(\Omega)$ and for all $q \in \mathbb{Z}^n_+$, $0 \leq |q| \leq \ell$,*

$$D^q(\mathcal{E}U)|_\Omega = U_q \quad \text{while} \quad \|\mathcal{E}U\|_\beta \leq c_\beta \|U\|_\beta,$$

where the C^β-norms are taken over the appropriate domains and the constant $c_\beta > 0$ depends only on n and β but not on Ω.

So that, $U \in C^\beta_{\mathrm{Wh}}(\Omega)$ may be regarded as the restriction of a function $\mathcal{E}U \in C^\beta_{\mathrm{Wh}}(\mathrm{R}^n)$ to the closed set Ω and the U_q do play the rôle of the partial derivatives of U_0. However, Definition 6.13 of $C^\beta_{\mathrm{Wh}}(\Omega)$ is completely intrinsic to Ω.

In the applications, the functions U^j, $j \in \mathrm{Z}_+$, and the limit function U^∞ in Lemma 6.14 may also depend on other variables. This dependence is analytic or smooth and sometimes periodic. Such variables should be treated as parameters. One can choose an extension operator $\mathcal{E}$ to preserve the analyticity (respectively smoothness) as well as periodicity with respect to these parameters.

Finally, note that Whitney-C^β functions defined on the whole space R^n can be approximated by entire real analytic functions $\mathrm{C}^n \to \mathrm{C}$. Let again $r_j = a\kappa^j$, $j \in \mathrm{Z}_+$, be a fixed geometric sequence with $a = r_0 > 0$ and $0 < \kappa < 1$. Consider the complex domains

$$W_j = \mathrm{R}^n + r_j = \{z \in \mathrm{C}^n \ : \ |\operatorname{Im} z| < r_j\}$$

(complex strips around R^n of width r_j).

Lemma 6.16 (Approximation Lemma) *Let $\beta > 0$ be an arbitrary number and $\{W_j\}_{j\geq 0}$ be as above. Then for each $U \in C^\beta_{\mathrm{Wh}}(\mathrm{R}^n)$ there exists a sequence $\{U^j\}_{j\geq 0}$ of entire real analytic functions, such that $U^0 \equiv 0$ while for all $j \geq 1$*

$$|U^j - U^{j-1}|_{W_j} \leq M_{\kappa,\beta} r_j^\beta \|U\|_\beta$$

for some constant $M_{\kappa,\beta} > 0$ (depending on n, κ, and β only) and

$$\|U - U^j\|_\alpha \to 0 \quad as \quad j \to +\infty$$

for all $\alpha < \beta$. Here $\|\cdot\|_\alpha$ and $\|\cdot\|_\beta$ are respectively the C^α- and C^β-norms on R^n.

Again, the functions U and U^j, $j \in \mathrm{Z}_+$, may also depend (analytically, smoothly, and/or periodically) on other variables to be treated as parameters.

Remark. The concept of Whitney-smoothness as well as Theorem 6.15 and Lemmas 6.14 and 6.16 admit straightforward generalizations to the case of the so called anisotropic differentiability (Pöschel [278]). This fact is of importance for the finite differentiable KAM theory. For some other generalizations of the Whitney Extension Theorem, see, e.g., [69] and references therein.

6.2 Conjectural new contexts

As we pointed out at the end of § 2.3.5, all our five "main" KAM Theorems 2.3–2.7 can be proven within a unified approach that treats each of the five contexts we have studied in this book as a special case (maybe, together with certain adaptations). This approach is applicable to other situations also, any of the latter can be considered as a new context. For instance, one can think of "richer" structures $\mathfrak{S}$ on the phase space, e.g., volume preserving, Hamiltonian, or reversible vector fields equivariant with respect to the action of some discrete Lie group. Nevertheless, there is no guarantee that those new contexts will be mathematically interesting. It seems, on the other hand, that, e.g., the Hamiltonian coisotropic context can be hardly included in the general scheme although it is of undoubtful mathematical importance and has physical applications. The KAM theory for this context was briefly discussed in § 1.4.2. Here we would like to invent some new contexts "intermediate" between the Hamiltonian coisotropic context and the volume preserving context with $p = 1$. Consider these two contexts again, supposing for simplicity that there are no external parameters.

Proposition 6.17 (A particular case of Theorem 2.27) *A typical globally divergence-free vector field on an N-dimensional manifold equipped with a volume element ($N \geq 2$) possesses Whitney-smooth 1-parameter families of Floquet Diophantine invariant analytic $(N-1)$-tori.*

Proposition 6.18 (Corollary of Theorems 1.5, 1.6, and 2.28) *A typical Hamiltonian vector field on an N-dimensional symplectic manifold ($N \geq 2$ being even) possesses, for each integer n in the range $N/2 \leq n \leq N-1$, Cantor-like $(N-n)$-parameter families of Floquet Diophantine coisotropic invariant analytic n-tori.*

In fact, these families are Whitney-smooth for $n = N/2$ (Theorem 2.28) and analytic for $n = N-1$ (Theorem 1.6). Conjecturally, the families of tori in Proposition 6.18 are Whitney-smooth for $N/2 < n < N-1$ also (see a discussion in [361] for the case $n = N-2$).

Remark. Of course, typical Hamiltonian vector fields with $N/2$ degrees of freedom possess also Floquet Diophantine invariant analytic n-tori for $0 \leq n \leq (N/2)-1$ (Theorem 2.28), but those tori are not coisotropic (they are isotropic) and they constitute n-parameter families rather than $(N-n)$-parameter ones. The torus dimension $n = N/2$ was always in this book included in the Hamiltonian isotropic context, but here we would prefer to consider it together with the dimensions $n > N/2$ pertaining to the Hamiltonian coisotropic context.

Comparing Propositions 6.17 and 6.18, one can conjecture that the KAM theory can be carried over to dynamical systems preserving a closed differential form *of an arbitrary degree.* Such generalization would be a "bridge" between the KAM theorems for Hamiltonian vector fields and those for globally divergence-free vector fields.

Let ζ be a closed differential ν-form on an N-dimensional manifold M where $2 \leq \nu \leq N$ and $\nu \equiv N \bmod 2$ (i.e., $N - \nu$ is even).

Definition 6.19 *The form ζ is said to be nondegenerate if for each point $u \in M$ and any nonzero vector $X \in T_uM$, there exist vectors $X_1, \ldots, X_{\nu-1} \in T_uM$ such that $\zeta(X, X_1, \ldots, X_{\nu-1}) \neq 0$.*

For the forms of degrees N and 2, this definition of nondegeneracy coincides with the standard definition.

Example 6.20 On even-dimensional manifolds, nondegenerate forms are exemplified by powers of a symplectic structure. Let (x, y) be local coordinates on a $2n$-dimensional manifold M, where $x \in \mathbb{R}^n$, $y \in \mathbb{R}^n$. The 2-form $\omega^2 = dy_1 \wedge dx_1 + \cdots + dy_n \wedge dx_n$ is a symplectic structure on M. Then for each integer m in the range $1 \leq m \leq n$ the $2m$-form $\zeta = (m!)^{-1}(\omega^2)^m$ is nondegenerate. Indeed, let X be a nonzero vector at some point of M. One can suppose without loss of generality that $dy_1(X) = a \neq 0$. Set $X_1 = \partial/\partial x_1$, $X_{2j} = \partial/\partial y_{j+1}$, $X_{2j+1} = \partial/\partial x_{j+1}$ $(1 \leq j \leq m-1)$, then $\zeta(X, X_1, \ldots, X_{2m-1}) = a$.

Example 6.21 Now consider odd-dimensional manifolds. Let (x, y, z) be local coordinates on a $(2n+1)$-dimensional manifold M, where $x \in \mathbb{R}^n$, $y \in \mathbb{R}^n$, $z \in \mathbb{R}$, and let $\omega^2 = dy_1 \wedge dx_1 + \cdots + dy_n \wedge dx_n$. Then for each integer m in the range $1 \le m \le n$ the $(2m+1)$-form $\zeta = (m!)^{-1}(\omega^2)^m \wedge dz$ is nondegenerate. Indeed, let X be a nonzero vector at some point of M. First suppose that $dz(X) = a \neq 0$. Set $X_{2j-1} = \partial/\partial y_j$, $X_{2j} = \partial/\partial x_j$ $(1 \le j \le m)$, then $\zeta(X, X_1, \ldots, X_{2m}) = a$. Now let $dz(X) = 0$. Then one can suppose without loss of generality that $dy_1(X) = a \neq 0$. Set $X_1 = \partial/\partial x_1$, $X_{2j} = \partial/\partial y_{j+1}$, $X_{2j+1} = \partial/\partial x_{j+1}$ $(1 \le j \le m-1)$, $X_{2m} = \partial/\partial z$, then $\zeta(X, X_1, \ldots, X_{2m}) = a$.

Definition 6.22 *A submanifold L of M is said to be ζ-coisotropic if for each point $u \in L$ and any vector $X \in T_uM$ not tangent to L, there exist vectors $X_1, \ldots, X_{\nu-1} \in T_uL$ such that $\zeta(X, X_1, \ldots, X_{\nu-1}) \neq 0$.*

In particular, if $L \subset M$ is a ζ-coisotropic submanifold then $\dim L \ge \nu - 1$. If $\nu = N$ and ζ is a volume element on M then any $(N-1)$-dimensional submanifold of M is ζ-coisotropic, whereas if $\nu = 2$ and ζ is a symplectic structure on M then the ζ-coisotropic submanifolds of M are just submanifolds coisotropic in the usual sense (see § 1.3.2).

Definition 6.23 *A vector field X on M is said to be ζ-preserving if the $(\nu-1)$-form $i_X\zeta$ is closed, and said to be globally ζ-preserving if this form is exact.*

Recall that the value of $i_X\zeta$ at the vectors $X_1, \ldots, X_{\nu-1}$ is equal to the value of ζ at the vectors $X, X_1, \ldots, X_{\nu-1}$. If ζ is a volume element $(\nu = N)$ then the globally ζ-preserving vector fields are just globally divergence-free vector fields, and if ζ is a symplectic structure $(\nu = 2)$ then the globally ζ-preserving vector fields are just Hamiltonian vector fields.

Conjecture 6.24 (Sevryuk [328]) *If the form ζ is nondegenerate then a typical globally ζ-preserving analytic vector field on M possesses, for each integer n in the range*

$$\tfrac{1}{2}(N+\nu) - 1 \le n \le N - 1, \tag{6.32}$$

Whitney-smooth $(N-n)$-parameter families of Floquet Diophantine ζ-coisotropic invariant analytic n-tori. The Floquet $(N-n) \times (N-n)$ matrix of each of these tori is zero.

For $\nu = N$ and $\nu = 2$, we arrive at Propositions 6.17 and 6.18, respectively.

If a globally ζ-preserving vector field in Conjecture 6.24 depends on s external parameters, we will obtain an $(N-n+s)$-parameter family of Floquet Diophantine ζ-coisotropic invariant n-tori.

Conjecture 6.24 can be further justified by the observation that in the *integrable* set-up, where the vector field in question and the form ζ are equivariant with respect to the free action of the standard n-torus $\mathbb{T}^n$, there is an *analytic* $(N-n)$-parameter family of ζ-coisotropic invariant analytic n-tori with parallel dynamics (otherwise speaking, the phase space turns out to be foliated into such tori). Indeed, consider the space $M = \mathbb{T}^n \times \mathbb{R}^{N-n}$ with coordinates $(x \in \mathbb{T}^n, y \in \mathbb{R}^{N-n})$ and a ν-form ζ on M with x-independent coefficients, each torus $\{y = const\}$ being ζ-coisotropic. Let

$$X = \omega(y)\frac{\partial}{\partial x} + f(y)\frac{\partial}{\partial y}$$

be a globally ζ-preserving integrable vector field on M, so that the $(\nu-1)$-form $\eta = i_X\zeta$ is exact. Prove that $f(y) \equiv 0$. Indeed, suppose that $f(y_0) \neq 0$ for some $y_0 \in \mathrm{R}^{N-n}$. Identify the tangent spaces $T_{(x_0,y_0)}(\mathrm{T}^n \times \{y_0\}) = U \cong \mathrm{R}^n$ to $\mathrm{T}^n \times \{y_0\}$ at all the points $x_0 \in \mathrm{T}^n$. We can treat $\mathbf{u} = \omega(y_0)\partial/\partial x$ as a vector in U and $\mathbf{v} = f(y_0)\partial/\partial y$ as a vector in $T_{y_0}\mathrm{R}^{N-n} \cong \mathrm{R}^{N-n}$. We know that $\mathbf{v} \neq 0$, the n-torus $\mathrm{T}^n \times \{y_0\}$ is ζ-coisotropic, and the form ζ has x_0-independent coefficients. Consequently, there are vectors $X_1^\sharp, \ldots, X_{\nu-1}^\sharp \in U$ such that $\eta(X_1^\sharp, \ldots, X_{\nu-1}^\sharp) = \zeta(\mathbf{u}+\mathbf{v}, X_1^\sharp, \ldots, X_{\nu-1}^\sharp) = a \neq 0$. On the other hand, since the form η is exact, so is its restriction $\eta|_{y_0}$ to the n-torus $\mathrm{T}^n \times \{y_0\}$. The form $\eta|_{y_0}$ is therefore a sum of terms of the form $[\partial F(x)/\partial x_{j_0}]dx_{j_1} \wedge \cdots \wedge dx_{j_{\nu-1}}$, and for any $\nu-1$ *constant* vector fields $X_1, \ldots, X_{\nu-1}$ on $\mathrm{T}^n \times \{y_0\}$ we will have

$$\int_{\mathrm{T}^n} \eta(X_1, \ldots, X_{\nu-1})dx = 0,$$

which contradicts the conclusion that $\eta(X_1^\sharp, \ldots, X_{\nu-1}^\sharp) = a \neq 0$ at every point $(x_0, y_0) \in \mathrm{T}^n \times \{y_0\}$ (cf. § 5.1.3). Thus, $f(y) \equiv 0$, and each torus $\{y = const\}$ is invariant under the flow of X and carries parallel dynamics with frequency vector $\omega(y)$.

Example 6.25 Let $n \geq 2$, $0 \leq p \leq (n-1)/2$, $N = 2(n-p)$ [n and p being integers], $M = \mathrm{T}^n \times \mathrm{R}^{n-2p}$ [so that $\dim M = N$], (φ, ψ) be coordinates on T^n with $\varphi \in \mathrm{T}^{n-2p}$, $\psi \in \mathrm{T}^{2p}$, and y be coordinates on R^{n-2p}. We wish to consider closed nondegenerate ν-forms on M of even degrees $\nu = 2m \geq 2$ satisfying inequalities (6.32):

$$n-p+m-1 \leq n \leq 2n-2p-1 \iff m \leq p+1.$$

Thus, $1 \leq m \leq p+1$. Let

$$\zeta = \frac{1}{m!}\left(\sum_{\iota=1}^{n-2p} dy_\iota \wedge d\varphi_\iota + \sum_{j=1}^{p} d\psi_j \wedge d\psi_{j+p}\right)^m.$$

This $2m$-form is nondegenerate (see Example 6.20). Moreover, each n-torus $\{y = const\}$ is ζ-coisotropic. Indeed, let $X \in T_{(\varphi_0,\psi_0,y_0)}M$ be not tangent to $T_{(\varphi_0,\psi_0,y_0)}(\mathrm{T}^n \times \{y_0\})$. We can suppose without loss of generality that $dy_1(X) = a \neq 0$. Set $X_1 = \partial/\partial\varphi_1$, $X_{2j} = \partial/\partial\psi_j$, $X_{2j+1} = \partial/\partial\psi_{j+p}$ $(1 \leq j \leq m-1 \leq p)$, then $\zeta(X, X_1, \ldots, X_{2m-1}) = a$. One can easily verify that any constant vector field on M of the form $\omega\,\partial/\partial\varphi$ is globally ζ-preserving.

Example 6.26 Let $n \geq 2$, $1 \leq p \leq n/2$, $N = 2(n-p)+1$ [n and p being integers], $M = \mathrm{T}^n \times \mathrm{R}^{n-2p+1}$ [so that $\dim M = N$], (φ, ψ, χ) be coordinates on T^n with $\varphi \in \mathrm{T}^{n-2p+1}$, $\psi \in \mathrm{T}^{2p-2}$, $\chi \in \mathrm{T}^1 = \mathrm{S}^1$, and y be coordinates on R^{n-2p+1}. We wish to consider closed nondegenerate ν-forms on M of odd degrees $\nu = 2m+1 \geq 3$ satisfying inequalities (6.32):

$$n-p+m \leq n \leq 2n-2p \iff m \leq p.$$

Thus, $1 \leq m \leq p$. Let

$$\zeta = \frac{1}{m!}\left(\sum_{\iota=1}^{n-2p+1} dy_\iota \wedge d\varphi_\iota + \sum_{j=1}^{p-1} d\psi_j \wedge d\psi_{j+p-1}\right)^m \wedge d\chi.$$

This $(2m+1)$-form is nondegenerate (see Example 6.21). Moreover, each n-torus $\{y = const\}$ is ζ-coisotropic. Indeed, let $X \in T_{(\varphi_0,\psi_0,\chi_0,y_0)}M$ be not tangent to $T_{(\varphi_0,\psi_0,\chi_0,y_0)}(\mathrm{T}^n \times$

$\{y_0\}$). We can suppose without loss of generality that $dy_1(X) = a \neq 0$. Set $X_1 = \partial/\partial\varphi_1$, $X_{2j} = \partial/\partial\psi_j$, $X_{2j+1} = \partial/\partial\psi_{j+p-1}$ $(1 \le j \le m-1 \le p-1)$, $X_{2m} = \partial/\partial\chi$, then $\zeta(X, X_1, \ldots, X_{2m}) = a$. One can easily verify that any constant vector field on M of the form $\omega\,\partial/\partial\varphi$ is globally ζ-preserving.

It would be also interesting to develop the KAM theory for vector fields globally preserving several forms $\zeta_1, \ldots, \zeta_l$ simultaneously.

One can hardly formulate an analogue of Conjecture 6.24 for diffeomorphisms (see § 5.1.3), in contrast to what is stated in [328].

6.3 The Bruno theory

The Bruno continuation theory describes invariant analytic surfaces passing through a Floquet quasi-periodic invariant analytic torus of an analytic family of analytic vector fields. Here we give a very brief summary of some concepts of this theory, and, in fact, the main purpose of the present section is to explain notations in the proofs of Theorems 3.1 and 3.4 in Section 3.1. For a detailed exposition of the theory, we refer the reader to original Bruno's book [72, Part II], see also [70].

Consider an analytic family $X = \{X^\mu\}_\mu$ of analytic vector fields on $M = \mathbb{T}^n \times \mathbb{R}^{m+p_++p_-}$, where $n, m, p_+, p_- \in \mathbb{Z}_+$ and $\mu \in \mathcal{O}_s(0)$, $s \in \mathbb{Z}_+$. Suppose that the system of differential equations determined by X has the form

$$\dot{x} = \omega + f(x, w, \mu), \qquad \dot{w} = \Omega w + W(x, w, \mu), \tag{6.33}$$

where $x \in \mathbb{T}^n$, $w \in \mathcal{O}_{m+p_++p_-}(0)$, $\omega \in \mathbb{R}^n$, $\Omega \in gl(m+p_++p_-, \mathbb{R})$, $f(x,w,\mu) = O(w,\mu)$, $W(x, w, \mu) = O_2(w) + O(\mu)$. In other words, the field X^0 possesses the Floquet invariant n-torus $\{w = 0\}$ with parallel dynamics, ω and Ω being the frequency vector and the Floquet matrix of this torus, respectively. Assume further that the torus $\{w = 0\}$ is quasi-periodic, i.e., the components of vector ω are rationally independent.

Let matrix Ω have p_+ eigenvalues $\Lambda_1^+, \ldots, \Lambda_{p_+}^+$ with positive real parts, p_- eigenvalues $\Lambda_1^-, \ldots, \Lambda_{p_-}^-$ with negative real parts, and m purely imaginary eigenvalues $\lambda_1, \ldots, \lambda_m$. Let $A \in gl(m+p_++p_-, \mathbb{C})$ be the Jordan normal form of Ω and $A = A^0 \oplus A^+ \oplus A^-$, where

$$A^0 \in gl(m, \mathbb{C}), \quad A^+ \in gl(p_+, \mathbb{C}), \quad A^- \in gl(p_-, \mathbb{C})$$

are Jordan normal matrices with diagonals

$$(\lambda_1, \ldots, \lambda_m), \quad (\Lambda_1^+, \ldots, \Lambda_{p_+}^+), \quad (\Lambda_1^-, \ldots, \Lambda_{p_-}^-),$$

respectively.

Our aim is to reduce the family X of vector fields to a "normal form" (to be more precise, to a "seminormal form") around the torus $\{w = 0\}$, cf. § 4.1.3. We will use coordinate changes *analytic* in x and *formal* in w and μ. Let $Z \in \mathbb{C}^L$ be some complex variable ranging in a neighborhood of the origin. Denote by $\mathfrak{R}_x[[Z]]^N$ the ring of formal power series

$$f = f(x, Z) = \sum_q f_q(x) Z^q, \quad q \in \mathbb{Z}_+^L, \tag{6.34}$$

whose coefficients $f_q : \mathrm{T}^n \to \mathrm{C}^N$ are analytic functions. A series (6.34) is called *convergent* if there exists a number $\epsilon_0 > 0$ such that the series

$$\sum_q |f_q(x)||Z_1|^{q_1} \cdots |Z_L|^{q_L}$$

converges for all complex x and Z in the domain $|\operatorname{Im} x| < \epsilon_0$, $|Z| < \epsilon_0$. Convergent series (6.34) constitute a subring $\mathfrak{R}_x^0[[Z]]^N$ of the ring $\mathfrak{R}_x[[Z]]^N$. If $N = 1$, the superscript [1] will be omitted. Our intention is to simplify the system (6.33) by means of coordinate transformations containing series in $\mathfrak{R}_x[[w, \mu]]$.

Consider a formal family of formal vector fields determining the system of differential equations

$$\begin{aligned}
\dot U &= \Upsilon(U,V,\mu) + \widehat\Upsilon(U,V,W^+,W^-,\mu)\\
\dot V &= \Xi(U,V,\mu) + \widehat\Xi(U,V,W^+,W^-,\mu)\\
\dot W^+ &= \mathfrak{W}^+(U,V,W^+,\mu) + \widehat{\mathfrak{W}}^+(U,V,W^+,W^-,\mu)\\
\dot W^- &= \mathfrak{W}^-(U,V,W^-,\mu) + \widehat{\mathfrak{W}}^-(U,V,W^+,W^-,\mu),
\end{aligned} \tag{6.35}$$

where $U \in \mathrm{T}^n$, $V \in \mathrm{C}^m$, $W^+ \in \mathrm{C}^{p_+}$, $W^- \in \mathrm{C}^{p_-}$, the series Υ, $\widehat\Upsilon$, Ξ, $\widehat\Xi$, $\mathfrak{W}^+$, $\widehat{\mathfrak{W}}^+$, $\mathfrak{W}^-$, $\widehat{\mathfrak{W}}^-$ in the right-hand sides belong to the rings $\mathfrak{R}_U[[V, W^+, W^-, \mu]]^N$ with appropriate dimensions $N = n, m, p_+, p_-$. Moreover, suppose that

$$\begin{aligned}
\Upsilon(U,V,\mu) &= \omega + O(V,\mu),\\
\Xi(U,V,\mu) &= A^0 V + O(V,\mu)O(V),\\
\mathfrak{W}^+(U,V,W^+,\mu) &= A^+W^+ + O(V,W^+,\mu)O(W^+),\\
\mathfrak{W}^-(U,V,W^-,\mu) &= A^-W^- + O(V,W^-,\mu)O(W^-),
\end{aligned} \tag{6.36}$$

and

$$\begin{aligned}
\widehat\Upsilon(U,V,W^+,W^-,\mu) &= O(W^+)O(W^-),\\
\widehat\Xi(U,V,W^+,W^-,\mu) &= O(W^+)O(W^-),\\
\widehat{\mathfrak{W}}^+(U,V,W^+,W^-,\mu) &= O(W^+)O(W^-),\\
\widehat{\mathfrak{W}}^-(U,V,W^+,W^-,\mu) &= O(W^+)O(W^-).
\end{aligned} \tag{6.37}$$

Here the symbol $O(a)O(b)$ [a and b being some complex multidimensional variables] denotes formal series in a, b (and, maybe, some other variables c) containing terms $C_{\alpha\beta\gamma}a^\alpha b^\beta c^\gamma$ with $|\alpha| \ge 1$, $|\beta| \ge 1$ only [α, β, γ being integer vectors with non-negative components].

Introduce also the following notation. Let $L \in \mathrm{N}$ and $1 \le j \le L$. By $\mathrm{N}_j^{(L)} \subset \mathrm{Z}^L$ we will denote the set of integer vectors

$$\mathrm{N}_j^{(L)} = \left\{ q \in \mathrm{Z}^L \ : \ q_1 \ge 0, \ldots, q_{j-1} \ge 0, q_j \ge -1, q_{j+1} \ge 0, \ldots, q_L \ge 0, \ \ q_1 + \cdots + q_L \ge 0 \right\}.$$

Set also $\mathrm{N}^{(L)} = \mathrm{N}_1^{(L)} \cup \cdots \cup \mathrm{N}_L^{(L)}$.

The series Υ, Ξ, $\mathfrak{W}^+$, $\mathfrak{W}^-$ can be expanded in the following way:

$$\begin{aligned}
\Upsilon_j(U,V,\mu) &= \sum_{k,q,\nu} \Upsilon_{jkq\nu} V^q \mu^\nu e^{i(k,U)},\\
1 \le j &\le n, \ \ q \in \mathrm{Z}_+^m;
\end{aligned} \tag{6.38}$$

$$\Xi_j(U,V,\mu) = V_j\Psi_j(U,V,\mu) = V_j \sum_{k,q,\nu} \Psi_{jkq\nu} V^q \mu^\nu e^{i\langle k,U\rangle},$$

$$1 \le j \le m, \quad q \in \mathbf{N}_j^{(m)}; \tag{6.39}$$

$$\mathfrak{W}_j^+(U,V,W^+,\mu) = W_j^+ \mathfrak{H}_j^+(U,V,W^+,\mu) = W_j^+ \sum_{k,q,r,\nu} \mathfrak{H}_{jkqr\nu}^+ V^q (W^+)^r \mu^\nu e^{i\langle k,U\rangle},$$

$$1 \le j \le p_+, \quad q \in \mathbf{Z}_+^m, \quad r \in \mathbf{N}_j^{(p_+)}; \tag{6.40}$$

$$\mathfrak{W}_j^-(U,V,W^-,\mu) = W_j^- \mathfrak{H}_j^-(U,V,W^-,\mu) = W_j^- \sum_{k,q,r,\nu} \mathfrak{H}_{jkqr\nu}^- V^q (W^-)^r \mu^\nu e^{i\langle k,U\rangle},$$

$$1 \le j \le p_-, \quad q \in \mathbf{Z}_+^m, \quad r \in \mathbf{N}_j^{(p_-)}. \tag{6.41}$$

Here $k \in \mathbf{Z}^n$ and $\nu \in \mathbf{Z}_+^s$ everywhere. Note that Ψ_j, $\mathfrak{H}_j^+$, and $\mathfrak{H}_j^-$ are *not*, generally speaking, formal power series in V, W^+, W^-, μ because they may contain terms where one of the variables is in power -1. These entities can be treated, however, as formal Laurent series in V, W^+, W^-, μ.

Definition 6.27 *A system (6.35) of differential equations with properties (6.36), (6.37) is called a Bruno seminormal form if in the Laurent–Fourier expansions (6.38)–(6.41),*

$$\begin{aligned} \Upsilon_{jkq\nu} &= 0 \quad \text{if} \quad i\langle k,\omega\rangle + \langle q,\lambda\rangle \ne 0, \\ \Psi_{jkq\nu} &= 0 \quad \text{if} \quad i\langle k,\omega\rangle + \langle q,\lambda\rangle \ne 0, \\ \mathfrak{H}_{jkqr\nu}^+ &= 0 \quad \text{if} \quad \langle r, \operatorname{Re}\Lambda^+\rangle \ne 0, \\ \mathfrak{H}_{jkqr\nu}^- &= 0 \quad \text{if} \quad \langle r, \operatorname{Re}\Lambda^-\rangle \ne 0. \end{aligned} \tag{6.42}$$

If this is the case, the system of differential equations

$$\dot U = \Upsilon(U,V,\mu), \qquad \dot V = \Xi(U,V,\mu) \tag{6.43}$$

obtained from system (6.35) by setting $W^+ = 0$, $W^- = 0$, is called a Bruno normal form. The corresponding family of vector fields is also called a Bruno normal form.

For any fixed integer vector $q \in \mathbf{N}^{(m)}$ define

$$\gamma(q) = \liminf_{|k|\to+\infty} \frac{\log|i\langle k,\omega\rangle + \langle q,\lambda\rangle|}{|k|}, \tag{6.44}$$

where the lim inf is taken over the integer vectors $k \in \mathbf{Z}^n \setminus \{0\}$ such that $i\langle k,\omega\rangle + \langle q,\lambda\rangle \ne 0$.

Theorem 6.28 *If $\gamma(q) \ge 0$ for all $q \in \mathbf{N}^{(m)}$ then system (6.33) can be reduced to Bruno seminormal form (6.35) by an invertible formal transformation*

$$x = U + \mathcal{U}(U,V,W^+,W^-,\mu), \quad w = \mathcal{W}(U,V,W^+,W^-,\mu) \tag{6.45}$$

where $\mathcal{U} \in \mathfrak{R}_U[[V,W^+,W^-,\mu]]^n$, $\mathcal{W} \in \mathfrak{R}_U[[V,W^+,W^-,\mu]]^{m+p_++p_-}$,

$$\begin{aligned} &\mathcal{U}(U,V,W^+,W^-,\mu) = O(V,W^+,W^-,\mu), \\ &\mathcal{W}(U,V,W^+,W^-,\mu) = S(V,W^+,W^-)^t + O_2(V,W^+,W^-) + O(\mu), \end{aligned}$$

$(V,W^+,W^-)^t$ is the vector $(V_1,\dots,V_m,W_1^+,\dots,W_{p_+}^+,W_1^-,\dots,W_{p_-}^-)$ treated as a column-vector, and $S \in GL(m+p_++p_-,\mathbf{C})$ is a matrix reducing the Floquet matrix Ω of the invariant n-torus $\{w=0\}$ of original vector field X^0 to the Jordan normal form: $S^{-1}\Omega S = A = A^0 \oplus A^+ \oplus A^-$.

Now for Bruno seminormal form (6.35), define the following formal sets:

$$\begin{aligned} \mathcal{A}_W &= \{(U,V,W^+,W^-,\mu) \ : \ \Pi(\Upsilon,\Psi) = \Pi(\omega,\lambda)\} \cap \{W^+ = 0 \text{ or } W^- = 0\}, \\ \mathcal{A} &= \{(U,V,0,0,\mu) \ : \ \Pi(\Upsilon,\Psi) = \Pi(\omega,\lambda)\}, \\ \mathcal{B}_W &= \{(U,V,W^+,W^-,\mu) \in \mathcal{A}_W \ : \ B^{n+m} = 0\}, \\ \mathcal{B} &= \{(U,V,0,0,\mu) \in \mathcal{A} \ : \ B^{n+m} = 0\}, \end{aligned} \tag{6.46}$$

where $B = B(U,V,\mu)$ is the following $(n+m) \times (n+m)$ matrix:

$$B = \begin{pmatrix} \partial\Upsilon/\partial U & \partial\Upsilon/\partial V \\ \partial\Xi/\partial U & D^0\partial\Psi/\partial V \end{pmatrix} \tag{6.47}$$

(the so called *Bruno matrix*), series Ψ are defined by (6.39), D^0 is the diagonal $m \times m$ matrix $diag(V_1, \ldots, V_m)$, and $\Pi : \mathbf{C}^d \setminus \{0\} \to \mathbf{CP}^{d-1}$ for $d \in \mathbf{N}$ denotes the natural projection of the d-dimensional space $\mathbf{C}^d \setminus \{0\}$ onto the $(d-1)$-dimensional complex projective space. Note that the condition $B^{n+m} = 0$ means just that matrix B is nilpotent.

One can verify that all the four sets (6.46) are invariant under system (6.35).

Each of the sets (6.46) can be considered as the set of solutions of some system of equations $f_1 = \ldots = f_L = 0$ where $f_j \in \mathfrak{R}_U[[V,W^+,W^-,\mu]]$, $1 \le j \le L$. The series $f_1, \ldots, f_L$ generate some ideal in the ring $\mathfrak{R}_U[[V,W^+,W^-,\mu]]$. The ideals corresponding to the sets $\mathcal{A}_W$, $\mathcal{A}$, $\mathcal{B}_W$, $\mathcal{B}$ will be denoted by $\mathcal{I}_W$, $\mathcal{I}$, $\mathcal{J}_W$, $\mathcal{J}$, respectively. Note that $\mathcal{I}_W \subset \mathcal{I} \subset \mathcal{J}$ and $\mathcal{I}_W \subset \mathcal{J}_W \subset \mathcal{J}$, because $\mathcal{B} \subset \mathcal{A} \subset \mathcal{A}_W$ and $\mathcal{B} \subset \mathcal{B}_W \subset \mathcal{A}_W$.

Finally, introduce the following quantities:

$$\alpha_l = \min_{k,q} |i\langle k,\omega\rangle + \langle q,\lambda\rangle|, \quad l \in \mathbf{N},$$

where the minimum is taken over all the pairs $(k \in \mathbf{Z}^n, q \in \mathbf{N}^{(m)})$ such that $i\langle k,\omega\rangle + \langle q,\lambda\rangle \neq 0$ and $|k| + q_1 + \cdots + q_m < 2^l$;

$$\beta = \sum_{l=1}^{\infty} \frac{\log \alpha_l}{2^l}.$$

The so called *Bruno condition* β consists in that

$$\beta > -\infty. \tag{6.48}$$

If this condition is satisfied then $\gamma(q) \ge 0$ for all $q \in \mathbf{N}^{(m)}$ [$\gamma(q)$ being defined by (6.44)], and one can apply Theorem 6.28 to the original system (6.33).

Theorem 6.29 *If the Bruno condition β (6.48) is satisfied, then each of the two ideals $\mathcal{J}_W$ and $\mathcal{J}$ in the ring $\mathfrak{R}_U[[V,W^+,W^-,\mu]]$ has a basis consisting of series in $\mathfrak{R}^0_U[[V,W^+,W^-,\mu]]$ (i.e., of convergent series). Moreover, there exists a seminormalizing transformation (6.45) reducing the original system (6.33) to a Bruno seminormal form (6.35) and possessing the following property: there are convergent series*

$$\begin{aligned} \widetilde{\mathcal{U}}(U,V,W^+,W^-,\mu) &\in \mathfrak{R}^0_U[[V,W^+,W^-,\mu]]^n, \\ \widetilde{\mathcal{W}}(U,V,W^+,W^-,\mu) &\in \mathfrak{R}^0_U[[V,W^+,W^-,\mu]]^{m+p_++p_-} \end{aligned}$$

with the property

$$\widetilde{\mathcal{U}}(U,V,W^+,W^-,\mu) = O(V,W^+,W^-,\mu),$$
$$\widetilde{\mathcal{W}}(U,V,W^+,W^-,\mu) = S(V,W^+,W^-)^t + O_2(V,W^+,W^-) + O(\mu),$$

and such that

$$\widetilde{\mathcal{U}}_j - \mathcal{U}_j \in \mathcal{J}_W, \qquad 1 \le j \le n,$$
$$\widetilde{\mathcal{W}}_j - \mathcal{W}_j \in \mathcal{J}_W, \qquad 1 \le j \le m + p_+ + p_-.$$

One often formulates this theorem in the following way: if the Bruno condition β (6.48) is satisfied, then there exists a seminormalizing transformation (6.45) analytic on the set $\mathcal{B}_W$, the latter set being itself analytic.

6.4 A proof of the Diophantine approximation lemma

In this section, we prove the theorem on Diophantine approximations of dependent quantities (Lemma 2.13) as stated in § 2.5.1, following the general lines of Bakhtin's argument [25].

Choose an arbitrary point $w^0 \in \Gamma$. The key observation in the proof is that there exists $c = c(w^0) > 0$ such that there are *no* integer vectors $k \in \mathrm{Z}^n$ subject to the inequalities

$$\Xi^Q_{\mathfrak{G}}(w^0) \,/\, \rho^Q_{\mathfrak{F}}(w^0) < \|k\| \le \left[\Xi^Q_{\mathfrak{G}}(w^0) \,/\, \rho^Q_{\mathfrak{F}}(w^0)\right] + c.$$

This means that if inequality

$$\langle k, \mathfrak{F}(w^0)\rangle \ne \mathfrak{G}(w^0), \quad k \in \mathrm{Z}^n$$

[cf. (2.87)] holds for

$$0 < \|k\| \le \Xi^Q_{\mathfrak{G}}(w^0) \,/\, \rho^Q_{\mathfrak{F}}(w^0)$$

then it holds for

$$0 < \|k\| \le \left[\Xi^Q_{\mathfrak{G}}(w^0) \,/\, \rho^Q_{\mathfrak{F}}(w^0)\right] + c.$$

Fix positive numbers $a < \rho^Q_{\mathfrak{F}}(w^0)$ and $b > \Xi^Q_{\mathfrak{G}}(w^0)$ such that

$$b/a = \left[\Xi^Q_{\mathfrak{G}}(w^0) \,/\, \rho^Q_{\mathfrak{F}}(w^0)\right] + c.$$

Introduce the notation

$$S = \min_{\|k\| > b/a} \|k\| > b/a$$

($k \in \mathrm{Z}^n$). Finally, fix a positive number

$$\sigma < \min_{0 < \|k\| \le b/a} |\langle k, \mathfrak{F}(w^0)\rangle - \mathfrak{G}(w^0)|$$

($k \in \mathrm{Z}^n$). Now one can choose a closed ball $\Gamma(w^0) \subset W$ centered at w^0 and a number $\delta^* = \delta^*(w^0) > 0$ such that for any C^Q mappings $\widetilde{\mathfrak{F}} : W \to \mathrm{R}^n$ and $\widetilde{\mathfrak{G}} : W \to \mathrm{R}$ satisfying conditions (2.88), the following inequalities are valid:

$$\rho^Q_{\widetilde{\mathfrak{F}}}(w^0) \ge a, \tag{6.49}$$

$$\max_{w\in\Gamma(w^0)} \Xi^Q_{\tilde{\mathfrak{G}}}(w) \le b, \tag{6.50}$$

$$\min_{w\in\Gamma(w^0)} \left|\left\langle k, \tilde{\mathfrak{F}}(w)\right\rangle - \tilde{\mathfrak{G}}(w)\right| \ge \sigma \quad \text{for all} \quad k \in Z^n,\ 0 < \|k\| \le b/a, \tag{6.51}$$

and

$$\max_{w\in\Gamma(w^0)} \max_{\|e\|=1} \max_{j=0}^{Q} \max_{\|u\|=1} \left|\sum_{|q|=j} \left\langle D^q\tilde{\mathfrak{F}}(w) - D^q\tilde{\mathfrak{F}}(w^0), e\right\rangle u^q\right| \le \frac{aS-b}{2S} \tag{6.52}$$

where $e \in R^n$, $u \in R^d$, $q \in Z^d_+$.

From now on, $\tilde{\mathfrak{F}}$ and $\tilde{\mathfrak{G}}$ will be fixed. Let also $k \in Z^n$ be fixed and $\|k\| > b/a$ (so that $\|k\| \ge S$). Consider a vector $u \in R^d$ with $\|u\| = 1$ and an integer j in the range $0 \le j \le Q$ such that

$$\left|\sum_{|q|=j} \left\langle D^q\tilde{\mathfrak{F}}(w^0), \frac{k}{\|k\|}\right\rangle u^q\right| \ge a.$$

The existence of u and j satisfying this condition is ensured by (6.49). Then for each $w \in \Gamma(w^0)$

$$\left|\sum_{|q|=j} \left[\left\langle D^q\tilde{\mathfrak{F}}(w), k\right\rangle - D^q\tilde{\mathfrak{G}}(w)\right] \frac{u^q}{\|k\|}\right| \ge a - \frac{aS-b}{2S} - \frac{b}{S} = \frac{aS-b}{2S},$$

here we have used (6.50) and (6.52). Let s be an arbitrary segment in $\Gamma(w^0)$ parallel to u. The jth derivative of the restriction of the function

$$\left\langle k, \tilde{\mathfrak{F}}(w)\right\rangle - \tilde{\mathfrak{G}}(w)$$

to s is separated from zero by the number

$$\frac{(aS-b)\|k\|}{2S} \ge \frac{aS-b}{2}.$$

Consequently, for any $\gamma > 0$ the (one-dimensional) Lebesgue measure of the set of those points $w \in s$ for which

$$\left|\left\langle k, \tilde{\mathfrak{F}}(w)\right\rangle - \tilde{\mathfrak{G}}(w)\right| < \gamma|k|^{-\tau} \tag{6.53}$$

does not exceed

$$C(j)\left(\frac{2S\gamma|k|^{-\tau}}{(aS-b)\|k\|}\right)^{1/j}$$

for $j \ge 1$ and is equal to 0 for $j = 0$ and $\gamma \le (aS-b)/2$. Here $C(j)$ is a certain universal positive constant. According to [24], one can take

$$C(j) = 4(j!/2)^{1/j}, \tag{6.54}$$

and this value is unimprovable. In fact, Bakhtin [24] proved the following statement:

Lemma 6.30 *If the jth derivative ($j \in \mathrm{N}$) of a C^j function $F : \mathrm{R} \to \mathrm{R}$ is no less than $\theta > 0$ everywhere, then*

$$\mathrm{meas}_1\{x \in \mathrm{R} : |F(x)| \le A\} \le C(j)(A/\theta)^{1/j}, \tag{6.55}$$

$C(j)$ being given by (6.54).

The simple proof of this lemma is based on the following property of Chebyshev polynomials: *on the space of all the polynomials $f(x)$ of degree j with real coefficients and coefficient 1 at x^j, the minimal value of the functional*

$$\max_{0 \le x \le X} |f(x)|$$

($X > 0$ is fixed) is attained at the first kind Chebyshev polynomial

$$\frac{X^j}{2^{2j-1}} \cos\left(j \arccos\left(\frac{2x - X}{X}\right)\right)$$

(see, e.g., [276]). For this polynomial F, inequality (6.55) becomes an equality with $\theta = j!$, $A = X^j 2^{1-2j}$ and $\mathrm{meas}_1\{x \in \mathrm{R} : |F(x)| \le A\} = \mathrm{meas}_1\{0 \le x \le X\} = X$.

Now the (d-dimensional) Lebesgue measure of the set of points $w \in \Gamma(w^0)$ subject to inequality (6.53) does not exceed

$$G_{d-1} r^{d-1} C(j) \left(\frac{2S\gamma|k|^{-\tau}}{(aS - b)\|k\|}\right)^{1/j}$$

for $j \ge 1$ and is equal to 0 for $j = 0$ and $\gamma \le (aS - b)/2$. Here r is the radius of $\Gamma(w^0)$ and G_{d-1} denotes the volume of the $(d-1)$-dimensional unit ball.[2]

Let

$$\gamma \le \min\left\{\frac{aS - b}{2}, \sigma, 1\right\}.$$

Then

$$\left|\left\langle k, \tilde{\mathfrak{F}}(w)\right\rangle - \tilde{\mathfrak{G}}(w)\right| \ge \gamma |k|^{-\tau}$$

for all $w \in \Gamma(w^0)$, $k \in \mathrm{Z}^n$, $0 < \|k\| \le b/a$ due to (6.51) and

$$\mathrm{meas}_d\left\{w \in \Gamma(w^0) : \left|\left\langle k, \tilde{\mathfrak{F}}(w)\right\rangle - \tilde{\mathfrak{G}}(w)\right| < \gamma|k|^{-\tau}\right\} \le K\gamma^{1/Q}|k|^{-(\tau+1)/Q}$$

for any $k \in \mathrm{Z}^n$, $\|k\| > b/a$ where $K = K(d, n, Q, a, b, S, r)$ is a certain constant. One observes that

$$\sum_k |k|^{-(\tau+1)/Q} < +\infty, \quad k \in \mathrm{Z}^n \setminus \{0\},$$

since $(\tau+1)/Q > n$. Therefore, the Lebesgue measure of the set of those points $w \in \Gamma(w^0)$, for which inequality (6.53) holds for at least one integer vector $k \in \mathrm{Z}^n \setminus \{0\}$, tends to 0 as $\gamma \downarrow 0$ uniformly in functions $\tilde{\mathfrak{F}}$ and $\tilde{\mathfrak{G}}$ satisfying (2.88). To complete the proof, it suffices to note that the compact set Γ can be covered by a *finite* collection of balls $\Gamma(w^0)$, $w^0 \in \Gamma$. □

[2] $G_d = \pi^{d/2}/\Gamma(\frac{d}{2}+1)$ where now Γ is the gamma function [44].

Bibliography

[1] R.H. Abraham and J.E. Marsden. *Foundations of Mechanics (2nd edition).* Benjamin/Cummings, 1978.

[2] J.F. Adams. *Lectures on Lie Groups.* Benjamin, New York, 1969.

[3] V.I. Arnol'd. On the stability of an equilibrium point of a Hamiltonian system of ordinary differential equations in the general elliptic case. *Sov. Math. Dokl.*, 2(2): 247–249, 1961.

[4] V.I. Arnol'd. Proof of a theorem by A.N. Kolmogorov on the persistence of quasi-periodic motions under small perturbations of the Hamiltonian. *Russian Math. Surveys*, 18(5): 9–36, 1963.

[5] V.I. Arnol'd. Small denominators and problems of stability of motion in classical and celestial mechanics. *Russian Math. Surveys*, 18(6): 85–191, 1963. [Corrigenda (in Russian): *Uspekhi Mat. Nauk*, 23(6): 216, 1968].

[6] V.I. Arnol'd. On Liouville's theorem concerning integrable problems of dynamics. *Sibirsk. Mat. Zh.*, 4(2): 471–474, 1963 [in Russian].

[7] V.I. Arnol'd. On the instability of dynamical systems with many degrees of freedom. *Sov. Math. Dokl.*, 5(3): 581–585, 1964.

[8] V.I. Arnol'd. Small divisors I. On mappings of the circle onto itself. *Amer. Math. Soc. Transl., Ser. 2*, 46: 213–284, 1965. [Russian original: *Izvest. Akad. Nauk SSSR, Ser. Mat.*, 25(1): 21–86, 1961; Corrigenda: *ibid.*, 28(2): 479–480, 1964].

[9] V.I. Arnol'd. On matrices depending on parameters. *Russian Math. Surveys*, 26(2): 29–43, 1971.

[10] V.I. Arnol'd. Lectures on bifurcations and versal families. *Russian Math. Surveys*, 27(5): 54–123, 1972.

[11] V.I. Arnol'd. Reversible systems. In R.Z. Sagdeev, editor, *Nonlinear and Turbulent Processes in Physics* (Kiev, 1983), *Vol. 3*, pages 1161–1174. Harwood Academic, Chur, New York, 1984.

[12] V.I. Arnol'd. *Geometrical Methods in the Theory of Ordinary Differential Equations (2nd edition).* Springer-Verlag, 1988. [Russian original: Nauka, Moscow, 1978].

[13] V.I. Arnol'd. *Mathematical Methods of Classical Mechanics (2nd edition).* Springer-Verlag, 1989. [Russian original: Nauka, Moscow, 1974].

[14] V.I. Arnol'd. About A.N. Kolmogorov. In A.N. Shiryaev, editor, *Kolmogorov in Recollections*, pages 144–172. Nauka, Moscow, 1993 [in Russian].

[15] V.I. Arnol'd. Catastrophe theory. In V.I. Arnol'd, editor, *Encyclopædia of Mathematical Sciences, Vol. 5, Dynamical Systems V*, pages 207–264. Springer-Verlag, 1994.

[16] V.I. Arnol'd, V.S. Afraĭmovich, Yu.S. Il'yashenko, and L.P. Shil'nikov. Bifurcation theory. In V.I. Arnol'd, editor, *Encyclopædia of Mathematical Sciences, Vol. 5, Dynamical Systems V*, pages 1–205. Springer-Verlag, 1994.

[17] V.I. Arnol'd and A. Avez. *Ergodic Problems of Classical Mechanics*. Addison-Wesley, 1989. [French original: Gauthier-Villars, 1968].

[18] V.I. Arnol'd and A.B. Givental'. Symplectic geometry. In V.I. Arnol'd and S.P. Novikov, editors, *Encyclopædia of Mathematical Sciences, Vol. 4, Dynamical Systems IV*, pages 1–136. Springer-Verlag, 1990.

[19] V.I. Arnol'd and Yu.S. Il'yashenko. Ordinary differential equations. In D.V. Anosov and V.I. Arnol'd, editors, *Encyclopædia of Mathematical Sciences, Vol. 1, Dynamical Systems I*, pages 1–148. Springer-Verlag, 1988.

[20] V.I. Arnol'd, V.V. Kozlov, and A.I. Neĭshtadt. Mathematical aspects of classical and celestial mechanics. In V.I. Arnol'd, editor, *Encyclopædia of Mathematical Sciences, Vol. 3, Dynamical Systems III*, pages 1–291. Springer-Verlag, 1988 [2nd edition: 1993].

[21] V.I. Arnol'd and L.D. Meshalkin. A.N. Kolmogorov's seminar on selected topics in analysis (1958–59). *Uspekhi Mat. Nauk*, 15(1): 247–250, 1960 [in Russian].

[22] V.I. Arnol'd and M.B. Sevryuk. Oscillations and bifurcations in reversible systems. In R.Z. Sagdeev, editor, *Nonlinear Phenomena in Plasma Physics and Hydrodynamics*, pages 31–64. Mir, Moscow, 1986.

[23] V.I. Bakhtin. Averaging in multifrequency systems. *Funct. Anal. Appl.*, 20(2): 83–88, 1986.

[24] V.I. Bakhtin. A strengthened extremal property of Chebyshev polynomials. *Moscow Univ. Math. Bull.*, 42(2): 24–26, 1987.

[25] V.I. Bakhtin. Diophantine approximations on images of mappings. *Dokl. Akad. Nauk Beloruss. SSR*, 35(5): 398–400, 1991 [in Russian].

[26] È.G. Belaga. On the reducibility of a system of ordinary differential equations in a neighborhood of a conditionally periodic motion. *Sov. Math. Dokl.*, 3(2): 360–364, 1962.

[27] G.R. Belitskiĭ. *Normal Forms, Invariants, and Local Mappings*. Naukova Dumka, Kiev, 1979 [in Russian].

[28] A.A. Bel'kovich. On the existence of invariant tori for a periodic Hamiltonian system having some of characteristic indices with non-zero real part. *Diff. Equat.*, 23(2): 133–138, 1987.

[29] G. Benettin, L. Galgani, and A. Giorgilli. A proof of Nekhoroshev's theorem for the stability times in nearly integrable Hamiltonian systems. *Celest. Mech.*, 37(1): 1–25, 1985.

[30] G. Benettin, L. Galgani, A. Giorgilli, and J.-M. Strelcyn. A proof of Kolmogorov's theorem on invariant tori using canonical transformations defined by the Lie method. *Nuovo Cimento B*, 79(2): 201–223, 1984.

[31] G. Benettin and G. Gallavotti. Stability of motions near resonances in quasi-integrable Hamiltonian systems. *J. Stat. Phys.*, 44(3–4): 293–338, 1986.

[32] P. Bergé, Y. Pomeau, and C. Vidal. *Order within Chaos. Towards a Deterministic Approach to Turbulence.* John Wiley, New York, 1986. [French original: Hermann, 1984].

[33] D. Bernstein and A. Katok. Birkhoff periodic orbits for small perturbations of completely integrable Hamiltonian systems with convex Hamiltonians. *Invent. Math.*, 88(2): 225–241, 1987.

[34] A. Bhowal, T.K. Roy, and A. Lahiri. Small-angle Krein collisions in a family of four-dimensional reversible maps. *Phys. Rev. E*, 47(6): 3932–3940, 1993.

[35] Yu.N. Bibikov. On the existence of invariant tori in a neighborhood of the equilibrium state of a system of differential equations. *Sov. Math. Dokl.*, 10(2): 261–265, 1969.

[36] Yu.N. Bibikov. The existence of conditionally periodic solutions of systems of differential equations. *Diff. Equat.*, 7(8): 1021–1027, 1971.

[37] Yu.N. Bibikov. A sharpening of a theorem of Moser. *Sov. Math. Dokl.*, 14(6): 1769–1773, 1973.

[38] Yu.N. Bibikov. *Local Theory of Nonlinear Analytic Ordinary Differential Equations.* Lect. Notes Math., Vol. 702. Springer-Verlag, 1979.

[39] Yu.N. Bibikov. *Multifrequency Nonlinear Oscillations and their Bifurcations.* Leningrad Univ. Press, 1991 [in Russian].

[40] Yu.N. Bibikov and V.A. Pliss. On the existence of invariant tori in a neighborhood of the zero solution of a system of ordinary differential equations. *Diff. Equat.*, 3(11): 967–976, 1967.

[41] G.D. Birkhoff. *Dynamical Systems.* AMS Colloquium Publications, Vol. IX, 1927 [revised edition: 1966].

[42] N.N. Bogolyubov, Yu.A. Mitropol'skiĭ, and A.M. Samoĭlenko. *Methods of Accelerated Convergence in Nonlinear Mechanics.* Springer-Verlag, 1976. [Russian original: Naukova Dumka, Kiev, 1969].

[43] J.-B. Bost. Tores invariants des systèmes dynamiques hamiltoniens (d'après Kolmogorov, Arnol'd, Moser, Rüssmann, Zehnder, Herman, Pöschel, ...). In *Séminaire Bourbaki, Vol. 639, 1984–85*, pages 113–157. Astérisque, 133–134, 1986.

[44] N. Bourbaki. *Éléments de Mathématique. Fasc. XXI. Livre V: Intégration. Chapitre V (2ième édition).* Hermann, 1967.

[45] N. Bourbaki. *Elements of Mathematics. Algebra I. Chapters 1–3.* Springer-Verlag, 1989. [French original: Hermann, 1970].

[46] B.L.J. Braaksma and H.W. Broer. Quasi-periodic flow near a codimension one singularity of a divergence free vector field in dimension four. In *Bifurcation, Théorie Ergodique et Applications* (Dijon, 1981), pages 74–142. Astérisque, 98–99, 1982.

[47] B.L.J. Braaksma and H.W. Broer. On a quasi-periodic Hopf bifurcation. *Ann. Institut Henri Poincaré, Analyse non linéaire*, 4(2): 115–168, 1987.

[48] B.L.J. Braaksma, H.W. Broer, and G.B. Huitema. Towards a quasi-periodic bifurcation theory. *Mem. Amer. Math. Soc.*, 83(421): 83–170, 1990.

[49] H. Brands, J.S.W. Lamb, and I. Hoveijn. Periodic orbits in k-symmetric dynamical systems. *Physica D*, 84(3–4): 460–475, 1995.

[50] G.E. Bredon. *Introduction to Compact Transformation Groups.* Acad. Press, 1972.

[51] T.J. Bridges. Symplecticity, reversibility and elliptic operators. In H.W. Broer, S.A. van Gils, I. Hoveijn, and F. Takens, editors, *Nonlinear Dynamical Systems and Chaos* (Proceedings of the *Dynamical Systems Conference*, Groningen, December 1995), pages 1–20. Birkhäuser, Basel, 1996.

[52] H.W. Broer. Formal normal form theorems for vector fields and some consequences for bifurcations in the volume preserving case. In D.A. Rand and L.S. Young, editors, *Dynamical Systems and Turbulence*, Lect. Notes Math., Vol. 898, pages 54–74. Springer-Verlag, 1981.

[53] H.W. Broer. Quasi-periodic flow near a codimension one singularity of a divergence free vector field in dimension three. In D.A. Rand and L.S. Young, editors, *Dynamical Systems and Turbulence*, Lect. Notes Math., Vol. 898, pages 75–89. Springer-Verlag, 1981.

[54] H.W. Broer. Quasi-periodicity in local bifurcation theory. In C.P. Bruter, A. Aragnol, and A. Lichnerowicz, editors, *Bifurcation Theory, Mechanics and Physics*, pages 177–208. Reidel, 1983. [Reprinted from *Nieuw Arch. Wisk.* 4, 1(1): 1–32, 1983].

[55] H.W. Broer. On some quasi-periodic bifurcations. *Delft Progress Report*, 12(1): 79–96, 1988.

[56] H.W. Broer. Quasi-periodic bifurcations, applications. In *Proceedings of the XIth Congress on Differential Equations and Applications / First Congress on Applied Mathematics* (Málaga, 1989), pages 3–21. University of Málaga, 1990.

[57] H.W. Broer. KAM theory: multi-periodicity in conservative and dissipative systems. *Nieuw Arch. Wisk.* 4, 14(1): 65–79, 1996.

[58] H.W. Broer, F. Dumortier, S.J. van Strien, and F. Takens. *Structures in Dynamics (Finite Dimensional Deterministic Studies).* Studies Math. Phys., Vol. 2. North-Holland, Elsevier, 1991.

[59] H.W. Broer and G.B. Huitema. A proof of the isoenergetic KAM-theorem from the "ordinary" one. *J. Differ. Eq.*, 90(1): 52–60, 1991.

[60] H.W. Broer and G.B. Huitema. Unfoldings of quasi-periodic tori in reversible systems. *J. Dynam. Differ. Eq.*, 7(1): 191–212, 1995.

[61] H.W. Broer, G.B. Huitema, and M.B. Sevryuk. Families of quasi-periodic motions in dynamical systems depending on parameters. In H.W. Broer, S.A. van Gils, I. Hoveijn, and F. Takens, editors, *Nonlinear Dynamical Systems and Chaos* (Proceedings of the *Dynamical Systems Conference*, Groningen, December 1995), pages 171–211. Birkhäuser, Basel, 1996.

[62] H.W. Broer, G.B. Huitema, and F. Takens. Unfoldings of quasi-periodic tori. *Mem. Amer. Math. Soc.*, 83(421): 1–81, 1990.

[63] H.W. Broer and F. Takens. Formally symmetric normal forms and genericity. In U. Kirchgraber and H.-O. Walther, editors, *Dynamics Reported, Vol. 2*, pages 39–59. Wiley, Chichester, 1989.

[64] H.W. Broer and F. Takens. Mixed spectra and rotational symmetry. *Arch. Rat. Mech. Anal.*, 124(1): 13–42, 1993.

[65] H.W. Broer and F.M. Tangerman. From a differentiable to a real analytic perturbation theory, applications to the Kupka–Smale theorems. *Ergod. Th. and Dynam. Syst.*, 6(3): 345–362, 1986.

[66] H.W. Broer and G. Vegter. Bifurcational aspects of parametric resonance. In C.K.R.T. Jones, U. Kirchgraber, and H.-O. Walther, editors, *Dynamics Reported (New Series), Vol. 1*, pages 1–53. Springer-Verlag, 1992.

[67] I.U. Bronstein and A.Ya. Kopanskiĭ. *Smooth Invariant Manifolds and Normal Forms.* World Scientific, River Edge, 1994. [Russian original: Shtiintsa, Kishinëv, 1992].

[68] I.U. Bronstein and A.Ya. Kopanskiĭ. Normal forms of vector fields satisfying certain geometric conditions. In H.W. Broer, S.A. van Gils, I. Hoveijn, and F. Takens, editors, *Nonlinear Dynamical Systems and Chaos* (Proceedings of the *Dynamical Systems Conference*, Groningen, December 1995), pages 79–101. Birkhäuser, Basel, 1996.

[69] Yu.A. Brudnyĭ and P.A. Shvartsman. Generalizations of Whitney's extension theorem. *Intern. Math. Res. Notices (electronic)*, (3): 129 *ff.* ($\approx$ 11 pp.), 1994.

[70] A.D. Bruno. Analytical form of differential equations, I and II. *Trans. Moscow Math. Soc.*, 25: 131–288, 1971 and 26: 199–239, 1972.

[71] A.D. Bruno. The normal form of a Hamiltonian system. *Russian Math. Surveys*, 43(1): 25–66, 1988.

[72] A.D. Bruno. *Local Methods in Nonlinear Differential Equations.* Springer-Verlag, 1989. [Russian original of Part I: Nauka, Moscow, 1979. Russian original of Part II: Akad. Nauk SSSR Inst. Prikl. Mat., preprints 97 and 98, 1974].

[73] A.D. Bruno. Normalization of a Hamiltonian system near an invariant cycle or torus. *Russian Math. Surveys*, 44(2): 53–89, 1989.

[74] A.D. Bruno. On conditions for nondegeneracy in Kolmogorov's theorem. *Sov. Math. Dokl.*, 45(1): 221–225, 1992.

[75] A.D. Bruno. *The Restricted 3-Body Problem: Plane Periodic Orbits.* Walter de Gruyter, Berlin, 1994. [Russian original: Nauka, Moscow, 1990].

[76] P. Brunovský. On one-parameter families of diffeomorphisms. *Comment. Math. Universitatis Carolinae*, 11(3): 559–582, 1970.

[77] P. Brunovský. On one-parameter families of diffeomorphisms II: Generic branching in higher dimensions. *Comment. Math. Universitatis Carolinae*, 12(4): 765–784, 1971.

[78] A. Celletti and L. Chierchia. Construction of analytic KAM surfaces and effective stability bounds. *Commun. Math. Phys.*, 118(1): 119–161, 1988.

[79] A. Celletti and L. Chierchia. Invariant curves for area-preserving twist maps far from integrable. *J. Stat. Phys.*, 65(3–4): 617–643, 1991.

[80] A. Celletti and C. Froeschlé. On the determination of the stochasticity threshold of invariant curves. *Intern. J. Bifurcation and Chaos*, 5(6): 1713–1719, 1995.

[81] A. Chenciner. Bifurcations de points fixes elliptiques, I. Courbes invariantes. *Publ. Math. I.H.E.S.*, 61: 67–127, 1985.

[82] A. Chenciner. Bifurcations de points fixes elliptiques, II. Orbites périodiques et ensembles de Cantor invariants. *Invent. Math.*, 80(1): 81–106, 1985.

[83] A. Chenciner. La dynamique au voisinage d'un point fixe elliptique conservatif: de Poincaré et Birkhoff à Aubry et Mather. In *Séminaire Bourbaki, Vol. 622, 1983–84*, pages 147–170. Astérisque, 121–122, 1985.

[84] A. Chenciner. Bifurcations de points fixes elliptiques, III. Orbites périodiques de "petites" périodes et élimination résonnante des couples de courbes invariantes. *Publ. Math. I.H.E.S.*, 66: 5–91, 1988.

[85] A. Chenciner and G. Iooss. Bifurcations de tores invariants. *Arch. Rat. Mech. Anal.*, 69(2): 109–198, 1979.

[86] A. Chenciner and G. Iooss. Persistance et bifurcation de tores invariants. *Arch. Rat. Mech. Anal.*, 71(4): 301–306, 1979.

[87] Ch.-Q. Cheng. Birkhoff–Kolmogorov–Arnol'd–Moser tori in convex Hamiltonian systems. *Commun. Math. Phys.*, 177(3): 529–559, 1996.

[88] Ch.-Q. Cheng and Y.-S. Sun. Existence of invariant tori in three-dimensional measure-preserving mappings. *Celest. Mech. Dynam. Astronom.*, 47(3): 275–292, 1990.

[89] Ch.-Q. Cheng and Y.-S. Sun. Existence of KAM tori in degenerate Hamiltonian systems. *J. Differ. Eq.*, 114(1): 288–335, 1994.

[90] L. Chierchia. On the stability problem for nearly-integrable Hamiltonian systems. In S.B. Kuksin, V.F. Lazutkin, and J. Pöschel, editors, *Seminar on Dynamical Systems* (St. Petersburg, 1991), pages 35–46. Birkhäuser, Basel, 1994.

[91] L. Chierchia and C. Falcolini. A direct proof of a theorem by Kolmogorov in Hamiltonian systems. *Ann. Sc. Norm. Super. Pisa, Sci. Fis. Mat., IV Ser.*, 21(4): 541–593, 1994.

[92] L. Chierchia and C. Falcolini. Compensations in small divisor problems. *Commun. Math. Phys.*, 175(1): 135–160, 1996.

[93] L. Chierchia and G. Gallavotti. Smooth prime integrals for quasi-integrable Hamiltonian systems. *Nuovo Cimento B*, 67(2): 277–295, 1982.

[94] L. Chierchia and G. Gallavotti. Drift and diffusion in phase space. *Ann. Institut Henri Poincaré, Physique théorique*, 60(1): 1–144, 1994.

[95] B.V. Chirikov. A universal instability of many-dimensional oscillator systems. *Phys. Rep.*, 52(5): 263–379, 1979.

[96] B.V. Chirikov and V.V. Vecheslavov. KAM integrability. In P.H. Rabinowitz and E. Zehnder, editors, *Analysis, et cetera* (research papers published in honor of J. Moser's 60th birthday), pages 219–236. Acad. Press, 1990.

[97] S.-N. Chow, K. Lu, and Y.-Q. Shen. Normal form and linearization for quasiperiodic systems. *Trans. Amer. Math. Soc.*, 331(1): 361–376, 1992.

[98] J.D. Crawford. Introduction to bifurcation theory. *Rev. Modern Phys.*, 63(4): 991–1037, 1991.

[99] R. Cushman and J.-C. van der Meer. The Hamiltonian Hopf bifurcation in the Lagrange top. In C. Albert, editor, Proceedings of the *5th Colloque International: Géométrie Symplectique et Mécanique*, Lect. Notes Math., Vol. 1416, pages 26–38. Springer-Verlag, 1990.

[100] H.H. de Jong. The measure of invariant subtori in a row of coupled pendulums *(in preparation)*.

[101] A. Delshams and R. de la Llave. Existence of quasi-periodic orbits and lack of transport for volume preserving transformations and flows. 1991 *(preprint)*.

[102] A. Delshams and P. Gutiérrez. Estimates on invariant tori near an elliptic equilibrium point of a Hamiltonian system. *J. Differ. Eq. (to appear)*.

[103] A. Delshams and P. Gutiérrez. Effective stability and KAM theory. *J. Differ. Eq.*, 128(2): 415–490, 1996.

[104] A. Delshams and P. Gutiérrez. Exponentially small estimates for KAM theorem near an elliptic equilibrium point. In C. Simó, editor, *Hamiltonian Systems with Three or More Degrees of Freedom* (S'Agaró, Spain, 1995), NATO ASI Series C: Math. Physics. Kluwer Academic, Dordrecht, 1996 *(to appear)*.

[105] R.L. Devaney. Reversible diffeomorphisms and flows. *Trans. Amer. Math. Soc.*, 218: 89–113, 1976.

[106] R.L. Devaney. Blue sky catastrophes in reversible and Hamiltonian systems. *Indiana Univ. Math. J.*, 26(2): 247–263, 1977.

[107] R. Douady. Une démonstration directe de l'équivalence des théorèmes de tores invariants pour difféomorphismes et champs de vecteurs. *C. R. Acad. Sci. Paris, Série I*, 295(2): 201–204, 1982.

[108] H. Dulac. Solutions d'un système d'équations différentielles dans le voisinage des valeurs singulières. *Bull. Soc. Math. France*, 40: 324–383, 1912.

[109] L.H. Eliasson. Perturbations of stable invariant tori for Hamiltonian systems. *Ann. Sc. Norm. Super. Pisa, Cl. Sci., IV Ser.*, 15(1): 115–147, 1988.

[110] L.H. Eliasson. Absolutely convergent series expansions for quasi-periodic motions. University of Stockholm, 1988 *(report 2–88)*.

[111] L.H. Eliasson. Hamiltonian systems with linear normal form near an invariant torus. In G. Turchetti, editor, *Nonlinear Dynamics* (Bologna, 1988), pages 11–29. World Sci. Publishing, Teaneck, NJ, 1989.

[112] L.H. Eliasson. Generalization of an estimate of small divisors by Siegel. In P.H. Rabinowitz and E. Zehnder, editors, *Analysis, et cetera* (research papers published in honor of J. Moser's 60th birthday), pages 283–299. Acad. Press, 1990.

[113] C. Falcolini. Compensations in small divisor problems. In C. Simó, editor, *Hamiltonian Systems with Three or More Degrees of Freedom* (S'Agaró, Spain, 1995), NATO ASI Series C: Math. Physics. Kluwer Academic, Dordrecht, 1996 *(to appear)*.

[114] H. Federer. *Geometric Measure Theory*. Springer-Verlag, 1969 [latest edition: 1996].

[115] N. Fenichel. Persistence and smoothness of invariant manifolds for flows. *Indiana Univ. Math. J.*, 21(3): 193–226, 1971.

[116] D. Flockerzi. Generalized bifurcation of higher dimensional tori. *J. Differ. Eq.*, 55(3): 346–367, 1984.

[117] M. Friedman. Quasi-periodic solutions of nonlinear ordinary differential equations with small damping. *Bull. Amer. Math. Soc.*, 73(3): 460–464, 1967.

[118] G. Gaeta. Normal forms of reversible dynamical systems. *Intern. J. Theor. Phys.*, 33(9): 1917–1928, 1994.

[119] D.M. Galin. On real matrices depending on parameters. *Uspekhi Mat. Nauk*, 27(1): 241–242, 1972 [in Russian].

[120] D.M. Galin. Versal deformations of linear Hamiltonian systems. *Amer. Math. Soc. Transl., Ser. 2*, 118: 1–12, 1982. [Russian original: *Trudy Sem. im. I.G. Petrovskogo*, 1: 63–74, 1975].

[121] G. Gallavotti. Twistless KAM tori. *Commun. Math. Phys.*, 164(1): 145–156, 1994.

[122] G. Gallavotti. Twistless KAM tori, quasi flat homoclinic intersections, and other cancellations in the perturbation series of certain completely integrable Hamiltonian systems. A review. *Rev. Math. Phys.*, 6(3): 343–411, 1994.

[123] G. Gallavotti and G. Gentile. Majorant series convergence for twistless KAM tori. *Ergod. Th. and Dynam. Syst.*, 15(5): 857–869, 1995.

[124] G. Gallavotti, G. Gentile, and V. Mastropietro. Field theory and KAM tori. *Math. Phys. Electron. J. (electronic)*, 1(5): ≈ 13 pp., 1995.

[125] G. Gentile. A proof of existence of whiskered tori with quasi-flat homoclinic intersections in a class of almost integrable Hamiltonian systems. *Forum Math.*, 7(6): 709–753, 1995.

[126] G. Gentile. Whiskered tori with prefixed frequencies and Lyapunov spectrum. *Dynam. Stability Syst.*, 10(3): 269–308, 1995.

[127] G. Gentile and V. Mastropietro. Tree expansion and multiscale analysis for KAM tori. *Nonlinearity*, 8(6): 1159–1178, 1995.

[128] G. Gentile and V. Mastropietro. Convergence of the Lindstedt series for KAM tori. 1995 *(preprint)*.

[129] G. Gentile and V. Mastropietro. KAM theorem revisited. *Physica D*, 90(3): 225–234, 1996.

[130] G. Gentile and V. Mastropietro. Methods for the analysis of the Lindstedt series for KAM tori and renormalizability in classical mechanics. A review with some applications. *Rev. Math. Phys.*, 8(3): 393–444, 1996.

[131] A. Giorgilli, A. Delshams, E. Fontich, L. Galgani, and C. Simó. Effective stability for a Hamiltonian system near an elliptic equilibrium point, with an application to the restricted three body problem. *J. Differ. Eq.*, 77(1): 167–198, 1989.

[132] A. Giorgilli and L. Galgani. Rigorous estimates for the series expansions of Hamiltonian perturbation theory. *Celest. Mech.*, 37(2): 95–112, 1985.

[133] A. Giorgilli and U. Locatelli. Kolmogorov theorem and classical perturbation theory. 1995 *(preprint)*.

[134] A. Giorgilli and A. Morbidelli. Invariant KAM tori and global stability for Hamiltonian systems. 1995 *(preprint)*.

[135] A.F. Golubchikov. On the structure of automorphisms of complex simple Lie groups. *Dokl. Akad. Nauk SSSR*, 77(1): 7–9, 1951 [in Russian].

[136] M. Golubitsky, J.E. Marsden, I. Stewart, and M. Dellnitz. The constrained Liapunov–Schmidt procedure and periodic orbits. In W.F. Langford and W. Nagata, editors, *Normal Forms and Homoclinic Chaos* (Waterloo, Canada, 1992), Fields Institute Communications, Vol. 4, pages 81–127. Amer. Math. Soc., Providence, RI, 1995.

[137] S.M. Graff. On the conservation of hyperbolic invariant tori for Hamiltonian systems. *J. Differ. Eq.*, 15(1): 1–69, 1974.

[138] S.M. Graff. Invariant tori for a class of Hamiltonian differential equations. In T.M. Rassias, editor, *Global Analysis – Analysis on Manifolds*, Teubner-Texte Math., Vol. 57, pages 111–125. Teubner-Verlag, Leipzig, 1983.

[139] J.M. Greene. A method for determining a stochastic transition. *J. Math. Phys.*, 20(6): 1183–1201, 1979.

[140] J.M. Greene. The status of KAM theory from a physicist's point of view. In G. Brown and A. Opie, editors, *Chaos in Australia* (Sydney, 1990), pages 8–23. World Scientific, River Edge, 1993.

[141] G. Haller and S. Wiggins. N-pulse homoclinic orbits in perturbations of resonant Hamiltonian systems. *Arch. Rat. Mech. Anal.*, 130(1): 25–101, 1995.

[142] G. Haller and S. Wiggins. Whiskered tori and chaos in resonant Hamiltonian normal forms. In W.F. Langford and W. Nagata, editors, *Normal Forms and Homoclinic Chaos* (Waterloo, Canada, 1992), Fields Institute Communications, Vol. 4, pages 129–149. Amer. Math. Soc., Providence, RI, 1995.

[143] H. Hanßmann. The quasi-periodic centre-saddle bifurcation *(in preparation)*.

[144] H. Hanßmann. Normal forms for perturbations of the Euler top. In W.F. Langford and W. Nagata, editors, *Normal Forms and Homoclinic Chaos* (Waterloo, Canada, 1992), Fields Institute Communications, Vol. 4, pages 151–173. Amer. Math. Soc., Providence, RI, 1995.

[145] H. Hanßmann. *Quasi-periodic motions of a rigid body – a case study on perturbations of superintegrable systems.* PhD thesis, University of Groningen, 1995.

[146] H. Hanßmann. Equivariant perturbations of the Euler top. In H.W. Broer, S.A. van Gils, I. Hoveijn, and F. Takens, editors, *Nonlinear Dynamical Systems and Chaos* (Proceedings of the *Dynamical Systems Conference*, Groningen, December 1995), pages 227–252. Birkhäuser, Basel, 1996.

[147] F. Hausdorff. *Set Theory (2nd edition).* Chelsea Publishing Co., New York, 1962. [German original: Veit (Leipzig), 1914; de Gruyter (Berlin), 1927].

[148] M.R. Herman. Construction d'un difféomorphisme minimal d'entropie topologique non nulle. *Ergod. Th. and Dynam. Syst.*, 1(1): 65–76, 1981.

[149] M.R. Herman. Une méthode pour minorer les exposants de Lyapounov et quelques exemples montrant le caractère local d'un théorème d'Arnol'd et de Moser sur le tore de dimension 2. *Comment. Math. Helvetici*, 58(3): 453–502, 1983.

[150] M.R. Herman. *Sur les courbes invariantes par les difféomorphismes de l'anneau, Vol. 1 et 2.* Astérisque, 103–104, 1983 and 144, 1986.

[151] M.R. Herman. Simple proofs of local conjugacy theorems for diffeomorphisms of the circle with almost every rotation number. *Bol. Soc. Bras. Mat.*, 16(1): 45–83, 1985.

[152] M.R. Herman. Existence et non existence de tores invariants par des difféomorphismes symplectiques. École Polytechnique (Palaiseau, France), 1988 *(preprint)*.

[153] M.R. Herman. *Talk held on the International Conference on Dynamical Systems* (Lyons), 1990.

[154] M.R. Herman. Inégalités "a priori" pour des tores lagrangiens invariants par des difféomorphismes symplectiques. *Publ. Math. I.H.E.S.*, 70: 47–101, 1990.

[155] M.R. Herman. Différentiabilité optimale et contre-exemples à la fermeture en topologie C^∞ des orbites récurrentes de flots hamiltoniens. *C. R. Acad. Sci. Paris, Série I*, 313(1): 49–51, 1991.

[156] M.R. Herman. Exemples de flots hamiltoniens dont aucune perturbation en topologie C^∞ n'a d'orbites périodiques sur un ouvert de surfaces d'énergies. *C. R. Acad. Sci. Paris, Série I*, 312(13): 989–994, 1991.

[157] M.W. Hirsch. *Differential Topology.* Springer-Verlag, 1976.

[158] M.W. Hirsch, C.C. Pugh, and M. Shub. *Invariant Manifolds.* Lect. Notes Math., Vol. 583. Springer-Verlag, 1977.

[159] P.J. Holmes and J.E. Marsden. Mel'nikov's method and Arnol'd diffusion for perturbations of integrable Hamiltonian systems. *J. Math. Phys.*, 23(4): 669–675, 1982.

[160] L. Hörmander. On the division of distributions by polynomials. *Ark. Mat.*, 3(6): 555–568, 1958.

[161] I. Hoveijn. Versal deformations and normal forms for reversible and Hamiltonian linear systems. *J. Differ. Eq.*, 126(2): 408–442, 1996.

[162] G.B. Huitema. *Unfoldings of quasi-periodic tori.* PhD thesis, University of Groningen, 1988.

[163] Yu.S. Il'yashenko. A steepness condition for analytic functions. *Russian Math. Surveys*, 41(1): 229–230, 1986.

[164] Yu.S. Il'yashenko. Stability of the equilibrium points in Hamiltonian systems with two degrees of freedom. L'Institut de Recherche Mathématique Avancée (Strasbourg, France), 1990 *(preprint 437/P-249)*.

[165] Yu.S. Il'yashenko and S.Yu. Yakovenko. Finitely-smooth normal forms of local families of diffeomorphisms and vector fields. *Russian Math. Surveys*, 46(1): 1–43, 1991.

[166] Yu.S. Il'yashenko and S.Yu. Yakovenko. Nonlinear Stokes phenomena in smooth classification problems. In Yu.S. Il'yashenko, editor, *Nonlinear Stokes Phenomena*, Advances Soviet Math., Vol. 14, pages 235–287. Amer. Math. Soc., Providence, RI, 1993.

[167] G. Iooss and J.E. Los. Quasi-genericity of bifurcations to high dimensional invariant tori for maps. *Commun. Math. Phys.*, 119(3): 453–500, 1988.

[168] R.A. Johnson and G.R. Sell. Smoothness of spectral subbundles and reducibility of quasi-periodic linear differential systems. *J. Differ. Eq.*, 41(2): 262–288, 1981.

[169] À. Jorba, R. de la Llave, and M. Zou. Lindstedt series for lower dimensional tori. In C. Simó, editor, *Hamiltonian Systems with Three or More Degrees of Freedom* (S'Agaró, Spain, 1995), NATO ASI Series C: Math. Physics. Kluwer Academic, Dordrecht, 1996 *(to appear)*.

[170] À. Jorba and C. Simó. On quasiperiodic perturbations of elliptic equilibrium points. *SIAM J. Math. Anal. (to appear)*.

[171] À. Jorba and C. Simó. On the reducibility of linear differential equations with quasiperiodic coefficients. *J. Differ. Eq.*, 98(1): 111–124, 1992.

[172] À. Jorba and J. Villanueva. On the persistence of lower dimensional invariant tori under quasiperiodic perturbations. 1996 *(preprint)*.

[173] Y. Katznelson. *An Introduction to Harmonic Analysis*. John Wiley, New York, 1968.

[174] A. Kelley. On the Liapounov subcenter manifold. *J. Math. Anal. Appl.*, 18(3): 472–478, 1967.

[175] U. Kirchgraber. A note on Liapunov's center theorem. *J. Math. Anal. Appl.*, 73(2): 568–570, 1980.

[176] J. Knobloch and A. Vanderbauwhede. Hopf bifurcation at k-fold resonances in reversible systems. TU Ilmenau, 1995 *(preprint No. M 16/95)*.

[177] J. Knobloch and A. Vanderbauwhede. Hopf bifurcation at k-fold resonances in conservative systems. In H.W. Broer, S.A. van Gils, I. Hoveijn, and F. Takens, editors, *Nonlinear Dynamical Systems and Chaos* (Proceedings of the *Dynamical Systems Conference*, Groningen, December 1995), pages 155–170. Birkhäuser, Basel, 1996.

[178] H. Koçak. Normal forms and versal deformations of linear Hamiltonian systems. *J. Differ. Eq.*, 51(3): 359–407, 1984.

[179] A.N. Kolmogorov. On the persistence of conditionally periodic motions under a small change of the Hamilton function. *Dokl. Akad. Nauk SSSR*, 98(4): 527–530, 1954 [in Russian]. [The English translation in G. Casati and J. Ford, editors, *Stochastic Behavior in Classical and Quantum Hamiltonian Systems*, Lect. Notes Phys., Vol. 93, pages 51–56. Springer-Verlag, 1979. Reprinted in: Bai Lin Hao, editor, *Chaos*, pages 81–86. World Scientific, 1984].

[180] A.N. Kolmogorov. The general theory of dynamical systems and classical mechanics. In *Proceedings of International Congress of Mathematicians* (Amsterdam, 1954), *Vol. 1*, pages 315–333. North-Holland, Amsterdam, 1957 [in Russian]. [Reprinted in: *International Mathematical Congress in Amsterdam, 1954 (Plenary Lectures)*, pages 187–208. Fizmatgiz, Moscow, 1961. The English translation as Appendix D in R.H.

Abraham. *Foundations of Mechanics*, pages 263–279. Benjamin, 1967. Reprinted as Appendix in R.H. Abraham and J.E. Marsden. *Foundations of Mechanics (2nd edition)*, pages 741–757. Benjamin/Cummings, 1978].

[181] V.V. Kozlov. *Qualitative Analysis Methods in Rigid Body Dynamics*. Moscow Univ. Press, 1980 [in Russian].

[182] V.V. Kozlov. Integrability and non-integrability in Hamiltonian mechanics. *Russian Math. Surveys*, 38(1): 1–76, 1983.

[183] V.V. Kozlov. *Symmetries, Topology, and Resonances in Hamiltonian Mechanics*. Springer-Verlag, 1996. [Russian original: Udmurt. Univ. Press, Izhevsk, 1995].

[184] S.B. Kuksin. An infinitesimal Liouville–Arnol'd theorem as a criterion of reducibility for variational Hamiltonian equations. *Chaos, Solitons & Fractals*, 2(3): 259–269, 1992.

[185] S.B. Kuksin. *Nearly Integrable Infinite-Dimensional Hamiltonian Systems*. Lect. Notes Math., Vol. 1556. Springer-Verlag, 1993.

[186] S.B. Kuksin. KAM theory for partial differential equations. In A. Joseph, F. Mignot, F. Murat, B. Prum, and R. Rentschler, editors, *Proceedings of the First European Congress of Mathematics* (Paris, 1992), *Invited Lectures, Vol. II*, Progress in Mathematics, Vol. 120, pages 123–157. Birkhäuser, Basel, 1994.

[187] A. Lahiri, A. Bhowal, and T.K. Roy. Fourth order resonant collisions of multipliers in reversible maps: Period-4 orbits and invariant curves. *Physica D*, 85(1-2): 10–24, 1995.

[188] A. Lahiri, A. Bhowal, T.K. Roy, and M.B. Sevryuk. Stability of invariant curves in four-dimensional reversible mappings near 1 : 1 resonance. *Physica D*, 63(1-2): 99–116, 1993.

[189] J.S.W. Lamb. Reversing symmetries in dynamical systems. *J. Phys. A: Math. Gen.*, 25(4): 925–937, 1992.

[190] J.S.W. Lamb. *Reversing symmetries in dynamical systems*. PhD thesis, University of Amsterdam, 1994.

[191] J.S.W. Lamb. Local bifurcations in k-symmetric dynamical systems. *Nonlinearity*, 9(2): 537–557, 1996.

[192] J.S.W. Lamb and M. Nicol. On symmetric ω-limit sets in reversible flows. In H.W. Broer, S.A. van Gils, I. Hoveijn, and F. Takens, editors, *Nonlinear Dynamical Systems and Chaos* (Proceedings of the *Dynamical Systems Conference*, Groningen, December 1995), pages 103–120. Birkhäuser, Basel, 1996.

[193] J.S.W. Lamb and G.R.W. Quispel. Reversing k-symmetries in dynamical systems. *Physica D*, 73(4): 277–304, 1994.

[194] J.S.W. Lamb and G.R.W. Quispel. Cyclic reversing k-symmetry groups. *Nonlinearity*, 8(6): 1005–1026, 1995.

[195] J.S.W. Lamb, J.A.G. Roberts, and H.W. Capel. Conditions for local (reversing) symmetries in dynamical systems. *Physica A*, 197(3): 379–422, 1993.

[196] P. Lancaster and M. Tismenetsky. *The Theory of Matrices (2nd edition)*. Academic, Orlando, Fla., 1985.

[197] V.F. Lazutkin. The existence of a continuum of closed invariant curves for a convex billiard. *Uspekhi Mat. Nauk*, 27(3): 201–202, 1972 [in Russian].

[198] V.F. Lazutkin. The existence of caustics for a billiard problem in a convex domain. *Math. USSR Izv.*, 7(1): 185–214, 1973.

[199] V.F. Lazutkin. Concerning Moser's theorem on invariant curves. In *Problems in the Dynamical Theory of Seismic Waves Propagation, Vol. 14*, pages 109–120. Nauka, Leningrad, 1974 [in Russian].

[200] V.F. Lazutkin. *Convex Billiard and Eigenfunctions of the Laplace Operator*. Leningrad Univ. Press, 1981 [in Russian].

[201] V.F. Lazutkin. *KAM Theory and Semiclassical Approximations to Eigenfunctions*. Springer-Verlag, 1993.

[202] A.J. Lichtenberg and M.A. Lieberman. *Regular and Chaotic Dynamics (2nd edition)*. Springer-Verlag, 1992.

[203] B.B. Lieberman. Quasi-periodic solutions of Hamiltonian systems. *J. Differ. Eq.*, 11(1): 109–137, 1972.

[204] P. Lochak. Canonical perturbation theory via simultaneous approximation. *Russian Math. Surveys*, 47(6): 57–133, 1992.

[205] P. Lochak. Hamiltonian perturbation theory: periodic orbits, resonances and intermittency. *Nonlinearity*, 6(6): 885–904, 1993.

[206] P. Lochak. Stability of Hamiltonian systems over exponentially long times: the near-linear case. In H.S. Dumas, K.R. Meyer, and D.S. Schmidt, editors, *Hamiltonian Dynamical Systems: History, Theory, and Applications* (Cincinnati, 1992), The IMA Volumes in Math. and Appl., Vol. 63, pages 221–229. Springer-Verlag, 1995.

[207] P. Lochak. Arnol'd diffusion; a compendium of remarks and questions. 1995 *(preprint)*.

[208] P. Lochak. Long time stability of many-dimensional Hamiltonian systems. In C. Simó, editor, *Hamiltonian Systems with Three or More Degrees of Freedom* (S'Agaró, Spain, 1995), NATO ASI Series C: Math. Physics. Kluwer Academic, Dordrecht, 1996 *(to appear)*.

[209] P. Lochak and A.I. Neĭshtadt. Estimates of stability time for nearly integrable systems with a quasiconvex Hamiltonian. *Chaos*, 2(4): 495–499, 1992.

[210] P. Lochak, A.I. Neĭshtadt, and L. Niederman. Stability of nearly integrable convex Hamiltonian systems over exponentially long times. In S.B. Kuksin, V.F. Lazutkin, and J. Pöschel, editors, *Seminar on Dynamical Systems* (St. Petersburg, 1991), pages 15–34. Birkhäuser, Basel, 1994.

[211] A.M. Lyapunov. *The General Problem of the Stability of Motion.* Taylor & Francis, London, 1992. [Russian original: Math. Soc. of Khar'kov, 1892. The French translation: *Problème Général de la Stabilité du Mouvement.* Ann. Math. Studies, Vol. 17. Princeton Univ. Press and Oxford Univ. Press, 1947].

[212] R.S. MacKay. Transition to chaos for area-preserving maps. In J.M. Jowett, M. Month, and S. Turner, editors, *Nonlinear Dynamics Aspects of Particle Accelerators*, Lect. Notes Phys., Vol. 247, pages 390–454. Springer-Verlag, 1986.

[213] R.S. MacKay. Converse KAM theory. In St. Pnevmatikos, T. Bountis, and Sp. Pnevmatikos, editors, *Singular Behavior and Nonlinear Dynamics, Vol. 1*, pages 109–113. World Sci. Publishing, Teaneck, NJ, 1989.

[214] R.S. MacKay. A criterion for nonexistence of invariant tori for Hamiltonian systems. *Physica D*, 36(1–2): 64–82, 1989.

[215] R.S. MacKay. *Renormalisation in Area-Preserving Maps.* World Scientific, River Edge, 1993.

[216] R.S. MacKay, J.D. Meiss, and I.C. Percival. Transport in Hamiltonian systems. *Physica D*, 13(1–2): 55–81, 1984.

[217] R.S. MacKay, J.D. Meiss, and J. Stark. Converse KAM theory for symplectic twist maps. *Nonlinearity*, 2(4): 555–570, 1989.

[218] R.S. MacKay and I.C. Percival. Converse KAM: theory and practice. *Commun. Math. Phys.*, 98(4): 469–512, 1985.

[219] L. Markus and K.R. Meyer. Generic Hamiltonian dynamical systems are neither integrable nor ergodic. *Mem. Amer. Math. Soc.*, (144): 1–52, 1974.

[220] J.E. Marsden and M. McCracken. *The Hopf Bifurcation and its Applications.* Springer-Verlag, 1976.

[221] W.S. Massey. *A Basic Course in Algebraic Topology.* Springer-Verlag, 1991.

[222] J.N. Mather. Nonexistence of invariant circles. *Ergod. Th. and Dynam. Syst.*, 4(2): 301–309, 1984.

[223] J.N. Mather. A criterion for the nonexistence of invariant circles. *Publ. Math. I.H.E.S.*, 63: 153–204, 1986.

[224] J.N. Mather and G. Forni. Action minimizing orbits in Hamiltonian systems. In S. Graffi, editor, *Transition to Chaos in Classical and Quantum Mechanics*, Lect. Notes Math., Vol. 1589, pages 92–186. Springer-Verlag, 1994.

[225] M.V. Matveev. Lyapunov stability of equilibrium points in reversible systems. *Math. Notes*, 57(1–2): 63–72, 1995.

[226] M.V. Matveev. *Stability of nonlinear reversible systems.* PhD thesis, Moscow State Aircraft Institute (Technical University), 1995 [in Russian].

[227] M.V. Matveev. Lyapunov and effective stability in reversible systems. 1996 *(preprint).*

[228] M.V. Matveev. Structure of the sets of invariant tori in the local KAM theory and problems of stability in reversible systems. In C. Simó, editor, *Hamiltonian Systems with Three or More Degrees of Freedom* (S'Agaró, Spain, 1995), NATO ASI Series C: Math. Physics. Kluwer Academic, Dordrecht, 1996 *(to appear).*

[229] J.D. Meiss. Symplectic maps, variational principles, and transport. *Rev. Modern Phys.*, 64(3): 795–848, 1992.

[230] I. Melbourne. Versal unfoldings of equivariant linear Hamiltonian vector fields. *Math. Proc. Camb. Phil. Soc.*, 114(3): 559–573, 1993.

[231] V.K. Mel'nikov. On some cases of conservation of conditionally periodic motions under a small change of the Hamilton function. *Sov. Math. Dokl.*, 6(6): 1592–1596, 1965.

[232] V.K. Mel'nikov. A family of conditionally periodic solutions of a Hamiltonian system. *Sov. Math. Dokl.*, 9(4): 882–886, 1968.

[233] K.R. Meyer. Hamiltonian systems with a discrete symmetry. *J. Differ. Eq.*, 41(2): 228–238, 1981.

[234] J. Milnor. *Morse Theory (5th printing).* Ann. Math. Studies, Vol. 51. Princeton Univ. Press, 1973.

[235] H.K. Moffatt. KAM-theory. *Bull. London Math. Soc.*, 22(1): 71–73, 1990.

[236] D. Montgomery and L. Zippin. *Topological Transformation Groups.* Interscience, New York, 1955.

[237] A. Morbidelli and A. Giorgilli. Quantitative perturbation theory by successive elimination of harmonics. *Celest. Mech. Dynam. Astronom.*, 55(2): 131–159, 1993.

[238] A. Morbidelli and A. Giorgilli. On a connection between KAM and Nekhoroshev's theorems. *Physica D*, 86(3): 514–516, 1995.

[239] A. Morbidelli and A. Giorgilli. Superexponential stability of KAM tori. *J. Stat. Phys.*, 78(5–6): 1607–1617, 1995.

[240] A. Morbidelli and A. Giorgilli. Sur un lien entre le théorème KAM et le théorème de Nekhoroshev. 1995 *(preprint).*

[241] A. Morbidelli and A. Giorgilli. Bounds on diffusion in phase space: Connections between KAM and Nekhoroshev's theorems and superexponential stability of invariant tori. In C. Simó, editor, *Hamiltonian Systems with Three or More Degrees of Freedom* (S'Agaró, Spain, 1995), NATO ASI Series C: Math. Physics. Kluwer Academic, Dordrecht, 1996 *(to appear).*

[242] F. Morgan. *Geometric Measure Theory: A Beginner's Guide*. Acad. Press, 1988.

[243] J. Moser. On invariant curves of area-preserving mappings of an annulus. *Nachr. Akad. Wiss. Göttingen, Math.-Phys. Kl. II*, (1): 1–20, 1962.

[244] J. Moser. On the theory of quasiperiodic motions. *SIAM Review*, 8(2): 145–172, 1966.

[245] J. Moser. A rapidly convergent iteration method and non-linear partial differential equations, I and II. *Ann. Sc. Norm. Super. Pisa, Sci. Fis. Mat., III Ser.*, 20(2): 265–315 and (3): 499–535, 1966.

[246] J. Moser. Convergent series expansions for quasi-periodic motions. *Math. Ann.*, 169(1): 136–176, 1967.

[247] J. Moser. Lectures on Hamiltonian systems. *Mem. Amer. Math. Soc.*, (81): 1–60, 1968 [3rd printing: 1989].

[248] J. Moser. *Stable and Random Motions in Dynamical Systems, with Special Emphasis on Celestial Mechanics*. Ann. Math. Studies, Vol. 77. Princeton Univ. Press, 1973.

[249] P. Moson. Quasi-periodic solutions of differential equations depending on parameters. *Z. Angew. Math. Mech.*, 65(4): 86–87, 1985.

[250] P. Moson. Quasiperiodic solutions of differential equations depending on parameters, I and II. *Vestnik Leningrad. Univ. Mat. Mekh. Astronom.*, (2): 16–22, 1986 and (3): 34–39, 1986 [in Russian].

[251] N.N. Nekhoroshev. On the behavior of Hamiltonian systems close to integrable ones. *Funct. Anal. Appl.*, 5(4): 338–339, 1971.

[252] N.N. Nekhoroshev. Action-angle variables and their generalizations. *Trans. Moscow Math. Soc.*, 26: 180–198, 1972.

[253] N.N. Nekhoroshev. Stable lower estimates for smooth mappings and the gradients of smooth functions. *Math. USSR Sbornik*, 19(3): 425–467, 1973.

[254] N.N. Nekhoroshev. An exponential estimate of the stability time for Hamiltonian systems close to integrable ones, I. *Russian Math. Surveys*, 32(6): 1–65, 1977.

[255] N.N. Nekhoroshev. An exponential estimate of the stability time for Hamiltonian systems close to integrable ones, II. In O.A. Oleĭnik, editor, *Topics in Modern Mathematics, Petrovskiĭ Seminar, No. 5*, pages 1–58. Consultant Bureau, New York, 1985. [Russian original: *Trudy Sem. im. I.G. Petrovskogo*, 5: 5–50, 1979].

[256] N.N. Nekhoroshev. The Poincaré–Lyapunov–Liouville–Arnol'd theorem. *Funct. Anal. Appl.*, 28(2): 128–129, 1994.

[257] A.I. Neĭshtadt. Estimates in the Kolmogorov theorem on the persistence of quasi-periodic motions. *J. Appl. Math. Mech.*, 45(6): 766–772, 1981.

[258] J.C. Oxtoby. *Measure and Category (2nd edition)*. Springer-Verlag, 1980.

[259] R.E.A.C. Paley and N. Wiener. *Fourier Transforms in the Complex Domain.* AMS Colloquium Publications, Vol. XIX, 1934.

[260] J. Palis and W.C. de Melo. *Geometric Theory of Dynamical Systems: An Introduction.* Springer-Verlag, 1982. [Portuguese original: IMPA, 1977].

[261] K.J. Palmer. Linearization of reversible systems. *J. Math. Anal. Appl.*, 60(3): 794–808, 1977.

[262] I.O. Parasyuk. Conservation of quasiperiodic motions in reversible multifrequency systems. *Dokl. Akad. Nauk Ukrain. SSR, Ser. A*, (9): 19–22, 1982 [in Russian].

[263] I.O. Parasyuk. Conservation of multidimensional invariant tori in Hamiltonian systems. *Ukrain. Math. J.*, 36(4): 380–385, 1984.

[264] I.O. Parasyuk. Coisotropic invariant tori of Hamiltonian systems in the quasiclassical theory of motion of a conduction electron. *Ukrain. Math. J.*, 42(3): 308–312, 1990.

[265] M.M. Peixoto. On an approximation theorem of Kupka and Smale. *J. Differ. Eq.*, 3(2): 214–227, 1967.

[266] I.C. Percival. A variational principle for invariant tori of fixed frequency. *J. Phys. A: Math. Gen.*, 12(3): L57–L60, 1979.

[267] I.C. Percival. Variational principles for invariant tori and cantori. In M. Month and J.C. Herrera, editors, *Nonlinear Dynamics and the Beam-Beam Interaction*, AIP Conference Proceedings, Vol. 57, pages 302–310. AIP Press, New York, 1980.

[268] I.C. Percival. Chaos in Hamiltonian systems. *Proc. Roy. Soc. London, Ser. A*, 413(1844): 131–143, 1987.

[269] I.C. Percival, R.S. MacKay, and J.D. Meiss. Transport in Hamiltonian systems. In R.Z. Sagdeev, editor, *Nonlinear and Turbulent Processes in Physics* (Kiev, 1983), *Vol. 3*, pages 1557-1572. Harwood Academic, Chur, New York, 1984.

[270] A.D. Perry and S. Wiggins. KAM tori are very sticky: rigorous lower bounds on the time to move away from an invariant Lagrangian torus with linear flow. *Physica D*, 71(1–2): 102–121, 1994.

[271] E. Piña and L. Jiménez Lara. On the symmetry lines of the standard mapping. *Physica D*, 26(1–3): 369–378, 1987.

[272] P.I. Plotnikov. Morse theory for quasi-periodic solutions of Hamiltonian systems. *Siberian Math. J.*, 35(3): 590–604, 1994.

[273] W. Pluschke. *Bifurcations of quasi-periodic solutions from fixed points of differentiable reversible systems.* PhD thesis, University of Stuttgart, 1989.

[274] H. Poincaré. *Sur les propriétés des fonctions définies par les équations aux dérivées partielles.* Thèse, Paris, 1879. [Reprinted in: *Œuvres de Henri Poincaré, Vol. 1*, Gauthier-Villars, 1928].

[275] H. Poincaré. *Les Méthodes Nouvelles de la Mécanique Céleste, I–III.* Dover Publications, 1957. [Original: Gauthier-Villars, 1892, 1893, 1899. The English translation: *New Methods of Celestial Mechanics.* AIP Press, Williston, 1992].

[276] G. Pólya and G. Szegö. *Problems and Theorems in Analysis, Vol. I and II.* Springer-Verlag, 1972 and 1976. [German original: Springer-Verlag, 1925, 1954, 1964].

[277] J. Pöschel. *Über invariante Tori in differenzierbaren Hamiltonschen Systemen.* Diplomarbeit, *Bonn. Math. Schr.*, Band 120, 1980.

[278] J. Pöschel. Integrability of Hamiltonian systems on Cantor sets. *Comm. Pure Appl. Math.*, 35(5): 653–696, 1982.

[279] J. Pöschel. On elliptic lower dimensional tori in Hamiltonian systems. *Math. Z.*, 202(4): 559–608, 1989.

[280] J. Pöschel. *On small divisors with spatial structure.* Habilitationsschrift, Rheinische Friedrich-Wilhelms-Universität, Bonn, 1989.

[281] J. Pöschel. Small divisors with spatial structure in infinite dimensional Hamiltonian systems. *Commun. Math. Phys.*, 127(2): 351–393, 1990.

[282] J. Pöschel. A lecture on the classical KAM theorem. ETH-Zürich, 1992 *(preprint)*.

[283] J. Pöschel. Nekhoroshev estimates for quasi-convex Hamiltonian systems. *Math. Z.*, 213(2): 187–216, 1993.

[284] T. Post, H.W. Capel, G.R.W. Quispel, and J.P. van der Weele. Bifurcations in two-dimensional reversible maps. *Physica A*, 164(3): 625–662, 1990.

[285] G.E. Prince, G.B. Byrnes, J. Sherring, and S.E. Godfrey. A generalization of the Liouville–Arnol'd theorem. *Math. Proc. Camb. Phil. Soc.*, 117(2): 353–370, 1995.

[286] A.S. Pyartli. Diophantine approximations on submanifolds of Euclidean space. *Funct. Anal. Appl.*, 3(4): 303–306, 1969.

[287] G.R.W. Quispel and M.B. Sevryuk. KAM theorems for the product of two involutions of different types. *Chaos*, 3(4): 757–769, 1993.

[288] W.P. Reinhardt. Chaos and collisions: introductory concepts. In F.A. Gianturco, editor, *Collision Theory for Atoms and Molecules*, NATO ASI Series B: Physics, Vol. 196, pages 465–518. Plenum Press, 1989.

[289] K.V. Rerikh. Non-algebraic integrability of the Chew–Low reversible dynamical system of the Cremona type and the relation with the 7th Hilbert problem (non-resonant case). *Physica D*, 82(1–2): 60–78, 1995.

[290] R.J. Rimmer. Generic bifurcations for involutory area-preserving maps. *Mem. Amer. Math. Soc.*, 41(272): 1–165, 1983.

[291] J.A.G. Roberts and M. Baake. Trace maps as 3D reversible dynamical systems with an invariant. *J. Stat. Phys.*, 74(3–4): 829–888, 1994.

[292] J.A.G. Roberts and G.R.W. Quispel. Chaos and time-reversal symmetry. Order and chaos in reversible dynamical systems. *Phys. Rep.*, 216(2–3): 63–177, 1992.

[293] R.C. Robinson. Generic properties of conservative systems, I and II. *Amer. J. Math.*, 92(3): 562–603 and (4): 897–906, 1970.

[294] T.K. Roy and A. Lahiri. Reversible Hopf bifurcation in four-dimensional maps. *Phys. Rev. A*, 44(8): 4937–4944, 1991.

[295] M. Rudnev and S. Wiggins. KAM theory near multiplicity one resonant surfaces in perturbations of a *priori* stable Hamiltonian systems. 1996 *(preprint)*.

[296] D. Ruelle. Strange attractors. *Math. Intelligencer*, 2(3): 126–137, 1980. [French original: *La Recherche* N° 108, Février 1980].

[297] D. Ruelle and F. Takens. On the nature of turbulence. *Commun. Math. Phys.*, 20(3): 167–192, 1971 and 23(4): 343–344, 1971.

[298] H. Rüssmann. Kleine Nenner I: Über invariante Kurven differenzierbarer Abbildungen eines Kreisringes. *Nachr. Akad. Wiss. Göttingen, Math.-Phys. Kl. II*, (5): 67–105, 1970.

[299] H. Rüssmann. Kleine Nenner II: Bemerkungen zur Newtonschen Methode. *Nachr. Akad. Wiss. Göttingen, Math.-Phys. Kl. II*, (1): 1–10, 1972.

[300] H. Rüssmann. Konvergente Reihenentwicklungen in der Störungstheorie der Himmelsmechanik. In K. Jacobs, editor, *Selecta Mathematica V (German)*, Heidelberger Taschenbücher, 201, pages 93–260. Springer-Verlag, 1979.

[301] H. Rüssmann. On the existence of invariant curves of twist mappings of an annulus. In J. Palis, editor, *Geometric Dynamics* (Proceedings of the *International Symposium on Dynamical Systems*, Rio de Janeiro, 1981), Lect. Notes Math., Vol. 1007, pages 677–718. Springer-Verlag, 1983.

[302] H. Rüssmann. Non-degeneracy in the perturbation theory of integrable dynamical systems. In M.M. Dodson and J.A.G. Vickers, editors, *Number Theory and Dynamical Systems*, London Math. Soc. Lect. Note Ser., Vol. 134, pages 5–18. Cambridge Univ. Press, 1989.

[303] H. Rüssmann. Nondegeneracy in the perturbation theory of integrable dynamical systems. In S. Albeverio, Ph. Blanchard, and D. Testard, editors, *Stochastics, Algebra and Analysis in Classical and Quantum Dynamics*, Math. and its Appl., Vol. 59, pages 211–223. Kluwer Academic, Dordrecht, 1990.

[304] H. Rüssmann. On twist-Hamiltonians. *Talk held on the Colloque international: Mécanique céleste et systèmes hamiltoniens* (Marseille), 1990.

[305] H. Rüssmann. On the frequencies of quasi periodic solutions of analytic nearly integrable Hamiltonian systems. In S.B. Kuksin, V.F. Lazutkin, and J. Pöschel, editors, *Seminar on Dynamical Systems* (St. Petersburg, 1991), pages 160–183. Birkhäuser, Basel, 1994.

[306] D. Salamon. The Kolmogorov–Arnol'd–Moser theorem. ETH-Zürich, 1986 *(preprint)*.

[307] D. Salamon and E. Zehnder. KAM theory in configuration space. *Comment. Math. Helvetici*, 64(1): 84–132, 1989.

[308] A.M. Samoĭlenko. *Elements of the Mathematical Theory of Multi-Frequency Oscillations.* Math. and its Appl., Vol. 71. Kluwer Academic, Dordrecht, 1991. [Russian original: Nauka, Moscow, 1987].

[309] V.S. Samovol. Linearization of systems of differential equations in a neighborhood of invariant toroidal manifolds. *Trans. Moscow Math. Soc.*, 38: 183–215, 1980.

[310] J. Scheurle. Bifurcation of a stationary solution of a dynamical system into n-dimensional tori of quasiperiodic solutions. In H.-O. Peitgen and H.-O. Walther, editors, *Functional Differential Equations and Approximation of Fixed Points*, Lect. Notes Math., Vol. 730, pages 442–454. Springer-Verlag, 1979.

[311] J. Scheurle. Quasi-periodic solutions of the plane three-body problem near Euler's orbits. *Celest. Mech.*, 28(1–2): 141–151, 1982.

[312] J. Scheurle. Bifurcation of quasi-periodic solutions from equilibrium points of reversible dynamical systems. *Arch. Rat. Mech. Anal.*, 97(2): 103–139, 1987.

[313] G.R. Sell. Bifurcation of higher dimensional tori. *Arch. Rat. Mech. Anal.*, 69(3): 199–230, 1979.

[314] G.R. Sell. Smooth linearization near a fixed point. *Amer. J. Math.*, 107(5): 1035–1091, 1985.

[315] M.B. Sevryuk. *Reversible Systems.* Lect. Notes Math., Vol. 1211. Springer-Verlag, 1986.

[316] M.B. Sevryuk. On invariant tori of reversible systems in a neighborhood of an equilibrium point. *Russian Math. Surveys*, 42(4): 147–148, 1987.

[317] M.B. Sevryuk. *Reversible dynamical systems.* PhD thesis, Moscow State University, 1987 [in Russian].

[318] M.B. Sevryuk. Invariant m-tori of reversible systems with the phase space of dimension greater than $2m$. *J. Soviet Math.*, 51(3): 2374–2386, 1990. [Russian original: *Trudy Sem. im. I.G. Petrovskogo*, 14: 109–124, 1989].

[319] M.B. Sevryuk. On the dimensions of invariant tori in the KAM theory. In V.V. Kozlov, editor, *Mathematical Methods in Mechanics*, pages 82–88. Moscow Univ. Press, 1990 [in Russian].

[320] M.B. Sevryuk. Lower-dimensional tori in reversible systems. *Chaos*, 1(2): 160–167, 1991.

[321] M.B. Sevryuk. Linear reversible systems and their versal deformations. *J. Soviet Math.*, 60(5): 1663–1680, 1992. [Russian original: *Trudy Sem. im. I.G. Petrovskogo*, 15: 33–54, 1991].

[322] M.B. Sevryuk. Invariant tori of reversible systems in the presence of additional even coordinates. *Russ. Acad. Sci. Dokl. Math.*, 46(2): 286–289, 1993.

[323] M.B. Sevryuk. Invariant tori of reversible systems of intermediate dimensions. *Russ. Acad. Sci. Dokl. Math.*, 47(1): 129–133, 1993. [Corrigenda (in Russian): *Dokl. Akad. Nauk*, 346(4): 576, 1996].

[324] M.B. Sevryuk. New cases of quasiperiodic motions in reversible systems. *Chaos*, 3(2): 211–214, 1993.

[325] M.B. Sevryuk. New results in the reversible KAM theory. In S.B. Kuksin, V.F. Lazutkin, and J. Pöschel, editors, *Seminar on Dynamical Systems* (St. Petersburg, 1991), pages 184–199. Birkhäuser, Basel, 1994.

[326] M.B. Sevryuk. The iteration-approximation decoupling in the reversible KAM theory. *Chaos*, 5(3): 552–565, 1995.

[327] M.B. Sevryuk. KAM-stable Hamiltonians. *J. Dynam. Control Syst.*, 1(3): 351–366, 1995.

[328] M.B. Sevryuk. Some problems of the KAM theory: quasi-periodic motions in typical systems. *Russian Math. Surveys*, 50(2): 341–353, 1995.

[329] M.B. Sevryuk. Invariant tori of Hamiltonian systems nondegenerate in the sense of Rüssmann. *Dokl. Akad. Nauk*, 346(5): 590–593, 1996 [in Russian, to be translated into English in *Russ. Acad. Sci. Dokl. Math.*].

[330] M.B. Sevryuk. Excitation of elliptic normal modes of invariant tori in Hamiltonian systems. 1996 *(preprint)*.

[331] M.B. Sevryuk. The lack-of-parameters problem in the KAM theory revisited. In C. Simó, editor, *Hamiltonian Systems with Three or More Degrees of Freedom* (S'Agaró, Spain, 1995), NATO ASI Series C: Math. Physics. Kluwer Academic, Dordrecht, 1996 *(to appear)*.

[332] M.B. Sevryuk and A. Lahiri. Bifurcations of families of invariant curves in four-dimensional reversible mappings. *Phys. Lett. A*, 154(3–4): 104–110, 1991.

[333] Ch.-W. Shih. Normal forms and versal deformations of linear involutive dynamical systems. *Chinese J. Math.*, 21(4): 333–347, 1993.

[334] C.L. Siegel. Über die Existenz einer Normalform analytischer HAMILTONscher Differentialgleichungen in der Nähe einer Gleichgewichtslösung. *Math. Ann.*, 128(2): 144–170, 1954.

[335] C.L. Siegel and J.K. Moser. *Lectures on Celestial Mechanics*. Springer-Verlag, 1971.

[336] J. Sotomayor. Generic bifurcations of dynamical systems. In M.M. Peixoto, editor, *Dynamical Systems*, pages 561–582. Acad. Press, 1973.

[337] V.G. Sprindžuk. *Metric Theory of Diophantine Approximations*. John Wiley, New York, 1979. [Russian original: Nauka, Moscow, 1977].

[338] E.M. Stein. *Singular Integrals and Differentiability Properties of Functions.* Princeton Math. Series, No. 30. Princeton Univ. Press, 1970.

[339] J.J. Stoker. *Nonlinear Vibrations in Mechanical and Electrical Systems.* John Wiley, New York, 1950 [reprinted: 1992].

[340] N.V. Svanidze. Small perturbations of an integrable dynamical system with an integral invariant. *Proc. Steklov Inst. Math.*, 2: 127–151, 1981.

[341] F. Takens. Singularities of vector fields. *Publ. Math. I.H.E.S.*, 43: 47–100, 1974.

[342] D.V. Treshchëv. A mechanism for the destruction of resonant tori in Hamiltonian systems. *Math. USSR Sbornik*, 68(1): 181–203, 1991.

[343] D.V. Treshchëv. Hyperbolic tori and asymptotic surfaces in Hamiltonian systems. *Russian J. Math. Phys.*, 2(1): 93–110, 1994.

[344] J.-C. van der Meer. *The Hamiltonian Hopf Bifurcation.* Lect. Notes Math., Vol. 1160. Springer-Verlag, 1985.

[345] J.-C. van der Meer. Bifurcation at nonsemisimple 1 : −1 resonance. *Z. Angew. Math. Phys.*, 37(3): 425–437, 1986.

[346] J.-C. van der Meer. Hamiltonian Hopf bifurcation with symmetry. *Nonlinearity*, 3(4): 1041–1056, 1990.

[347] S.J. van Strien. Center manifolds are not C^∞. *Math. Z.*, 166(2): 143–145, 1979.

[348] A. Vanderbauwhede. Families of periodic solutions for autonomous systems. In A.R. Bednarek and L. Cesari, editors, *Dynamical Systems, Part II*, pages 427–446. Acad. Press, 1982.

[349] A. Vanderbauwhede. *Local Bifurcation and Symmetry.* Research Notes Math., Vol. 75. Pitman, Boston, 1982.

[350] A. Vanderbauwhede. Hopf bifurcation for equivariant conservative and time-reversible systems. *Proc. Roy. Soc. Edinburgh, Ser. A*, 116(1–2): 103–128, 1990.

[351] P. Veerman and P. Holmes. The existence of arbitrarily many distinct periodic orbits in a two degree of freedom Hamiltonian system. *Physica D*, 14(2): 177–192, 1985.

[352] F.O.O. Wagener. Quasi-periodic stability of invariant circles of an unfolded skew Hopf bifurcation *(in preparation)*.

[353] Y.-H. Wan. Versal deformations of infinitesimally symplectic transformations with antisymplectic involutions. In M. Roberts and I. Stewart, editors, *Singularity Theory and its Applications, Part II*, Lect. Notes Math., Vol. 1463, pages 301–320. Springer-Verlag, 1991.

[354] A. Weinstein. *Lectures on Symplectic Manifolds.* Regional Conf. Ser. Math., no. 29. Amer. Math. Soc., Providence, RI, 1977 [2nd printing: 1979].

[355] H. Whitney. Analytic extensions of differentiable functions defined in closed sets. *Trans. Amer. Math. Soc.*, 36(1): 63–89, 1934.

[356] S. Wiggins. *Normally Hyperbolic Invariant Manifolds in Dynamical Systems.* Springer-Verlag, 1994.

[357] B.P. Wood, A.J. Lichtenberg, and M.A. Lieberman. Arnol'd and Arnol'd-like diffusion in many dimensions. *Physica D*, 71(1–2): 132–145, 1994.

[358] Zh. Xia. Existence of invariant tori in volume-preserving diffeomorphisms. *Ergod. Th. and Dynam. Syst.*, 12(3): 621–631, 1992.

[359] Zh. Xia. Existence of invariant tori for certain non-symplectic diffeomorphisms. In H.S. Dumas, K.R. Meyer, and D.S. Schmidt, editors, *Hamiltonian Dynamical Systems: History, Theory, and Applications* (Cincinnati, 1992), The IMA Volumes in Math. and Appl., Vol. 63, pages 373–385. Springer-Verlag, 1995.

[360] J. Xu, J. You, and Q. Qiu. Invariant tori for nearly integrable Hamiltonian systems with degeneracy. ETH-Zürich, 1994 *(preprint)*. [A revised and abridged version of this preprint has been submitted to *Math. Z.*].

[361] J.-C. Yoccoz. Travaux de Herman sur les tores invariants. In *Séminaire Bourbaki, Vol. 754, 1991–92*, pages 311–344. Astérisque, 206, 1992.

[362] G.M. Zaslavskiĭ and R.Z. Sagdeev. *Introduction to Nonlinear Physics. From the Pendulum to Turbulence and Chaos.* Nauka, Moscow, 1988 [in Russian. The English version: R.Z. Sagdeev, D.A. Usikov, and G.M. Zaslavsky. *Nonlinear Physics: From the Pendulum to Turbulence and Chaos.* Contemporary Concepts in Physics, Vol. 4. Harwood Academic Publ., 1988].

[363] E. Zehnder. An implicit function theorem for small divisor problems. *Bull. Amer. Math. Soc.*, 80(1): 174–179, 1974.

[364] E. Zehnder. Generalized implicit function theorems with applications to some small divisor problems, I and II. *Comm. Pure Appl. Math.*, 28(1): 91–140, 1975 and 29(1): 49–111, 1976.

Index

Printing and Binding: Druckpartner Rübelmann GmbH , Hemsbach

Lecture Notes in Mathematics

For information about Vols. 1–1459
please contact your bookseller or Springer-Verlag

Vol. 1460: G. Toscani, V. Boffi, S. Rionero (Eds.), Mathematical Aspects of Fluid Plasma Dynamics. Proceedings, 1988. V, 221 pages. 1991.

Vol. 1461: R. Gorenflo, S. Vessella, Abel Integral Equations. VII, 215 pages. 1991.

Vol. 1462: D. Mond, J. Montaldi (Eds.), Singularity Theory and its Applications. Warwick 1989, Part I. VIII, 405 pages. 1991.

Vol. 1463: R. Roberts, I. Stewart (Eds.), Singularity Theory and its Applications. Warwick 1989, Part II. VIII, 322 pages. 1991.

Vol. 1464: D. L. Burkholder, E. Pardoux, A. Sznitman, Ecole d'Eté de Probabilités de Saint- Flour XIX-1989. Editor: P. L. Hennequin. VI, 256 pages. 1991.

Vol. 1465: G. David, Wavelets and Singular Integrals on Curves and Surfaces. X, 107 pages. 1991.

Vol. 1466: W. Banaszczyk, Additive Subgroups of Topological Vector Spaces. VII, 178 pages. 1991.

Vol. 1467: W. M. Schmidt, Diophantine Approximations and Diophantine Equations. VIII, 217 pages. 1991.

Vol. 1468: J. Noguchi, T. Ohsawa (Eds.), Prospects in Complex Geometry. Proceedings, 1989. VII, 421 pages. 1991.

Vol. 1469: J. Lindenstrauss, V. D. Milman (Eds.), Geometric Aspects of Functional Analysis. Seminar 1989-90. XI, 191 pages. 1991.

Vol. 1470: E. Odell, H. Rosenthal (Eds.), Functional Analysis. Proceedings, 1987-89. VII, 199 pages. 1991.

Vol. 1471: A. A. Panchishkin, Non-Archimedean L-Functions of Siegel and Hilbert Modular Forms. VII, 157 pages. 1991.

Vol. 1472: T. T. Nielsen, Bose Algebras: The Complex and Real Wave Representations. V, 132 pages. 1991.

Vol. 1473: Y. Hino, S. Murakami, T. Naito, Functional Differential Equations with Infinite Delay. X, 317 pages. 1991.

Vol. 1474: S. Jackowski, B. Oliver, K. Pawałowski (Eds.), Algebraic Topology, Poznań 1989. Proceedings. VIII, 397 pages. 1991.

Vol. 1475: S. Busenberg, M. Martelli (Eds.), Delay Differential Equations and Dynamical Systems. Proceedings, 1990. VIII, 249 pages. 1991.

Vol. 1476: M. Bekkali, Topics in Set Theory. VII, 120 pages. 1991.

Vol. 1477: R. Jajte, Strong Limit Theorems in Noncommutative L_2-Spaces. X, 113 pages. 1991.

Vol. 1478: M.-P. Malliavin (Ed.), Topics in Invariant Theory. Seminar 1989-1990. VI, 272 pages. 1991.

Vol. 1479: S. Bloch, I. Dolgachev, W. Fulton (Eds.), Algebraic Geometry. Proceedings, 1989. VII, 300 pages. 1991.

Vol. 1480: F. Dumortier, R. Roussarie, J. Sotomayor, H. Żoładek, Bifurcations of Planar Vector Fields: Nilpotent Singularities and Abelian Integrals. VIII, 226 pages. 1991.

Vol. 1481: D. Ferus, U. Pinkall, U. Simon, B. Wegner (Eds.), Global Differential Geometry and Global Analysis. Proceedings, 1991. VIII, 283 pages. 1991.

Vol. 1482: J. Chabrowski, The Dirichlet Problem with L^2-Boundary Data for Elliptic Linear Equations. VI, 173 pages. 1991.

Vol. 1483: E. Reithmeier, Periodic Solutions of Nonlinear Dynamical Systems. VI, 171 pages. 1991.

Vol. 1484: H. Delfs, Homology of Locally Semialgebraic Spaces. IX, 136 pages. 1991.

Vol. 1485: J. Azéma, P. A. Meyer, M. Yor (Eds.), Séminaire de Probabilités XXV. VIII, 440 pages. 1991.

Vol. 1486: L. Arnold, H. Crauel, J.-P. Eckmann (Eds.), Lyapunov Exponents. Proceedings, 1990. VIII, 365 pages. 1991.

Vol. 1487: E. Freitag, Singular Modular Forms and Theta Relations. VI, 172 pages. 1991.

Vol. 1488: A. Carboni, M. C. Pedicchio, G. Rosolini (Eds.), Category Theory. Proceedings, 1990. VII, 494 pages. 1991.

Vol. 1489: A. Mielke, Hamiltonian and Lagrangian Flows on Center Manifolds. X, 140 pages. 1991.

Vol. 1490: K. Metsch, Linear Spaces with Few Lines. XIII, 196 pages. 1991.

Vol. 1491: E. Lluis-Puebla, J.-L. Loday, H. Gillet, C. Soulé, V. Snaith, Higher Algebraic K-Theory: an overview. IX, 164 pages. 1992.

Vol. 1492: K. R. Wicks, Fractals and Hyperspaces. VIII, 168 pages. 1991.

Vol. 1493: E. Benoît (Ed.), Dynamic Bifurcations. Proceedings, Luminy 1990. VII, 219 pages. 1991.

Vol. 1494: M.-T. Cheng, X.-W. Zhou, D.-G. Deng (Eds.), Harmonic Analysis. Proceedings, 1988. IX, 226 pages. 1991.

Vol. 1495: J. M. Bony, G. Grubb, L. Hörmander, H. Komatsu, J. Sjöstrand, Microlocal Analysis and Applications. Montecatini Terme, 1989. Editors: L. Cattabriga, L. Rodino. VII, 349 pages. 1991.

Vol. 1496: C. Foias, B. Francis, J. W. Helton, H. Kwakernaak, J. B. Pearson, H_∞-Control Theory. Como, 1990. Editors: E. Mosca, L. Pandolfi. VII, 336 pages. 1991.

Vol. 1497: G. T. Herman, A. K. Louis, F. Natterer (Eds.), Mathematical Methods in Tomography. Proceedings 1990. X, 268 pages. 1991.

Vol. 1498: R. Lang, Spectral Theory of Random Schrödinger Operators. X, 125 pages. 1991.

Vol. 1499: K. Taira, Boundary Value Problems and Markov Processes. IX, 132 pages. 1991.

Vol. 1500: J.-P. Serre, Lie Algebras and Lie Groups. VII, 168 pages. 1992.

Vol. 1501: A. De Masi, E. Presutti, Mathematical Methods for Hydrodynamic Limits. IX, 196 pages. 1991.

Vol. 1502: C. Simpson, Asymptotic Behavior of Monodromy. V, 139 pages. 1991.

Vol. 1503: S. Shokranian, The Selberg-Arthur Trace Formula (Lectures by J. Arthur). VII, 97 pages. 1991.

Vol. 1504: J. Cheeger, M. Gromov, C. Okonek, P. Pansu, Geometric Topology: Recent Developments. Editors: P. de Bartolomeis, F. Tricerri. VII, 197 pages. 1991.

Vol. 1505: K. Kajitani, T. Nishitani, The Hyperbolic Cauchy Problem. VII, 168 pages. 1991.

Vol. 1506: A. Buium, Differential Algebraic Groups of Finite Dimension. XV, 145 pages. 1992.

Vol. 1507: K. Hulek, T. Peternell, M. Schneider, F.-O. Schreyer (Eds.), Complex Algebraic Varieties. Proceedings, 1990. VII, 179 pages. 1992.

Vol. 1508: M. Vuorinen (Ed.), Quasiconformal Space Mappings. A Collection of Surveys 1960-1990. IX, 148 pages. 1992.

Vol. 1509: J. Aguadé, M. Castellet, F. R. Cohen (Eds.), Algebraic Topology - Homotopy and Group Cohomology. Proceedings, 1990. X, 330 pages. 1992.

Vol. 1510: P. P. Kulish (Ed.), Quantum Groups. Proceedings, 1990. XII, 398 pages. 1992.

Vol. 1511: B. S. Yadav, D. Singh (Eds.), Functional Analysis and Operator Theory. Proceedings, 1990. VIII, 223 pages. 1992.

Vol. 1512: L. M. Adleman, M.-D. A. Huang, Primality Testing and Abelian Varieties Over Finite Fields. VII, 142 pages. 1992.

Vol. 1513: L. S. Block, W. A. Coppel, Dynamics in One Dimension. VIII, 249 pages. 1992.

Vol. 1514: U. Krengel, K. Richter, V. Warstat (Eds.), Ergodic Theory and Related Topics III, Proceedings, 1990. VIII, 236 pages. 1992.

Vol. 1515: E. Ballico, F. Catanese, C. Ciliberto (Eds.), Classification of Irregular Varieties. Proceedings, 1990. VII, 149 pages. 1992.

Vol. 1516: R. A. Lorentz, Multivariate Birkhoff Interpolation. IX, 192 pages. 1992.

Vol. 1517: K. Keimel, W. Roth, Ordered Cones and Approximation. VI, 134 pages. 1992.

Vol. 1518: H. Stichtenoth, M. A. Tsfasman (Eds.), Coding Theory and Algebraic Geometry. Proceedings, 1991. VIII, 223 pages. 1992.

Vol. 1519: M. W. Short. The Primitive Soluble Permutation Groups of Degree less than 256. IX, 145 pages. 1992.

Vol. 1520: Yu. G. Borisovich, Yu. E. Gliklikh (Eds.), Global Analysis – Studies and Applications V. VII, 284 pages. 1992.

Vol. 1521: S. Busenberg, B. Forte, H. K. Kuiken, Mathematical Modelling of Industrial Process. Bari, 1990. Editors: V. Capasso, A. Fasano. VII, 162 pages. 1992.

Vol. 1522: J.-M. Delort, F. B. I. Transformation. VII, 101 pages. 1992.

Vol. 1523: W. Xue, Rings with Morita Duality. X, 168 pages. 1992.

Vol. 1524: M. Coste, L. Mahé, M.-F. Roy (Eds.), Real Algebraic Geometry. Proceedings, 1991. VIII, 418 pages. 1992.

Vol. 1525: C. Casacuberta, M. Castellet (Eds.), Mathematical Research Today and Tomorrow. VII, 112 pages. 1992.

Vol. 1526: J. Azéma, P. A. Meyer, M. Yor (Eds.), Séminaire de Probabilités XXVI. X, 633 pages. 1992.

Vol. 1527: M. I. Freidlin, J.-F. Le Gall, Ecole d'Eté de Probabilités de Saint-Flour XX – 1990. Editor: P. L. Hennequin. VIII, 244 pages. 1992.

Vol. 1528: G. Isac, Complementarity Problems. VI, 297 pages. 1992.

Vol. 1529: J. van Neerven, The Adjoint of a Semigroup of Linear Operators. X, 195 pages. 1992.

Vol. 1530: J. G. Heywood, K. Masuda, R. Rautmann, S. A. Solonnikov (Eds.), The Navier-Stokes Equations II – Theory and Numerical Methods. IX, 322 pages. 1992.

Vol. 1531: M. Stoer, Design of Survivable Networks. IV, 206 pages. 1992.

Vol. 1532: J. F. Colombeau, Multiplication of Distributions. X, 184 pages. 1992.

Vol. 1533: P. Jipsen, H. Rose, Varieties of Lattices. X, 162 pages. 1992.

Vol. 1534: C. Greither, Cyclic Galois Extensions of Commutative Rings. X, 145 pages. 1992.

Vol. 1535: A. B. Evans, Orthomorphism Graphs of Groups. VIII, 114 pages. 1992.

Vol. 1536: M. K. Kwong, A. Zettl, Norm Inequalities for Derivatives and Differences. VII, 150 pages. 1992.

Vol. 1537: P. Fitzpatrick, M. Martelli, J. Mawhin, R. Nussbaum, Topological Methods for Ordinary Differential Equations. Montecatini Terme, 1991. Editors: M. Furi, P. Zecca. VII, 218 pages. 1993.

Vol. 1538: P.-A. Meyer, Quantum Probability for Probabilists. X, 287 pages. 1993.

Vol. 1539: M. Coornaert, A. Papadopoulos, Symbolic Dynamics and Hyperbolic Groups. VIII, 138 pages. 1993.

Vol. 1540: H. Komatsu (Ed.), Functional Analysis and Related Topics, 1991. Proceedings. XXI, 413 pages. 1993.

Vol. 1541: D. A. Dawson, B. Maisonneuve, J. Spencer, Ecole d' Eté de Probabilités de Saint-Flour XXI - 1991. Editor: P. L. Hennequin. VIII, 356 pages. 1993.

Vol. 1542: J.Fröhlich, Th.Kerler, Quantum Groups, Quantum Categories and Quantum Field Theory. VII, 431 pages. 1993.

Vol. 1543: A. L. Dontchev, T. Zolezzi, Well-Posed Optimization Problems. XII, 421 pages. 1993.

Vol. 1544: M.Schürmann, White Noise on Bialgebras. VII, 146 pages. 1993.

Vol. 1545: J. Morgan, K. O'Grady, Differential Topology of Complex Surfaces. VIII, 224 pages. 1993.

Vol. 1546: V. V. Kalashnikov, V. M. Zolotarev (Eds.), Stability Problems for Stochastic Models. Proceedings, 1991. VIII, 229 pages. 1993.

Vol. 1547: P. Harmand, D. Werner, W. Werner, M-ideals in Banach Spaces and Banach Algebras. VIII, 387 pages. 1993.

Vol. 1548: T. Urabe, Dynkin Graphs and Quadrilateral Singularities. VI, 233 pages. 1993.

Vol. 1549: G. Vainikko, Multidimensional Weakly Singular Integral Equations. XI, 159 pages. 1993.

Vol. 1550: A. A. Gonchar, E. B. Saff (Eds.), Methods of Approximation Theory in Complex Analysis and Mathematical Physics IV, 222 pages, 1993.

Vol. 1551: L. Arkeryd, P. L. Lions, P.A. Markowich, S.R. S. Varadhan. Nonequilibrium Problems in Many-Particle Systems. Montecatini, 1992. Editors: C. Cercignani, M. Pulvirenti. VII, 158 pages 1993.

Vol. 1552: J. Hilgert, K.-H. Neeb, Lie Semigroups and their Applications. XII, 315 pages. 1993.

Vol. 1553: J.-L- Colliot-Thélène, J. Kato, P. Vojta. Arithmetic Algebraic Geometry. Trento, 1991. Editor: E. Ballico. VII, 223 pages. 1993.

Vol. 1554: A. K. Lenstra, H. W. Lenstra, Jr. (Eds.), The Development of the Number Field Sieve. VIII, 131 pages. 1993.

Vol. 1555: O. Liess, Conical Refraction and Higher Microlocalization. X, 389 pages. 1993.

Vol. 1556: S. B. Kuksin, Nearly Integrable Infinite-Dimensional Hamiltonian Systems. XXVII, 101 pages. 1993.

Vol. 1557: J. Azéma, P. A. Meyer, M. Yor (Eds.), Séminaire de Probabilités XXVII. VI, 327 pages. 1993.

Vol. 1558: T. J. Bridges, J. E. Furter, Singularity Theory and Equivariant Symplectic Maps. VI, 226 pages. 1993.

Vol. 1559: V. G. Sprindžuk, Classical Diophantine Equations. XII, 228 pages. 1993.

Vol. 1560: T. Bartsch, Topological Methods for Variational Problems with Symmetries. X, 152 pages. 1993.

Vol. 1561: I. S. Molchanov, Limit Theorems for Unions of Random Closed Sets. X, 157 pages. 1993.

Vol. 1562: G. Harder, Eisensteinkohomologie und die Konstruktion gemischter Motive. XX, 184 pages. 1993.

Vol. 1563: E. Fabes, M. Fukushima, L. Gross, C. Kenig, M. Röckner, D. W. Stroock, Dirichlet Forms. Varenna, 1992. Editors: G. Dell'Antonio, U. Mosco. VII, 245 pages. 1993.

Vol. 1564: J. Jorgenson, S. Lang, Basic Analysis of Regularized Series and Products. IX, 122 pages. 1993.

Vol. 1565: L. Boutet de Monvel, C. De Concini, C. Procesi, P. Schapira, M. Vergne. D-modules, Representation Theory, and Quantum Groups. Venezia, 1992. Editors: G. Zampieri, A. D'Agnolo. VII, 217 pages. 1993.

Vol. 1566: B. Edixhoven, J.-H. Evertse (Eds.), Diophantine Approximation and Abelian Varieties. XIII, 127 pages. 1993.

Vol. 1567: R. L. Dobrushin, S. Kusuoka, Statistical Mechanics and Fractals. VII, 98 pages. 1993.

Vol. 1568: F. Weisz, Martingale Hardy Spaces and their Application in Fourier Analysis. VIII, 217 pages. 1994.

Vol. 1569: V. Totik, Weighted Approximation with Varying Weight. VI, 117 pages. 1994.

Vol. 1570: R. deLaubenfels, Existence Families, Functional Calculi and Evolution Equations. XV, 234 pages. 1994.

Vol. 1571: S. Yu. Pilyugin, The Space of Dynamical Systems with the C^0-Topology. X, 188 pages. 1994.

Vol. 1572: L. Göttsche, Hilbert Schemes of Zero-Dimensional Subschemes of Smooth Varieties. IX, 196 pages. 1994.

Vol. 1573: V. P. Havin, N. K. Nikolski (Eds.), Linear and Complex Analysis – Problem Book 3 – Part I. XXII, 489 pages. 1994.

Vol. 1574: V. P. Havin, N. K. Nikolski (Eds.), Linear and Complex Analysis – Problem Book 3 – Part II. XXII, 507 pages. 1994.

Vol. 1575: M. Mitrea, Clifford Wavelets, Singular Integrals, and Hardy Spaces. XI, 116 pages. 1994.

Vol. 1576: K. Kitahara, Spaces of Approximating Functions with Haar-Like Conditions. X, 110 pages. 1994.

Vol. 1577: N. Obata, White Noise Calculus and Fock Space. X, 183 pages. 1994.

Vol. 1578: J. Bernstein, V. Lunts, Equivariant Sheaves and Functors. V, 139 pages. 1994.

Vol. 1579: N. Kazamaki, Continuous Exponential Martingales and *BMO*. VII, 91 pages. 1994.

Vol. 1580: M. Milman, Extrapolation and Optimal Decompositions with Applications to Analysis. XI, 161 pages. 1994.

Vol. 1581: D. Bakry, R. D. Gill, S. A. Molchanov, Lectures on Probability Theory. Editor: P. Bernard. VIII, 420 pages. 1994.

Vol. 1582: W. Balser, From Divergent Power Series to Analytic Functions. X, 108 pages. 1994.

Vol. 1583: J. Azéma, P. A. Meyer, M. Yor (Eds.), Séminaire de Probabilités XXVIII. VI, 334 pages. 1994.

Vol. 1584: M. Brokate, N. Kenmochi, I. Müller, J. F. Rodriguez, C. Verdi, Phase Transitions and Hysteresis. Montecatini Terme, 1993. Editor: A. Visintin. VII. 291 pages. 1994.

Vol. 1585: G. Frey (Ed.), On Artin's Conjecture for Odd 2-dimensional Representations. VIII, 148 pages. 1994.

Vol. 1586: R. Nillsen, Difference Spaces and Invariant Linear Forms. XII, 186 pages. 1994.

Vol. 1587: N. Xi, Representations of Affine Hecke Algebras. VIII, 137 pages. 1994.

Vol. 1588: C. Scheiderer, Real and Étale Cohomology. XXIV, 273 pages. 1994.

Vol. 1589: J. Bellissard, M. Degli Esposti, G. Forni, S. Graffi, S. Isola, J. N. Mather, Transition to Chaos in Classical and Quantum Mechanics. Montecatini Terme, 1991. Editor: S. Graffi. VII, 192 pages. 1994.

Vol. 1590: P. M. Soardi, Potential Theory on Infinite Networks. VIII, 187 pages. 1994.

Vol. 1591: M. Abate, G. Patrizio, Finsler Metrics – A Global Approach. IX, 180 pages. 1994.

Vol. 1592: K. W. Breitung, Asymptotic Approximations for Probability Integrals. IX, 146 pages. 1994.

Vol. 1593: J. Jorgenson & S. Lang, D. Goldfeld, Explicit Formulas for Regularized Products and Series. VIII, 154 pages. 1994.

Vol. 1594: M. Green, J. Murre, C. Voisin, Algebraic Cycles and Hodge Theory. Torino, 1993. Editors: A. Albano, F. Bardelli. VII, 275 pages. 1994.

Vol. 1595: R.D.M. Accola, Topics in the Theory of Riemann Surfaces. IX, 105 pages. 1994.

Vol. 1596: L. Heindorf, L. B. Shapiro, Nearly Projective Boolean Algebras. X, 202 pages. 1994.

Vol. 1597: B. Herzog, Kodaira-Spencer Maps in Local Algebra. XVII, 176 pages. 1994.

Vol. 1598: J. Berndt, F. Tricerri, L. Vanhecke, Generalized

Heisenberg Groups and Damek-Ricci Harmonic Spaces. VIII, 125 pages. 1995.

Vol. 1599: K. Johannson, Topology and Combinatorics of 3-Manifolds. XVIII, 446 pages. 1995.

Vol. 1600: W. Narkiewicz, Polynomial Mappings. VII, 130 pages. 1995.

Vol. 1601: A. Pott, Finite Geometry and Character Theory. VII, 181 pages. 1995.

Vol. 1602: J. Winkelmann, The Classification of Three-dimensional Homogeneous Complex Manifolds. XI, 230 pages. 1995.

Vol. 1603: V. Ene, Real Functions – Current Topics. XIII, 310 pages. 1995.

Vol. 1604: A. Huber, Mixed Motives and their Realization in Derived Categories. XV, 207 pages. 1995.

Vol. 1605: L. B. Wahlbin, Superconvergence in Galerkin Finite Element Methods. XI, 166 pages. 1995.

Vol. 1606: P.-D. Liu, M. Qian, Smooth Ergodic Theory of Random Dynamical Systems. XI, 221 pages. 1995.

Vol. 1607: G. Schwarz, Hodge Decomposition – A Method for Solving Boundary Value Problems. VII, 155 pages. 1995.

Vol. 1608: P. Biane, R. Durrett, Lectures on Probability Theory. Editor: P. Bernard. VII, 210 pages. 1995.

Vol. 1609: L. Arnold, C. Jones, K. Mischaikow, G. Raugel, Dynamical Systems. Montecatini Terme, 1994. Editor: R. Johnson. VIII, 329 pages. 1995.

Vol. 1610: A. S. Üstünel, An Introduction to Analysis on Wiener Space. X, 95 pages. 1995.

Vol. 1611: N. Knarr, Translation Planes. VI, 112 pages. 1995.

Vol. 1612: W. Kühnel, Tight Polyhedral Submanifolds and Tight Triangulations. VII, 122 pages. 1995.

Vol. 1613: J. Azéma, M. Emery, P. A. Meyer, M. Yor (Eds.), Séminaire de Probabilités XXIX. VI, 326 pages. 1995.

Vol. 1614: A. Koshelev, Regularity Problem for Quasilinear Elliptic and Parabolic Systems. XXI, 255 pages. 1995.

Vol. 1615: D. B. Massey, Lê Cycles and Hypersurface Singularities. XI, 131 pages. 1995.

Vol. 1616: I. Moerdijk, Classifying Spaces and Classifying Topoi. VII, 94 pages. 1995.

Vol. 1617: V. Yurinsky, Sums and Gaussian Vectors. XI, 305 pages. 1995.

Vol. 1618: G. Pisier, Similarity Problems and Completely Bounded Maps. VII, 156 pages. 1996.

Vol. 1619: E. Landvogt, A Compactification of the Bruhat-Tits Building. VII, 152 pages. 1996.

Vol. 1620: R. Donagi, B. Dubrovin, E. Frenkel, E. Previato, Integrable Systems and Quantum Groups. Montecatini Terme, 1993. Editors:M. Francaviglia, S. Greco. VIII, 488 pages. 1996.

Vol. 1621: H. Bass, M. V. Otero-Espinar, D. N. Rockmore, C. P. L. Tresser, Cyclic Renormalization and Auto-morphism Groups of Rooted Trees. XXI, 136 pages. 1996.

Vol. 1622: E. D. Farjoun, Cellular Spaces, Null Spaces and Homotopy Localization. XIV, 199 pages. 1996.

Vol. 1623: H.P. Yap, Total Colourings of Graphs. VIII, 131 pages. 1996.

Vol. 1624: V. Brînzănescu, Holomorphic Vector Bundles over Compact Complex Surfaces. X, 170 pages. 1996.

Vol.1625: S. Lang, Topics in Cohomology of Groups. VII, 226 pages. 1996.

Vol. 1626: J. Azéma, M. Emery, M. Yor (Eds.), Séminaire de Probabilités XXX. VIII, 382 pages. 1996.

Vol. 1627: C. Graham, Th. G. Kurtz, S. Méléard, Ph. E. Protter, M. Pulvirenti, D. Talay, Probabilistic Models for Nonlinear Partial Differential Equations. Montecatini Terme, 1995. Editors: D. Talay, L. Tubaro. X, 301 pages. 1996.

Vol. 1628: P.-H. Zieschang, An Algebraic Approach to Association Schemes. XII, 189 pages. 1996.

Vol. 1629: J. D. Moore, Lectures on Seiberg-Witten Invariants. VII, 105 pages. 1996.

Vol. 1630: D. Neuenschwander, Probabilities on the Heisenberg Group: Limit Theorems and Brownian Motion. VIII, 139 pages. 1996.

Vol. 1631: K. Nishioka, Mahler Functions and Transcendence.VIII, 185 pages.1996.

Vol. 1632: A. Kushkuley, Z. Balanov, Geometric Methods in Degree Theory for Equivariant Maps. VII, 136 pages. 1996.

Vol.1633: H. Aikawa, M. Essén, Potential Theory – Selected Topics. IX, 200 pages.1996.

Vol. 1634: J. Xu, Flat Covers of Modules. IX, 161 pages. 1996.

Vol. 1635: E. Hebey, Sobolev Spaces on Riemannian Manifolds. X, 116 pages. 1996.

Vol. 1636: M. A. Marshall, Spaces of Orderings and Abstract Real Spectra. VI, 190 pages. 1996.

Vol. 1637: B. Hunt, The Geometry of some special Arithmetic Quotients. XIII, 332 pages. 1996.

Vol. 1638: P. Vanhaecke, Integrable Systems in the realm of Algebraic Geometry. VIII, 218 pages. 1996.

Vol. 1639: K. Dekimpe, Almost-Bieberbach Groups: Affine and Polynomial Structures. X, 259 pages. 1996.

Vol. 1640: G. Boillat, C. M. Dafermos, P. D. Lax, T. P. Liu, Recent Mathematical Methods in Nonlinear Wave Propagation. Montecatini Terme, 1994. Editor: T. Ruggeri. VII, 142 pages. 1996.

Vol. 1641: P. Abramenko, Twin Buildings and Applications to S-Arithmetic Groups. IX, 123 pages. 1996.

Vol. 1642: M. Puschnigg, Asymptotic Cyclic Cohomology. XXII, 138 pages. 1996.

Vol. 1643: J. Richter-Gebert, Realization Spaces of Polytopes. XI, 187 pages. 1996.

Vol. 1644: A. Adler, S. Ramanan, Moduli of Abelian Varieties. VI, 196 pages. 1996.

Vol. 1645: H. W. Broer, G. B. Huitema, M. B. Sevryuk, Quasi-Periodic Motions in Families of Dynamical Systems. XI, 195 pages. 1996.

Vol. 1646: J.-P. Demailly, T. Peternell, G. Tian, A. N. Tyurin, Transcendental Methods in Algebraic Geometry. Cetraro, 1994. Editors: F. Catanese, C. Ciliberto. VII, 257 pages. 1996.

Vol. 1647: D. Dias, P. Le Barz, Configuration Spaces over Hilbert Schemes and Applications. VII. 143 pages. 1996.

Vol. 1648: R. Dobrushin, P. Groeneboom, M. Ledoux, Lectures on Probability Theory and Statistics. Editor: P. Bernard. VIII, 300 pages. 1996.

4. Lecture Notes are printed by photo-offset from the master-copy delivered in camera-ready form by the authors. Springer-Verlag provides technical instructions for the preparation of manuscripts. Macro packages in TEX, LATEX2e, LATEX2.09 are available from Springer's web-pages at

http://www.springer.de/math/authors/b-tex.html.

Careful preparation of the manuscripts will help keep production time short and ensure satisfactory appearance of the finished book.

The actual production of a Lecture Notes volume takes approximately 12 weeks.

5. Authors receive a total of 50 free copies of their volume, but no royalties. They are entitled to a discount of 33.3 % on the price of Springer books purchase for their personal use, if ordering directly from Springer-Verlag.

Commitment to publish is made by letter of intent rather than by signing a formal contract. Springer-Verlag secures the copyright for each volume. Authors are free to reuse material contained in their LNM volumes in later publications: A brief written (or e-mail) request for formal permission is sufficient.

Addresses:

Professor F. Takens, Mathematisch Instituut,
Rijksuniversiteit Groningen, Postbus 800,
9700 AV Groningen, The Netherlands
E-mail: F.Takens@math.rug.nl

Professor B. Teissier
Université Paris 7
UFR de Mathématiques
Equipe Géométrie et Dynamique
Case 7012
2 place Jussieu
75251 Paris Cedex 05
E-mail: Teissier@math.jussieu.fr

Springer-Verlag, Mathematics Editorial, Tiergartenstr. 17,
D-69121 Heidelberg, Germany,
Tel.: *49 (6221) 487-701
Fax: *49 (6221) 487-355
E-mail: lnm@Springer.de